Astronomy: Structure of the Universe

Third Edition

Astronomy: Structure of the Universe

Third Edition

A. E. Roy, Ph.D., F.R.A.S., F.R.S.E., F.B.I.S.,
Titular Professor of Astronomy, University of Glasgow

D. Clarke, Ph.D., M.Inst.P., F.R.A.S.,
Director of Glasgow University Observatory
Reader in Astronomy, University of Glasgow

Adam Hilger, Bristol and New York

First published 1977
Reprinted 1978
Second edition 1982
Third edition 1989
Reprinted 1990

To our students
who have learned almost as much from us
as we have from them

British Library Cataloguing in Publication Data
Roy, A. E. (Archie Edmiston) 1924–
Astronomy.—3rd ed.
1. Astronomical bodies
I. Title II. Clarke, D. (David) 1936–
523

ISBN 0-85274-082-4

US Library of Congress Cataloging-in-Publication Data
are available

Published under the Adam Hilger imprint by IOP Publishing Ltd
Techno House, Redcliffe Way, Bristol BS1 6NX, England
335 East 45th Street, New York, NY 10017-3483, USA

Printed in Great Britain by J W Arrowsmith Ltd, Bristol

Contents

Foreword

In our previous book (*Astronomy*: *Principles and Practice*, Adam Hilger 1977) we presented the principles and practice of astronomy. We now give a description of the universe as revealed by the application of the methods discussed in the first book.

Although the work has been published in two separate volumes, it was conceived as one book, written to fulfil the need of a preliminary *science course* or a *liberal arts course* at the first year university level. It is hoped that those polytechnics that now provide courses in astronomy will find it useful; in addition the serious amateur astronomer will find helpful discussions of topics in which he is interested.

The present volume contains three parts. These perform the following functions:

Part 1: Describes the Solar System.
Part 2: Describes the constituents of the Galaxy: stars and interstellar material.
Part 3: Deals with our own and other galaxies, closing with a discussion of cosmological ideas.

Problems are also given in the relevant places to test thoroughly the student's appreciation of the principles involved. Exercises suitable for laboratory or outdoor work are also provided. A small, carefully selected list of books is also included, the references given seeming to us to enlarge usefully on particular topics.

Figures and equations are given the number of the chapter in which they appear, followed by a number denoting their order in that chapter. SI units and those preferred by the *International Astronomical Union* are used wherever possible.

Many references occur in the second volume to topics discussed in the first; this is inevitable because of the design of the work. Nevertheless it is hoped that just as the first volume can be read with profit without reference to the second, so a useful first reading of the second volume can be made without recourse to the first volume.

Inevitably, some errors of fact and misprints will occur and the authors would be glad to hear of them.

Second edition. Very little need be added to the Foreword. A number of sections have been rewritten because of the increase in our knowledge of various topics due to the harvest of new facts gathered by spacecraft such as the Voyagers and by modern instrumentation. A few errors have been removed and we are grateful to those who have taken the trouble to notify us of them. Again our thanks go to Mrs M I Morris of the Department of Astronomy for much careful typing.

Third edition. Only few changes of substance have been made, most of the additions being related to advances of knowledge, particularly those made by spacecraft. Again some occasional errors have been rectified and we are grateful to those people who have informed us about them. Once again we thank Mrs M I Morris for her help in preparing the typescript for the improvements of this edition.

Acknowledgments

Teaching is a constant process of reiteration and regeneration. The response that a teacher has from his students provides a form of feedback, effecting an improvement in the content and method of presentation of succeeding courses. Without this intelligent response from students, courses would remain dull and static. Our first acknowledgment must be to our former students who have helped in giving shape to the contents of this book.

The presentation of many of the topics in the book has undoubtedly benefited from innumerable discussions with past and present members of the staff of the Department of Astronomy, Glasgow University. In particular we would like to thank Professor Peter A Sweet not only for helpful advice but also for providing some source material.

We would like to thank the Oxford University Press and Professor A Thom for permission to include Figure 5.13, redrawn from Figure 3.5 of his work *Megalithic Sites in Britain*; the Controller of Her Majesty's Stationery Office for permission to reproduce from the *Explanatory Supplement to the Astronomical Ephemeris* much of the astronomical data in Tables 6.3, 6.4, 6.5, 6.6 and 6.7; the Council for the British Astronomical Association for permission to use material from *The Handbook of the BAA* for 1970 for Tables 6.1, 6.6 and 6.9; Methuen and Co. Ltd, London, for permission to use material of Table I from *The Nature of Comets* by N B Richter; Gall and Inglis, Edinburgh, for permission to present in Chapter 7 an extract from *Norton's Star Atlas.*

We also acknowledge with thanks the following sources of photographs: Sacramento Peak Observatories, Hale Observatories, National Aeronautics and Space Administration, Jet Propulsion Laboratories, The Royal Astronomical Society, Lowell Observatory and Mt. Wilson and Palomar Observatories. The source of each photograph is included in its caption.

The authors also acknowledge with sincere gratitude their debt to Mrs Margaret I Morris, MA, FRAS, of the Department of Astronomy, Glasgow University; not only did she type the first draft of the manuscript but she also read it critically, suggesting a large number of improvements in the mode of presentation that the authors have been happy to incorporate. Our thanks are also due to Mrs L Williamson, of the same Department, for much additional typing.

Archie E Roy
David Clarke

Part 1
The Solar System

Chapters 1–6

PROGRAMME: The description of the planets and other bodies of the Solar System is introduced by presenting the range and type of measurements which are applied to their study; much of the material is based on the most recent ground-based and space-vehicle measurements. Special attention is given to descriptions of the Sun, Earth, Moon, the Earth–Moon system and eclipses. Finally, the overall features of the Solar System are discussed and its creation and development considered.

1
The Solar System

1.1 Introduction

The Solar System is a tiny part of the Universe. The bodies in the Solar System, with the exception of most comets, are all contained in a volume of space of diameter about one thirty-thousandth the distance to the nearest star. The perturbing effects of the stars on the orbits of the Solar System members are therefore negligible, the System being effectively isolated as far as its internal movements are concerned.

The Sun, a typical star, dominates the System in size and mass. The largest planet, Jupiter, has a diameter of only one-tenth that of the Sun; its mass is but one-thousandth. Because of this predominance in mass, the planets, asteroids, comets and meteors move to a high degree of approximation as if they were attracted solely by the Sun, their orbits being ellipses of various sizes about the Sun and with the Sun at one focus. The eccentricities of these orbits have a range of values. In all planetary cases, the eccentricities are small, being less than 0·1, except those of the orbits of Mercury and Pluto, where they have values of 0·206 and 0·250 respectively.

Eccentricities of the asteroid orbits are in general more pronounced while those of cometary and meteor orbits can approach unity.

The planes of the planetary orbits contain the Sun's centre and are inclined to the Earth's orbital plane by a few degrees at most. Again it is seen that Mercury and Pluto are the exceptions; their orbital inclinations are 7° and 17° 10′ respectively. The Solar System is therefore two-dimensional to a first approximation. Indeed it would be difficult to draw distinctly an accurate diagram displaying the angular separation of most of the orbits from each other, so small are these angles.

We can however make a plan view diagram of the System, the distances being drawn to scale, the planes of the planetary orbits being rotated on to the Earth's orbital plane (Fig. 1.1).

Also inserted in the diagram is the asteroid belt, a region between the orbits of Jupiter and Mars occupied by the orbits of thousands of minor planets, the largest of which, Ceres, is only 770 km in diameter.

It will be noticed that the minimum distance of Pluto from the Sun is less than that of Neptune from the Sun. If it were not for the mutual inclination of their orbital planes, the probability of a collision of these planets would be markedly increased.

The symbols for the planets are also given in Fig. 1.1. The symbol for the Sun, namely ⊙, has not been used to avoid confusion in the diagram.

It is seen then that the planetary orbits are almost circular and coplanar. A further regularity is that the direction of movement of the planets in their orbits is the same though the rates vary within each orbit and from orbit to orbit, in accordance with Kepler's laws.

Asteroids, comets and meteors also move in accordance with Kepler's laws. No example of an asteroid moving in a retrograde orbit is known but, although most comets and meteors move about the Sun in the direction in which the planets revolve, there are a number of exceptions, the most famous of which is Halley's comet.

Each planet, except Mercury and Venus, is attended by one or more satellites. Saturn, in addition to having seventeen moons, possesses a ring

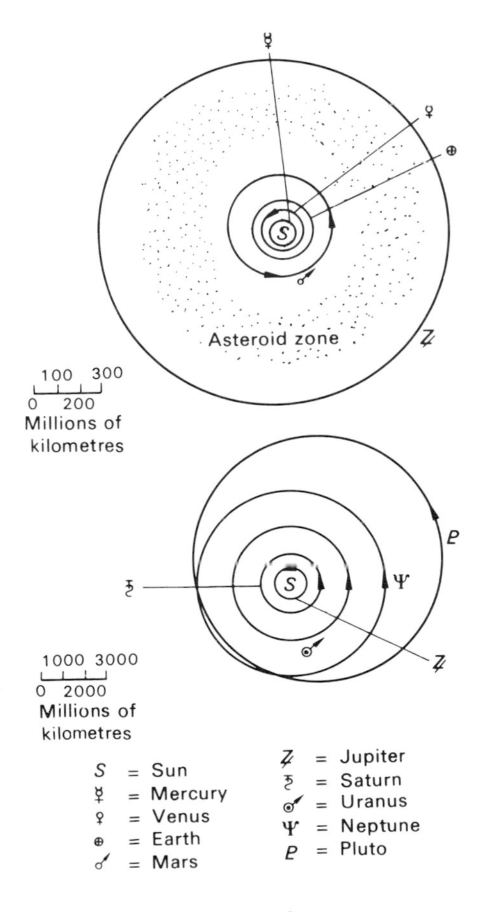

Fig. 1.1 A plan of the Solar System

system composed of millions of tiny satellites, of uncertain nature, moving in coplanar, almost circular orbits about the planet. The Earth has one large satellite, the Moon, almost one-eightieth its mass. Mars has two small satellites with diameters less than 16 km. Jupiter has at least sixteen moons, four of which (the Galilean satellites) are large, being of the order of size of the Moon. The other twelve are much smaller, ranging in diameter from about 160 km downwards. Saturn's seventeen moons include Titan, almost as large as the planet Mercury, and bodies very much smaller. Uranus has fifteen known satellites, all revolving in almost circular, coplanar orbits. Neptune's two moons are of widely different mass. One, Triton, must be as large as Titan; the other, Nereid, is much smaller and moves in a highly-eccentric orbit about the planet. Pluto's moon, Charon, is the largest in the Solar System in terms of its planet's size so that the Pluto–Charon system is essentially a double planet.

In addition to their satellites, Jupiter and Uranus possess ring systems.

Most of the satellites revolve about the planets in elliptic orbits of small eccentricity and in almost coplanar systems, though the mean plane of one planet's system of satellites may be very different from that in another's. All but a few of the satellites revolve about their planets in the direction in which the planets move about the Sun.

In general, the planets and satellites rotate about axes fixed within them, the direction of rotation for most being in the same direction in which the planets revolve about the Sun. The periods of rotation of some satellites, including the Moon, are equal to their periods of revolution about their primaries so that they keep the same face turned towards the planet they belong to.

Surface conditions of planets and satellites vary greatly from body to body as does the structure and composition of these bodies. All these factors depend upon the past history of the individual body, its mass and radius, its distance from the Sun, and so forth. The ways by which the physical properties are determined are discussed below.

Further data about the Solar System as an entity will be given at the end of Chapter 6.

1.2 The Basic Measurements and Their Interpretation

1.2.1 Introduction All the information obtained about the bodies of the Solar System has been collected by astronomers at the telescope or by instruments carried by balloons, artificial satellites or space probes. The observations in essence are all those which can possibly be made, involving the recording of the position of the source, its brightness and the polarization of its light over the spectral ranges available for measurement, all as a function of time. In most cases, the bodies of the Solar System are large enough in angular size to enable the light from selected areas to be studied separately. Variations in brightness over the face of the body in some cases reveal structural details which may be studied by visual use of a telescope or recorded by photography and mapped.

The general analysis of the light which the planets send us allows us in fact to build up pictures of their physical conditions and their apparent chemical composition. Interpretation of the basic measurements gives us a means of knowing a planet's size, mass, temperature, etc.

From visual and photographic observations with the aid of telescopes, the planets reveal themselves as being globes with surface features, some of which may be only transient, lasting a few hours to a few months. In the cases of the Earth and Mars, some surface features vary according to the orbital position and are thus said to be seasonal. Most of the planets have some kind of gaseous atmosphere; some atmospheres are thin, allowing observation of the permanent solid surface; others are optically thick, any features that are seen corresponding to structure within the atmosphere. Thus, we can see immediately that all the planets are different and that they all deserve individual detailed study.

1.2.2 Planetary Orbits and Distances By careful observations of the positions of the planets, it has been possible to determine the elements of their orbits very accurately. A preliminary orbit can be obtained for any body from three positional measurements. Six numbers—the orbital elements—have to be found in order to define not only the size, shape and orientation of the orbit but also the planet's position within it at any time. Measurements of both the planet's right ascension and declination, taken at three different times, are therefore sufficient in principle to enable the elements to be found. This branch of astronomy, called **orbit determination and improvement**, has been used by space scientists in recent years in guiding their spacecraft to Mercury, Venus, Mars, Jupiter and Saturn.

One of the orbital elements is of course the **mean distance** of the planet from the Sun. In the first instance this distance is determined in relation to the Astronomical Unit and then converted to units more common to scientific measure on the Earth, such as the kilometre. The most recent and accurate determination of the Astronomical Unit has been obtained directly from radar experiments with Venus and confirmed by tracking artificial Martian satellites.

1.2.3 Physical Sizes Determinations of the angular sizes of many of the bodies of the Solar System can be made directly on the disks that are presented. This may be done using a telescope and a micrometer eyepiece or it may be performed by taking a photograph on a large plate scale and subsequently making measurements by microscope in the laboratory. In the special case of Jupiter's larger satellites where a measurable disk is not normally seen in the telescope, measurements of their angular sizes were made by Michelson using his interferometric technique. Such measurements have of course been outdated by the successes of the Voyager space missions to the outer planets.

A series of measurements, taken over periods of a few days to a few months, reveals that the apparent sizes of the bodies vary, this being a reflection of the fact that the Earth–planetary distances are constantly changing. From any one apparent angular size, α, knowledge of the object's distance, Δ, allows its physical size (diameter $= d$) to be determined. Thus

$$d = \alpha \Delta. \tag{1.1}$$

It will be appreciated that any instrument is capable of measuring only to a certain degree of accuracy. The uncertainty in any determination, as a fraction of the quantity being measured, normally depends on the value of that quantity. In the case of measuring the diameters of planets, it is obvious that the accuracy of the values obtained depends on the angular sizes of the disks that are presented. The diameter of Jupiter is thus known very accurately while that of Pluto is very uncertain.

Indeed in some cases where the accuracy is good, departures from sphericity may also be measured. In the case of Jupiter, for example, even a cursory visual inspection using a small telescope reveals that Jupiter's disk appears to be oblate.

The flattening of a planetary body is obtained from a comparison of its polar diameter, d_P, with its equatorial diameter, d_E. The degree of flattening, f, is then defined by the relationship:

$$f = \left(1 - \frac{d_P}{d_E}\right). \tag{1.2}$$

1.2.4 Mass Determinations Measurements of the mass of any of the Solar System's bodies can best be obtained if the body happens to possess its own satellite. The method of determining the mass of a planet by this method in relation to the mass of the Sun is given in detail in Roy & Clarke, *Astronomy: Principles and Practice*, Section 12.6.7.

For bodies without satellites, determinations of their mass is effected by considering the interplay between themselves and their nearest neighbours within the Solar System. Generally the mutual perturbations are small and a long series of positional measurements is required before satisfactory answers can be obtained from the theories of these planets' motions. The accuracy of a mass determination from a study of the effects of perturbations is usually inferior to a determination by means of the satellite method.

With the advent of space techniques, it is now possible to provide artificial satellites for bodies without natural satellites, so allowing a mass determination to be made. This has already been done in the case of the Moon. Not only has an accurate mass determination been made possible, but also detailed measurement of the artificial satellite's orbit has allowed the distribution of the mass beneath the lunar surface to be investigated. Similar investigations have been made for Mars and Venus with the insertion of interplanetary probes into orbit about these planets.

Measurements of the deflections in the trajectories in the fly-pasts of the Mariner space probes have allowed an accurate mass to be assigned to Mercury.

1.2.5 Density Determinations From knowledge of the mass, M, of any body and its physical size, given by diameter d, a value for its mean density, $\bar{\rho}$, is immediately available. Thus

$$\bar{\rho} = 6M/\pi d^3. \tag{1.3}$$

It will be seen from Table 6.6, listing the physical parameters of the planets, that their mean values of density vary greatly.

Measurements of the densities of planetary atmospheres have been obtained by ground-based observers by two means. Sunlight which is incident on any particular planetary atmosphere is scattered and, from measurements of the brightness and polarization of this scattered light, values of gas densities may be obtained. Any planetary atmosphere also absorbs radiation, in some cases preventing visibility of the planet's surface or, in others, reducing the clarity of the ground surface detail that can be seen. From observations of the apparent brightness variations in the light from a satellite at the time of an eclipse by a planet or of the changes in brightness of a star as it is occulted by a planet, the density and structure of the planet's atmosphere may be inferred.

More direct measurements of planetary atmospheres can be obtained from space probes. The Mariner 6 and 7 vehicles of 1969 each carried short-wave radio transmitters. Monitoring of their signals around the time of the vehicles' occultation by Mars showed that the radio waves were refracted by the Martian atmosphere, thus yielding data on the density and pressure of the atmosphere. More accurate measurements have been made over longer periods by the instruments in Mariner 9 while in orbit about Mars. If an instrument package is inserted into a planetary atmosphere, as has been done with those of Venus and Mars, even more direct measurements are of course possible.

In particular the Viking landers on Mars have enabled regular meteorological reports to be received from the Martian surface.

1.2.6 Surface Gravity To a first order, the value of the acceleration due to gravity on the surface of a planet is given by the planet's mass and its radius. If m is the mass of a body and it is placed on a planet with surface gravity, g_P, then the force acting on it is mg_P. This force arises from the mutual gravitation between the body and the planet. If M_P is the mass of the planet and R_P its radius then the gravitational force is given by GM_Pm/R_P^2, where G is the universal constant of gravitation. Thus

$$mg_P = GM_Pm/R_P^2$$

and therefore

$$g_P = GM_P/R_P^2. \tag{1.4}$$

In the case of the Earth this may be written as

$$g_\oplus = GM_\oplus/R_\oplus^2$$

where the subscript $\oplus$ refers to the values of g, M and R for the Earth. Values of g_P may be related to the Earth's surface gravity by letting $g_\oplus$ equal unity. By doing this, the term G may be expressed as

$$G = R_\oplus^2/M_\oplus.$$

Substitution of this into Equation (1.4) gives

$$g_P = \frac{M_P}{M_\oplus}\frac{R_\oplus^2}{R_P^2}. \tag{1.5}$$

Thus from measurements of a planet's mass and its size, both in relation to the Earth, its surface gravity may be calculated by Equation (1.5), in units of the Earth's surface gravity.

It is well known that gravity over the Earth's surface varies as a result of small changes in the Earth's radius with latitude. Similar gravity changes occur over the surfaces of the planets, being more pronounced according to the degree of flattening.

If the planet has a short rotation period, the value of the surface gravity is reduced on account of the action of centrifugal force. The contribution of the centrifugal force varies with the latitude of the point considered on the surface, being a maximum at the equator and zero at the poles. Taking this force into account, Equation (1.4) may be rewritten as

$$g_P = \frac{GM_P}{R_P^2} - \frac{V^2}{R_P \cos \phi} \tag{1.6}$$

where V is the velocity of the planet's rotation at the surface at latitude ϕ where the surface gravity is being considered.

Now the velocity of the planet's surface at latitude ϕ is given by

$$V = 2\pi R_P \cos \phi / T$$

where T is the period of rotation. Substituting this in Equation (1.6) gives

$$g_P = \frac{GM_P}{R_P^2} - \frac{4\pi^2 R_P \cos \phi}{T^2}. \tag{1.7}$$

In the case of Jupiter, because of the pronounced flattening and short period of rotation, the surface gravity decreases by some 15% in going from the poles to the equator. For the Earth the change is much smaller, being of the order of 0·5%.

1.2.7 The Velocity of Escape The ability of a planet to retain an atmosphere is a function of the planet's mass and radius and its proximity to the Sun. The energy received from the Sun imparts kinetic energy to the molecules in the planet's atmosphere. As seen in Section 1.2.6, the planet's mass and radius dictate the acceleration due to gravity at the planet's surface; on them also depends a quantity known as the velocity of escape.

In Roy & Clarke, *op. cit.*, Section 12.6.4, the velocity, V, of a planet in its elliptical orbit about the Sun is given by Equation (12.29), namely

$$V^2 = \mu\left(\frac{2}{r} - \frac{1}{a}\right),$$

where r was the planet's heliocentric distance from the Sun. The quantity a was the semi-major axis of the orbit; μ was the product of G and $(M+m)$, where G is the universal constant of gravitation, and M and m were the masses of Sun and planet respectively.

Let M now be the planet's mass, while m is the mass of a molecule or atom in the planet's atmosphere. Then, since $M \gg m$

$$V^2 = GM\left(\frac{2}{r} - \frac{1}{a}\right),$$

with r being the distance of the molecule or atom from the planet's centre and a the semi-major axis of the molecule's orbit. If the molecule had a velocity just sufficient to put it into an orbit which enabled it to leave the planet and reach infinity, the orbit would be the limiting case of an ellipse with one focus at infinity. Then the semi-major axis a would tend to infinity so that we could write

$$V^2 = 2GM/r = 2\mu/r. \tag{1.8}$$

The velocity V calculated in this way is the so-called **velocity of escape**. Now circular velocity V_c at the same distance r is given by

$$V_c^2 = \mu/r.$$

Hence

$$V = \sqrt{2}\, V_c,$$

a useful relationship.

If we put r equal to the planet's radius, Equation (1.8) enables the velocity of escape for the planet to be found. Then if the Sun heats the planet's atmosphere to such an extent that the molecules have velocities of that order, they will escape from the planet into interplanetary space, resulting in the loss of the atmosphere by that planet. The kinetic theory of gases shows that the average velocity of molecules in a gas is directly proportional to the absolute temperature and inversely proportional to its mean molecular weight. Those gases with less massive molecules and atoms therefore have velocities much higher than those made of more massive particles and so tend to escape before the others.

Collisions among the molecules are always taking place so that some increase their speeds while others lose momentum. Table 1.1 shows the time it takes for a planet to lose *half* its atmosphere if the molecules of the gas forming that atmosphere have mean velocities x times the planet's velocity of escape.

Table 1.1

x	T
1/3	6 weeks
1/4	10^3 years
1/5	10^8 years

Knowledge of a planet's velocity of escape and its surface temperature is therefore important in deducing properties of its atmosphere, or whether in fact it can possess one at all.

1.2.8 Brightness Observations It is evident from simple consideration of the brightness changes of planets as they execute their orbits that the light we receive from them originates in the Sun and is reflected by them. Any light which a planet may generate itself is insignificantly small. The variations in brightness can be described by considering the planet's distance, r, from the Sun and its distance, Δ, from the Earth. Let us first consider the outer planets which, to all intents and purposes, present themselves to us as a whole disk.

The amount of light, I, received by a planet is inversely proportional to the square of its distance from the Sun. We may therefore write

$$I = K/r^2, \tag{1.9}$$

where K is a value related to the brightness of the Sun. A fraction, A, of this light is reflected by the planet towards the Earth and the amount of light, B, received at the Earth is given by

$$B = I(A/\Delta^2)$$

and hence,

$$B = KA/r^2\Delta^2.$$

The variations in the magnitude of a planet may be expressed by inserting B into Pogson's equation, relating brightness to magnitude, which gives

$$m = k - 2{\cdot}5 \log_{10} (KA/r^2\Delta^2)$$
$$= k' + 2{\cdot}5 \log_{10} r^2\Delta^2$$

where the constant k' incorporates k, A and K. Hence

$$m = k' + 5 \log_{10} r\Delta. \tag{1.10}$$

The term A is a property of the planet's apparent surface and designates its reflecting ability. It is related to the **albedo** of the planet, which is strictly defined as the ratio of the total light reflected from a sphere to the total light incident on it. The value of A in practice varies with the phase angle of the planet and in a different way for each planet (see Section 1.2.9). As a consequence of this, Equation (1.10) is not strictly correct and additional terms are required to express planetary magnitude variations.

For the cases of bodies lying outside the Earth's orbit, the values of the phase angle are limited to a small range and good agreement to observation can be obtained by using an expression of the form

$$m = k' + 5 \log_{10} r\Delta + C\phi, \tag{1.11}$$

where C is a constant for the particular planet or asteroid and ϕ is the phase angle. In the special case of Saturn, terms are also required to take account of the contribution to the total light by the ring system whose aspect to the Earth varies according to the positions of both the Earth and Saturn in their orbits.

For the inferior planets, the phase angle may take any value between 0° and 180° and account must also be taken of the fact that we do not see all of the illuminated face. This may be done by adding extra terms involving second and third powers of ϕ to Equation (1.11).

1.2.9 Planetary Spectra Further evidence that the main part of the light from the planets is reflected sunlight comes from observations of planetary spectra. These spectra resemble the solar spectrum to a good degree with the exception that extra absorption features are sometimes present, particularly in the infra-red. These extra features have led to the identification of some of the gases present in the planetary atmospheres.

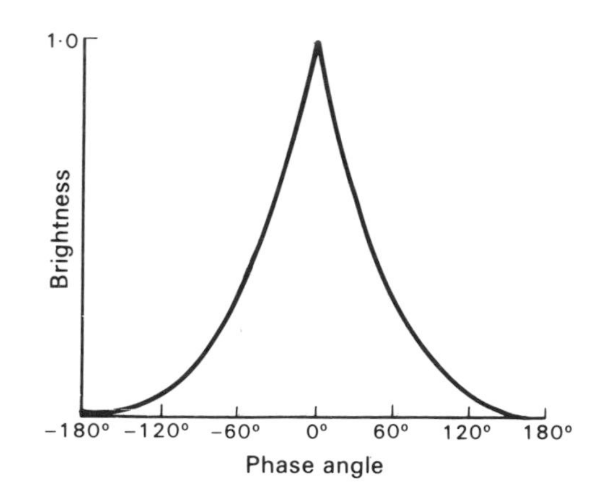

Fig. 1.2 Brightness–phase angle curve for the whole lunar disk

In the case of comets, emission bands appear in their spectra. Very small emission features in lunar spectra have been detected and these have been ascribed to the lunar material luminescing under the action of short-wave solar radiation or particle bombardment; emission lines have also been detected in spectra of Venus.

1.2.10 Planetary Scattering Properties We have already seen that the brightness of a planet changes according to the positions of the planet and the Earth in their orbits and that some account needs to be taken of the way the particular planet scatters the incident sunlight. By observing how the brightness of a planet varies and by removing not only the contributions to this caused by Sun–planet, planet–Earth distance changes but also the effects caused by the observer seeing only part of the fully illuminated face, it is possible to deduce the scattering laws. These laws describe the behaviour of a surface in the way it scatters light according to the phase angle. As an example of the curve relating the strength of scattered light with phase angle, that of the Moon is illustrated in Fig. 1.2. For some of the planets it is possible to isolate different areas of their surfaces, allowing studies to be made of regional differences in surface conditions.

Similar curves may be obtained by measuring how the polarization of the planet's light varies with phase angle. For polarization measurements, where one component in the received light is measured as a fraction of the whole, it is not necessary to make corrections for the distance of the planet or the aspect presented to the observer; the polarization measurements may be plotted directly as the phase angle changes. The polarization–phase angle curve for the light from the whole Moon is illustrated in Fig. 1.3 to serve as an example of the kind of curve that is obtained. Polarization–phase angle curves are usually more sensitive to the scattering laws than are those of brightness–phase angle. By comparison with laboratory samples and theoretical models, both types of curve supplement each other in helping to deduce the nature of any planetary surface or the particles within a planetary atmosphere.

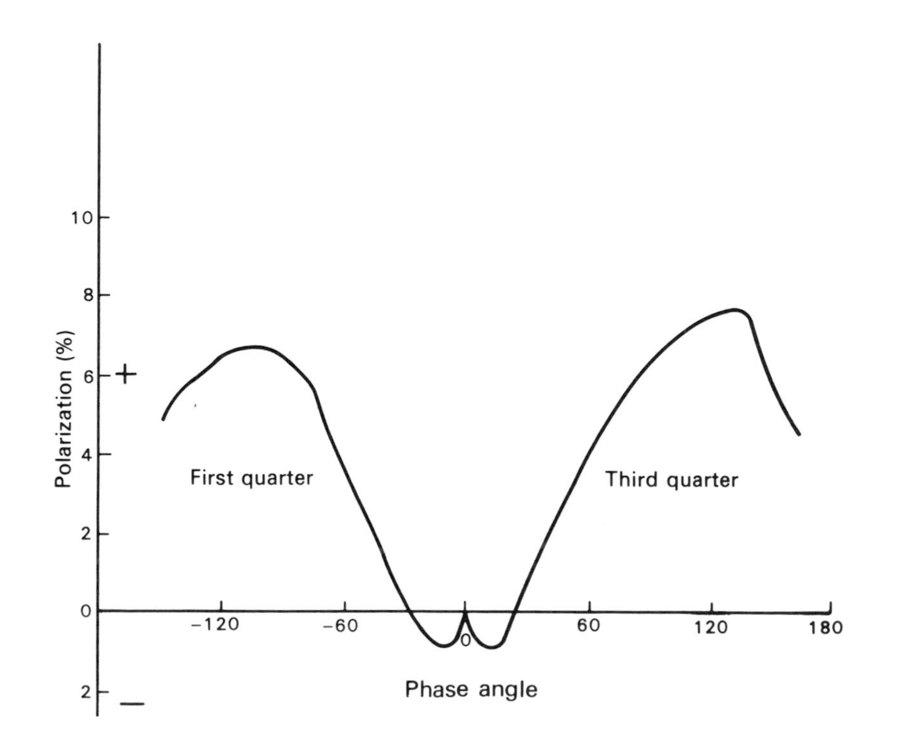

Fig. 1.3 The polarization–phase angle curve for the light from the whole lunar disk

1.2.11 Rotation Periods Some observed planetary features are permanent or semi-permanent and by timing their appearance, it is a simple matter to obtain values for the rotation period.

For bodies whose angular sizes do not permit inspection of surface detail, particularly asteroids, rotation periods have been obtained by observing regular brightness and polarization variations. It may also be mentioned here that other long-term variations in the brightnesses of planets are suspected, perhaps reflecting the level of solar activity. In effect this would mean that the term K in Equation (1.9) is variable.

In the special cases of Mercury and Venus, where apparent surface details are too faint to allow reliable positional measurements to be made of them, rotational periods have been obtained by using radar and measuring the Doppler shifts that the rotation imparts.

1.2.12 Temperature As the planets do not in any significant way generate heat of their own with the noteworthy exception of Jupiter, their temperatures are dictated by the amount of energy which they can trap from the Sun. This quantity varies inversely as the square of the planet's distance from the Sun. The inferior planets are obviously hotter than the Earth; the superior planets are obviously cooler.

An approximate value of the temperature of any planet may be obtained by considering its distance from the Sun, in relation to that of the Earth's distance, and applying Stefan's law.

If $T_\oplus$ and T_P are the temperatures of the Earth and any planet respectively, and $E_\oplus$ and E_P the total amount of energy that each body radiates per unit area, then according to Stefan's law we may write

$$E_\oplus = \sigma T_\oplus^4$$

and

$$E_P = \sigma T_P^4.$$

Dividing the first of these equations by the second we have

$$E_\oplus / E_P = T_\oplus^4 / T_P^4,$$

therefore

$$T_P^4 = \frac{E_P}{E_\oplus} T_\oplus^4. \tag{1.12}$$

Let us assume that the total amount of energy radiated per unit area is equal to the total amount of energy received per unit area from the Sun. Now the energy received by a planet is given by an expression similar to Equation (1.9), i.e. $E = K/r^2$. By using astronomical units to express the planetary distances, the energy received by the Earth is given by

$$E_\oplus = K/1^2 = K.$$

Therefore we may write the energy received by any planet as

$$E_P = E_\oplus / r^2.$$

Substituting this in Equation (1.12) we have

$$T_P^4 = \frac{E_\oplus}{r^2 E_\oplus} T_\oplus^4$$

and therefore

$$T_P = T_\oplus / (r^{1/2}). \tag{1.13}$$

As an example, consider the temperature of Saturn which, according to Table 6.4 is at a distance of 9·54 AU. By assuming the Earth's temperature, T_E, to be 300 K, Equation (1.13) gives a temperature for Saturn of

$$T_{♄} = 300/(9{\cdot}54^{1/2}) = 300/3{\cdot}09 = 97\ \mathrm{K}.$$

If each planet is considered in turn, we find that the range of temperatures runs from tens of degrees to hundreds of degrees. According to Planck's law, this suggests that the thermal radiation generated by any planet puts the peak of the energy distribution into the infra-red or beyond.

Actual temperatures are best measured by observing the spectral energy distribution in the infra-red or short-wave radio region of the spectrum. The values obtained from these techniques agree fairly well with those that are predicted by the simple method above.

1.2.13 Spacecraft Experiments In addition to the information above obtained by Earth-bound studies, knowledge of the Solar System bodies has been extended at an unprecedented rate since the advent of spacecraft. Such vehicles have carried various kinds of instrumentation including cameras, photometers, polarimeters, spectrometers, magnetometers, etc., the information being telemetered to Earth and received as radio signals. Much of the early work on the Earth environment was performed by rocket probes and satellites. Major results from these studies include the discovery of the van Allen belts, the nature of the Earth's magnetosphere and the detailed figure of our planet.

One achievement of great significance was the return of rock samples from the Apollo missions to the Moon. Analysis in the laboratory allowed study of the chemical abundances and the morphology of lunar materials. The results have given direct information on the anomalous abundances of elements such as titanium, on the lunar age and on the molten states of the ground materials in aeons past.

Planetary rendezvous have allowed the mapping of surface features with high spatial resolution. Not only has this been done in two dimensions using cameras, but it has also been extended to three dimensions. Satellites in orbit about the planets have determined the contours by on-board radar. In the case of Mars, the depths of its canyons have been measured using spectrometers to determine the number of atoms and hence the local atmospheric pressure between the orbiter and the sub-ground point. All of these studies have opened up new branches of study under the name of astrogeology.

Space probes that have landed on Venus and Mars have been able to transmit direct information about the planetary environments. Indeed, the equivalent of daily weather reports have been received from the Viking landing site on Mars. Planetary meteorology has been added to the studies of the Solar System.

Important too have been the discoveries and extension of our knowledge related to planetary satellites and the ring systems. The photographic images which revealed that Io is currently volcanically active were a remarkable revelation. And the story is not yet over. Future probes into the depths of the Solar System will continue to present many unexpected and unthought of surprises.

2 The Sun

2.1 The Sun's Appearance

The Sun is a gaseous sphere of diameter over one hundred times that of the Earth. Its volume is therefore some one and a third million times the Earth's although its mass is only one third of a million times the terrestrial mass. Dividing mass by volume, its mean density is found to be one quarter the Earth's mean density, or 1·4 times the density of water.

The surface temperature is about 6000 °C. The temperature increases with depth and theories of stellar structure estimate the Sun's central temperature to be some 13 000 000 °C at which temperature nuclear reactions take place, providing the source of its energy. It may be noted that each second the Sun emits 4×10^{26} joules. A discussion of the internal constitution of the Sun will be left to a later section when theories of stellar structure, stellar energy and stellar evolution are dealt with.

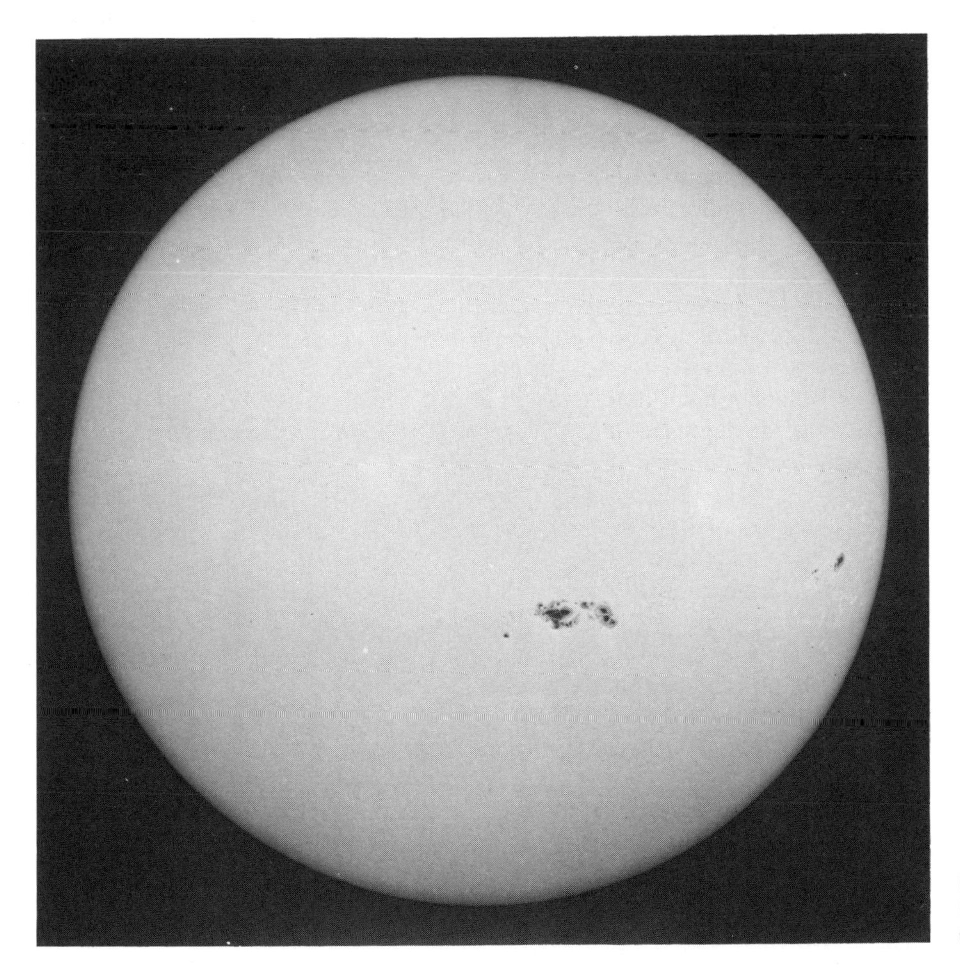

Fig. 2.1 A photoheliogram showing a large sunspot and the presence of faculae towards the limb

The **photosphere** is the name given to the sharp boundary from which comes the light producing the continuous part of the solar spectrum. Very bright localized areas, the **faculae**, are often present and **sunspots** may also be visible.

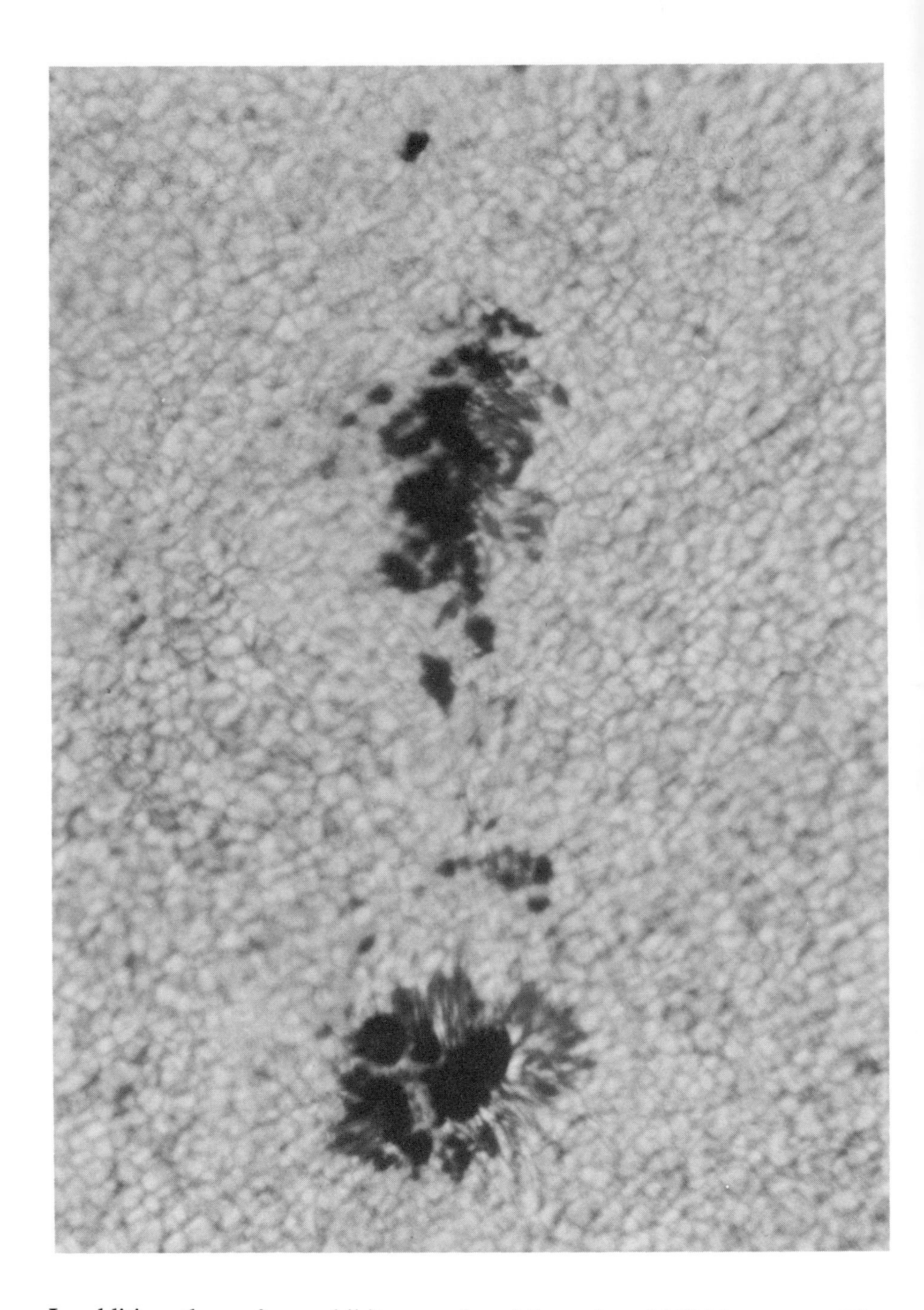

Fig. 2.2 Solar granulation and a small, growing group of spots and pores. (By courtesy of the Sacramento Peak Observatory, Air Force Cambridge Research Laboratories)

In addition, the surface exhibits **granules** of the order of 1600 km across. The constant changing of the granular pattern reveals a general state of turbulence on the solar surface. The granules are interpreted as being flow tubes, the hot material flowing outwards and the cooler material flowing inwards. The photosphere appears sharply bounded but in fact it is a layer some 400 km thick within which the opacity increases rapidly with depth.

2.2 The Sun's Atmosphere

Above the photosphere lies the **reversing layer**, about 300 km thick, in which has been detected more than 60 of the elements found on Earth, such as iron and calcium. This layer, at a lower temperature than the photosphere, produces most of the absorption lines in the solar spectrum.

A region of tenuous gas, called the **chromosphere**, lies above the reversing layer. It is some thousands of kilometres thick and consists chiefly of hydrogen, helium and calcium. It merges into the **solar corona** which is visible during total eclipses of the Sun. At such times the corona is seen to stretch as far as 8 000 000 km out from the Sun. The temperature of this very tenuous region is estimated to be 1 000 000 °C.

Other phenomena associated with the Sun's atmosphere (made up of the reversing layer and the chromosphere) are **prominences** and **flares**. Prominences appear in some cases like giant flames, taking all kinds of shapes and attaining heights of hundreds of thousands of kilometres. Others originate high in the chromosphere and are seen to stream downwards rather than upwards. The behaviour of the gas forming a prominence is controlled by a complicated interplay of magnetic fields, the solar gravitational field and solar radiation. The phenomena associated with prominences may be exhibited dramatically using time-lapse photography (see Figs. 2.3(*a*) to (*e*)).

Solar flares are associated with sunspot groups. If photographs of the Sun are taken in light of a monochromatic wavelength, the distribution of particular elements in the Sun's upper photosphere may be studied. Formerly, this was achieved by spreading the light out by spectrometer into a spectrum and using a slit to allow only a narrow region of it to fall on a photographic plate. If the solar image is then allowed to trail across the first slit of the spectrograph while the photographic plate is moved at the same speed across the second slit, an image of the Sun is built up in the light of that wavelength band. The same result is achieved nowadays by using light filters that allow only a narrow wavelength band to pass through them. Such a photograph is called a **spectroheliogram**. The regions about sunspots often become very much brighter than the surrounding areas, especially in the light of the Hα line of hydrogen (see Fig. 2.4).

These flares have a duration ranging from minutes to hours. From them ultra-violet radiation and electrically-charged particles are ejected, crossing space to affect markedly the Earth's upper atmosphere and magnetic field.

During a total solar eclipse, when the photosphere is hidden, the chromosphere gives a bright-line or emission spectrum as it should if it consists of hot, tenuous gas. Study of the patterns of lines and their strengths reveals the presence of hydrogen, helium and calcium and their relative abundances.

Recently the question of the possibility of there being various kinds of oscillation or pulsation in the Sun has been raised, opening up the field of solar seismology. Several sensitive techniques based on high resolution spectrometry and disk diameter measurements have been developed to explore the possible modes of oscillation. The most readily observed phenomenon is the five-minute oscillation which is detectable by Doppler spectroheliograms or their equivalent, this being related to the upper photosphere. Longer periods ($\sim$160 minute) oscillations have also been reported and a study of these raises the exciting prospect of performing solar seismology, probing the convection zone and the solar interior.

It is seen, therefore, that the Sun's surface and atmosphere always show varying degrees of activity. Variations averaged out over the whole disk would be small and if the Sun were to be viewed at interstellar distances, it would be difficult to detect its variability, but none the less it could be done by monitoring the overall brightness or the subtle changes in the cores of some of the spectral lines.

2.3 Sunspots

Projection of the Sun's image on a white surface usually shows the presence of dark spots on the Sun's surface. Some spots may be 100 000 km in diameter

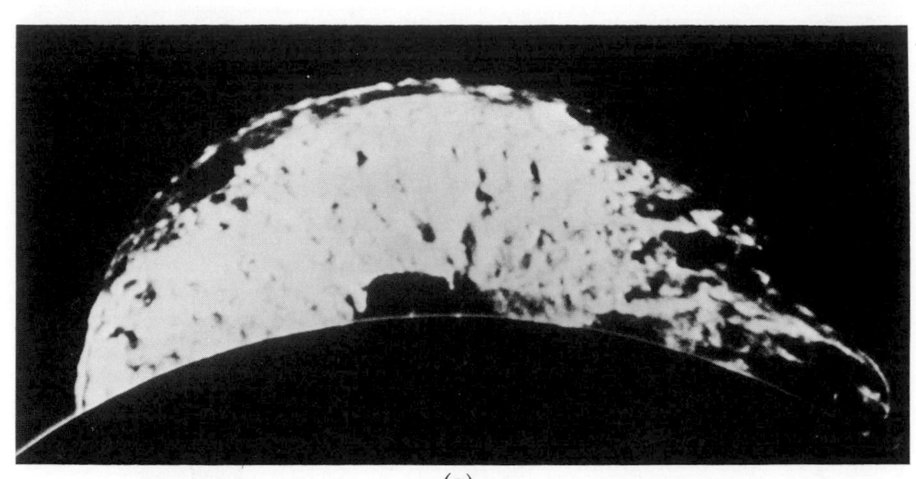

(a)

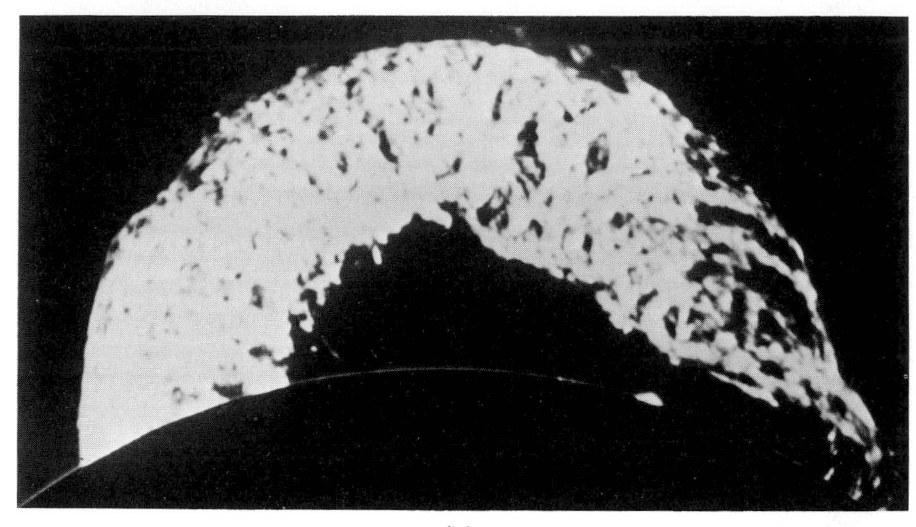

(b)

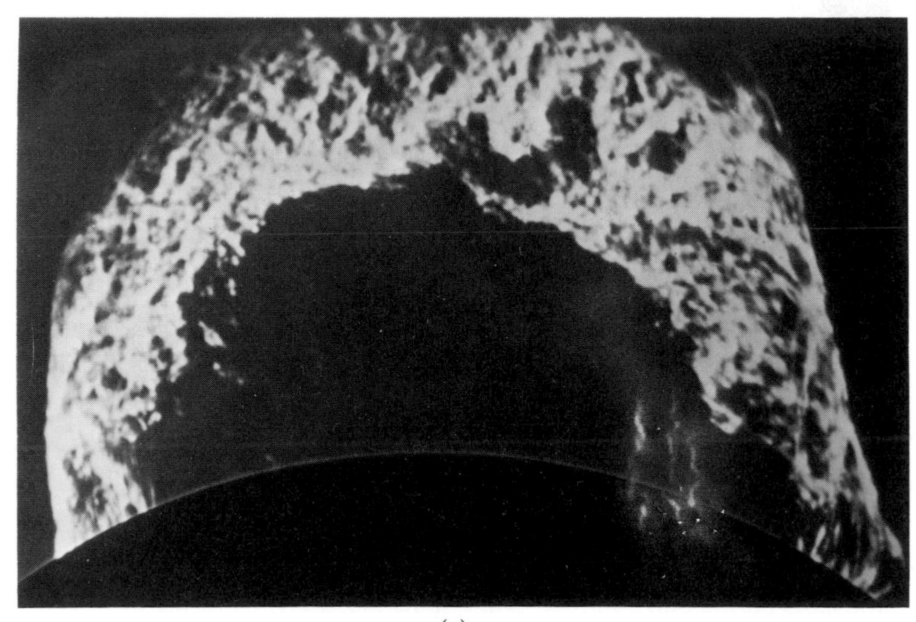

(c)

Fig. 2.3 The development of a solar prominence on June 4th, 1946. (*a*) $16^h\ 3^m$; (*b*) $16^h\ 36^m$; (*c*) $16^h\ 51^m$; (*d*) $17^h\ 3^m$; (*e*) $17^h\ 23^m$. (By courtesy of the Royal Astronomical Society)

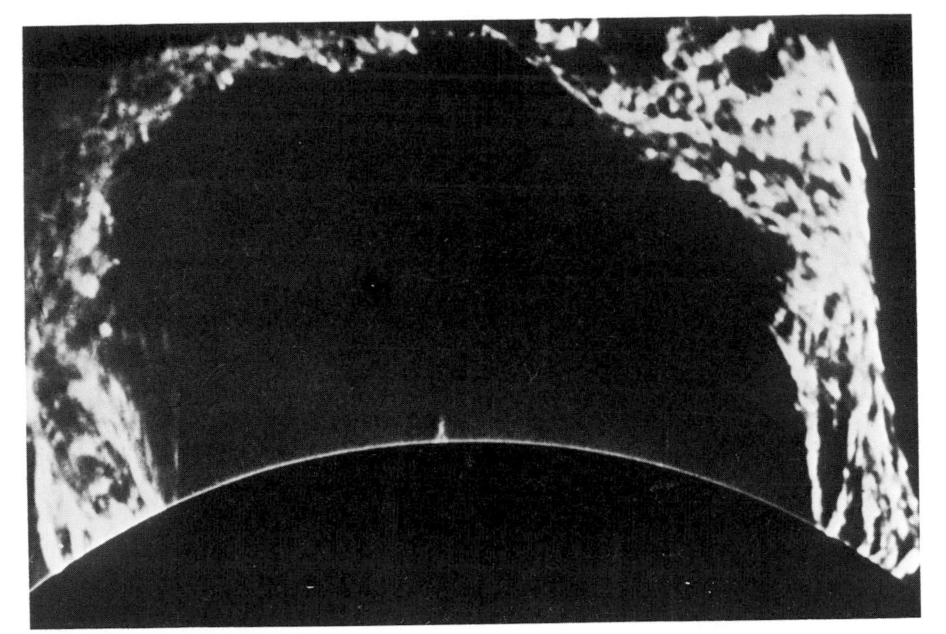

(d)

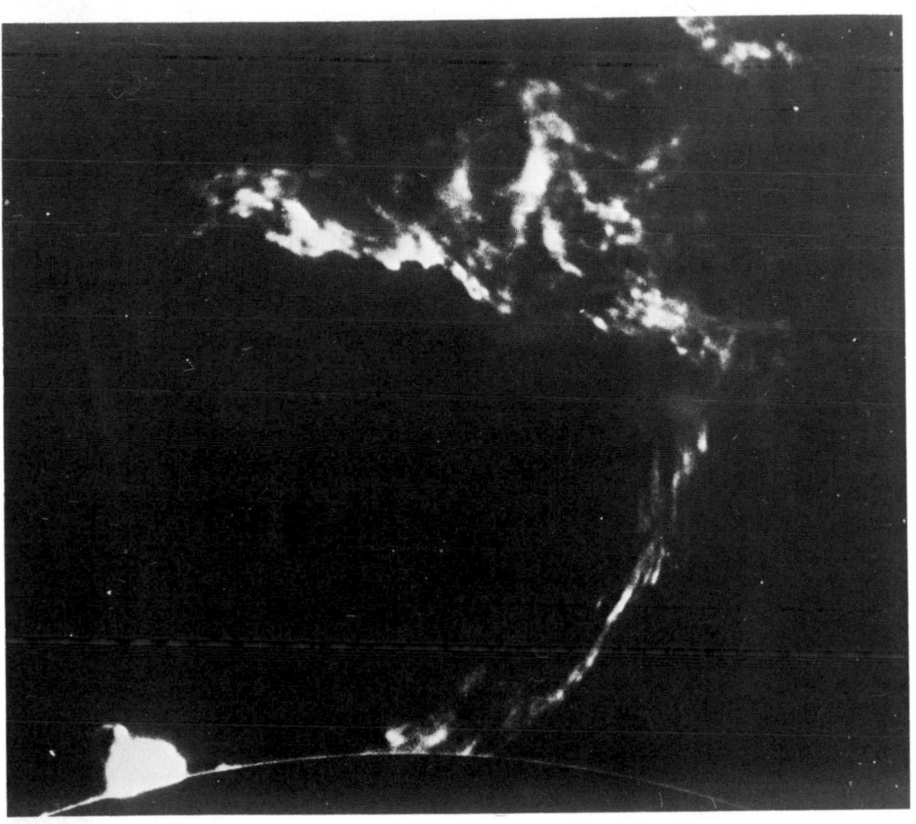

(e)

Fig. 2.4 An Hα spectroheliogram of a flare near its maximum

though most are much smaller than this. These spots change with time, their progression across the solar disk revealing the Sun's rotation. The rotational period of the Sun is a function of latitude, varying from 25 to 34 days, the period increasing as higher solar latitudes are reached. When a spot approaches the solar limb, it appears as a cavity on the solar surface. This phenomenon is known as the **Wilson effect**, named after the first professor of astronomy (1760) at the University of Glasgow.

Study of the paths traced out by spots at various times of the year shows that the Sun's equator is inclined at about 7° to the ecliptic. Confirmation of the Sun's rotation is obtained by the fact that whereas the lines in the solar spectrum of light from one edge of the Sun are displaced by Doppler shift towards the red, spectral lines from light emitted from the opposite edge or limb are displaced towards the blue.

Sunspots are dark only by comparison with the surrounding photosphere; temperatures, even in the dark central **umbra**, are of the order of 4500 °C. Surrounding the umbra is the **penumbra**, a region almost as bright as the photosphere, made up of twisted bright filaments aligned roughly radially to the spot centre. Spots usually occur in groups. The number in a particular group will vary with time. Generally speaking, a group develops into two parts. The leader–in the sense of the direction of the solar rotation–becomes dominated by one large spot; the follower likewise possesses a main spot with a number of minor ones.

Associated with a sunspot is a magnetic field. Its strength depends upon the size of the spot and ranges from about 100 gauss to several thousands of gauss. The Zeeman effect enables measurements to be made of the directions taken by the lines of force in various parts of the spot. In the centre of the umbra the lines are vertical; at the edge they are almost horizontal. There also exists an outward flow of gas from the umbra, revealed by a Doppler shift in the spectral lines of the penumbra, observed when the spot lies near the edge of the Sun's disk. This is known as the **Evershed effect**.

It may also be mentioned here that measurement of Zeeman-split spectrum lines allows the magnetic field to be determined at any point over the Sun's surface. By making a rastor scan, a picture of the Sun's general field over its surface, or a **magnetogram**, can be obtained. The general magnetic field of the Sun is typically a few gauss. Its pattern is similar to that of a dipole field, the poles being within a few degrees of the rotational poles.

An interesting feature of sunspots is that if the spots in the leading part of a group show positive magnetic polarity, the followers have negative polarity. What is more remarkable is that all other groups in that hemisphere of the Sun, say the northern hemisphere, show the same distribution of magnetic polarity while all groups in the southern hemisphere exhibit a reverse polarity, the leaders being negative, the followers being positive. This situation is maintained throughout a so-called sunspot cycle for members of that cycle.

The **sunspot cycle** was discovered by Schwabe in 1843 after twenty years of systematic solar observation. The average number of sunspots visible on a particular day shows a cyclic variation with a period of about 11 years. There is a suggestion that an 80 year period may also exist.

As a new cycle begins, small spots in latitudes as high as 30° or more appear in both hemispheres. The spots become larger, increasing in number, the latitudes in which they appear gradually moving towards the equator. By

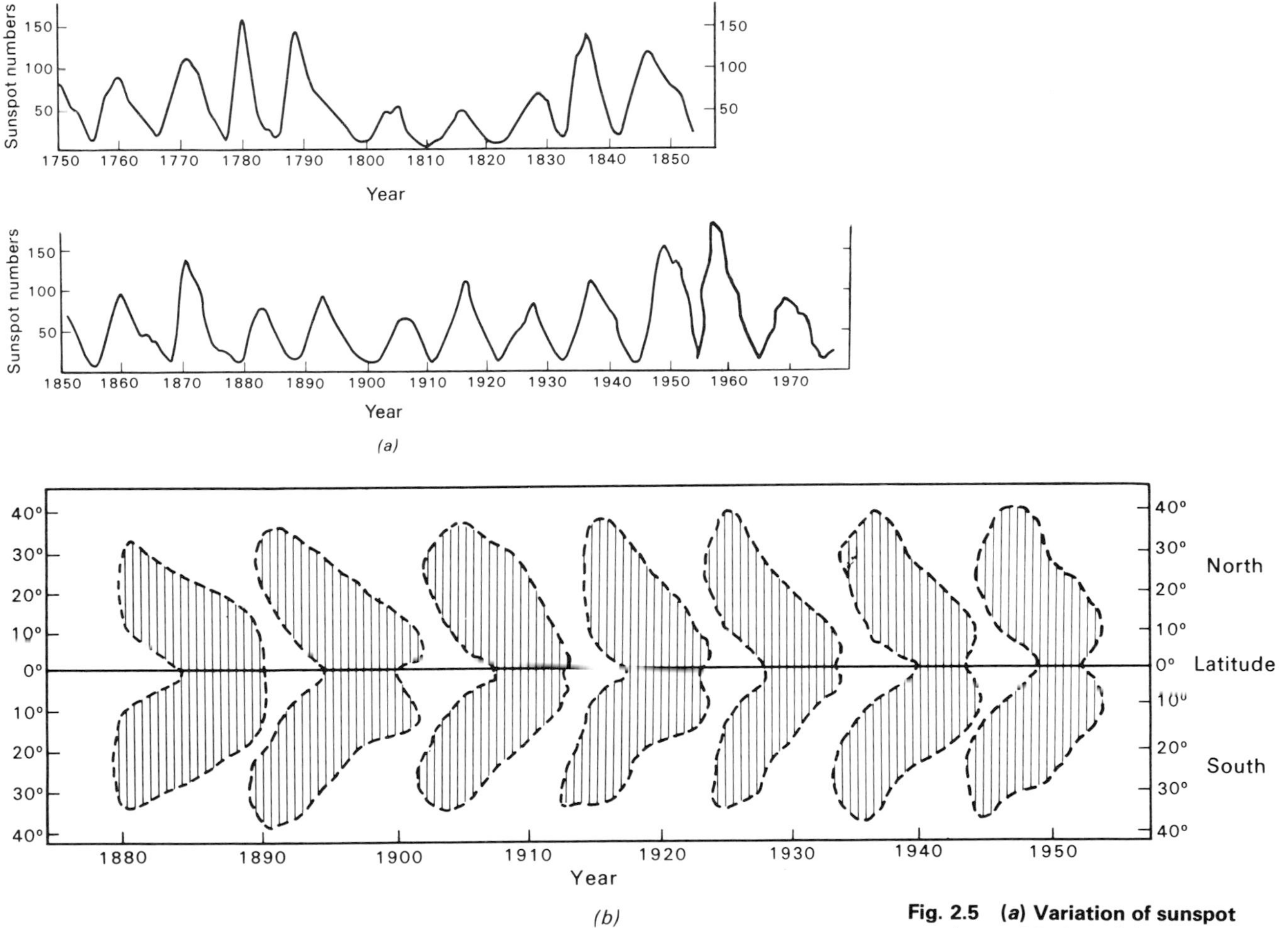

Fig. 2.5 ***(a)*** **Variation of sunspot numbers with time.** ***(b)*** **The butterfly diagram. The shaded areas show the regions in which sunspots appeared at the time specified between 1880 and 1950**

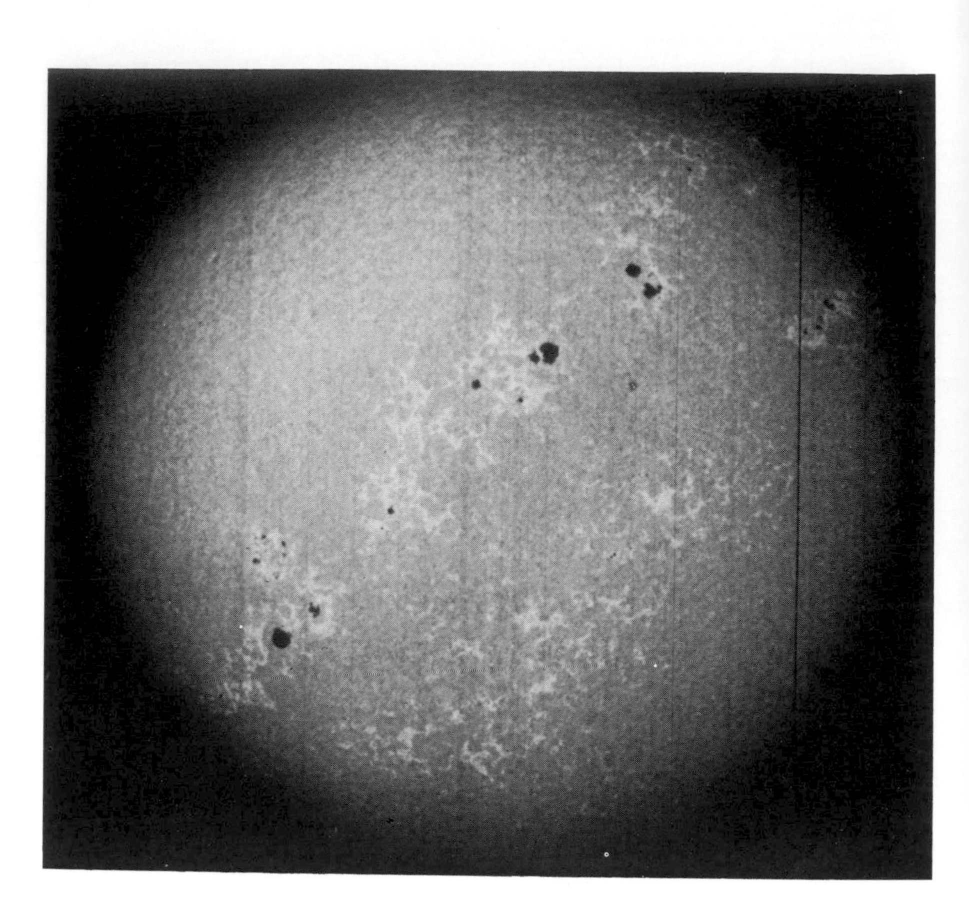

Fig. 2.6 Full disk spectroheliogram showing sunspots and faculae concentrated in two belts of activity in the northern and southern hemispheres

mid-cycle, that is at maximum activity, spots are most numerous around latitude 15°. Towards the end of a cycle, when very few spots of the old cycle are present, usually at about latitude 5°, the first spots of the next cycle appear around the higher latitude limits. In Fig. 2.5(*a*), a graph of the number of sunspots from 1755 to 1950 is shown, exhibiting the irregular nature of the cycle. Fig. 2.5(*b*), depicts the so-called "Butterfly Diagram", showing the variation of latitude in which sunspots appear with time. Fig. 2.6 illustrates the belts of activity in the two hemispheres at a particular phase of the cycle.

It may also be mentioned that there is a reversal of magnetic polarity with each new sunspot cycle. When the first sunspot groups of the new cycle appear about ±30°, those parts that had positive polarity in the old cycle now have negative polarity, while those formerly negative are now positive. If such magnetic phenomena are taken into account, the basic sunspot cycle is 22 rather than 11 years long.

Many theories attempting to explain sunspot phenomena have been put forward. A plausible theory proposed by Hale many years ago suggested that simple vortices in the hot gas sometimes sweep up to the surface and break, their ends forming a sunspot pair. The ends would rotate in opposite directions, like the ends of a whirlpool vortex formed on either side of an oar blade drawn through water. Because the gas is ionized and therefore electrically charged, the rotation would give rise to magnetic fields of opposite polarity. Apart from such attempts to explain the phenomena associated with sunspots, solar physicists have the problem of the Sun's general magnetic field to account for. There is a vast and ever-growing literature dealing with such phenomena, an understanding of the processes involved necessitating a mastery of the branch of physics called magnetohydrodynamics. It is not possible in this book to go further into these more recent developments except to sketch Babcock's model of the solar cycle, proposed in 1960 to account for many of the phenomena, including the 22-year solar cycle. In what follows we should remember that terms such as "lines of magnetic force", "flux tubes", "magnetic pressure" are essentially attempts to describe in understandable and pictorial everyday language concepts that are only adequately treated by the methods of magnetohydrodynamics.

At the beginning of a new 22-year solar cycle, the Sun's magnetic field resembles that of a bar magnet, the magnetic lines of force emerging from the surface around the south magnetic pole (around latitude 60° S) and re-entering near the north magnetic pole (around latitude 60° N). The continuations of the lines below the surface lie at a depth about one-tenth of the way to the centre (Fig. 2.7(*a*)). It has been noted already that the rotational period of the Sun varies from 25 to 34 days, the period increasing as higher solar latitudes are reached. This differential rotation distorts the lines of force within the Sun, each line being wound around the Sun many times until the situation *for a particular line of force* is as shown in Fig. 2.7(*b*). The number of lines of force crossing a given area will vary with latitude and the greater the number, the

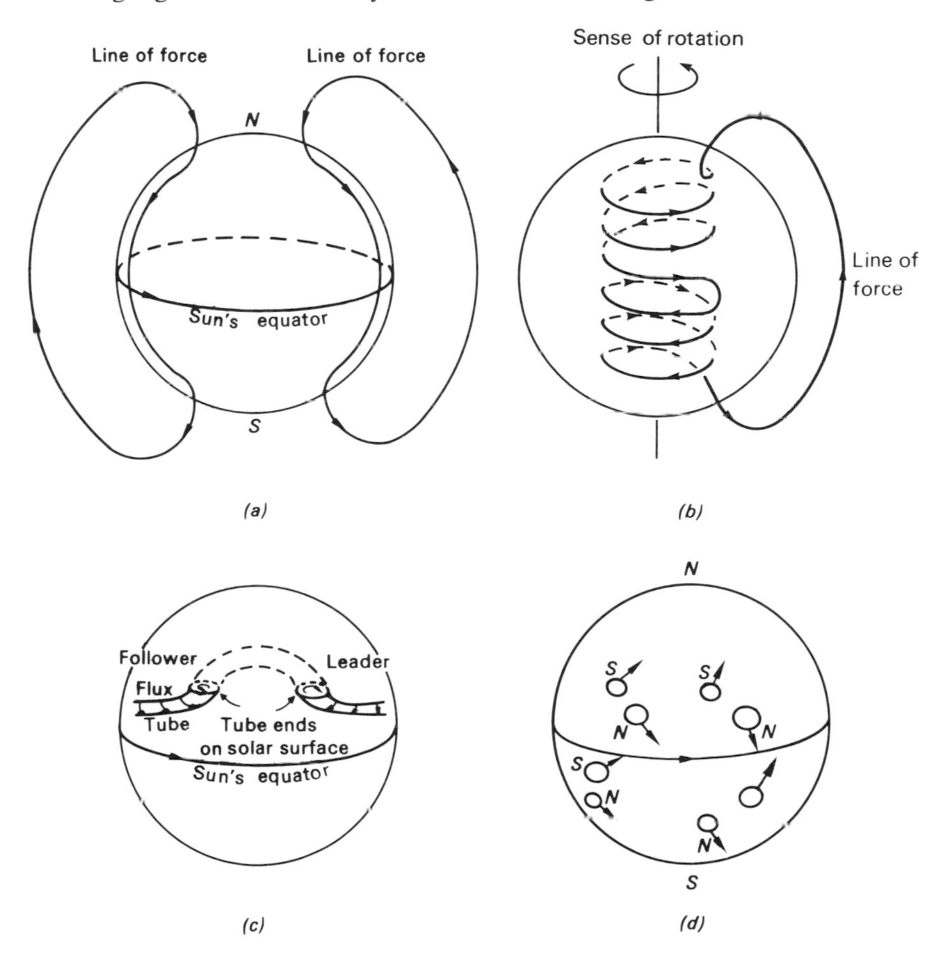

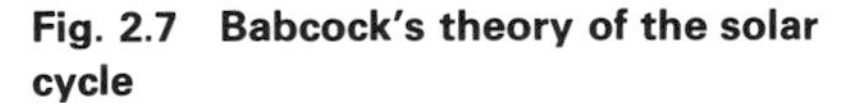
Fig. 2.7 Babcock's theory of the solar cycle

greater will become the magnetic pressure. Gas movement upwards by convection in the outer layers of the Sun will twist the lines of forces into magnetic "ropes". Such a "rope" or flux tube can be shown to experience magnetic forces tending to send a part of it up to the surface where it produces a sunspot pair of opposite polarities. It can also be shown that this process is first active at higher latitudes (around 35° N and 35° S), which is in accordance with our knowledge that the first spots of a new cycle usually occur in such latitudes (Fig. 2.7(*c*)).

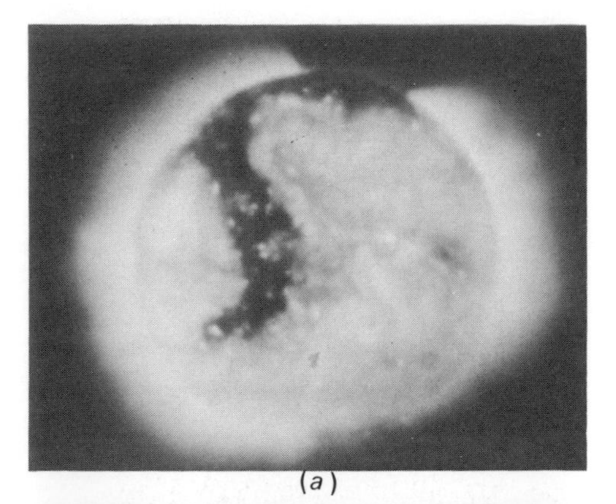

(*a*)

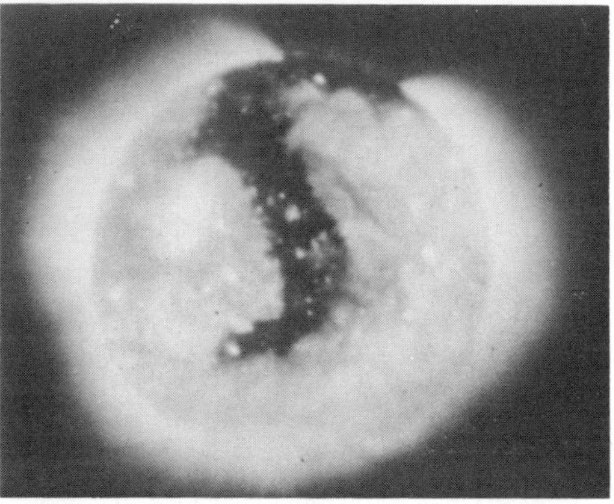

(*b*)

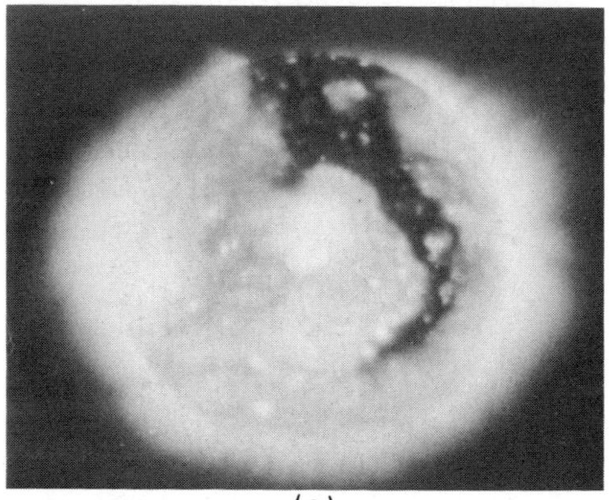

(*c*)

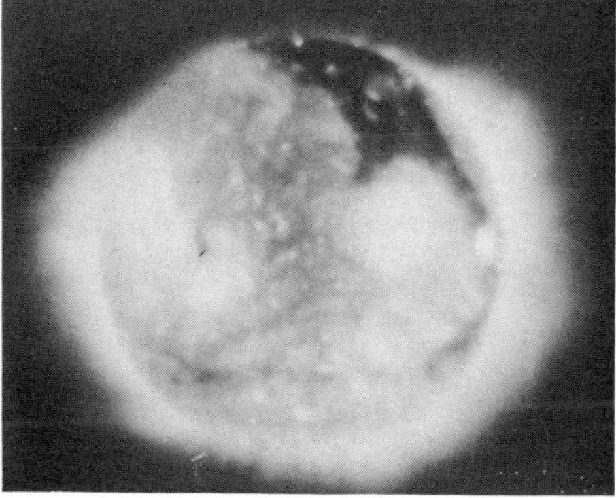

(*d*)

Fig. 2.8 X-ray photographs of the Sun showing the rotation of a coronal hole

As time goes on, lower and lower latitudes are affected. At sunspot maximum, the field strength of the magnetic forces is greatest over a wide band of latitude in each solar hemisphere. From then on, further mixing of lines tends to decrease the magnetic field strengths, the solar differential rotation forces the leader of each sunspot pair towards the equator, at the same time as its follower is forced towards the poles. It will be remembered that (i) all sunspot pair leaders in one hemisphere show the same polarity, their followers exhibiting the opposite polarity, (ii) the groups in the other hemisphere, while obeying this rule, show reverse polarity. Thus the magnetic forces action in sending all leaders in both hemispheres towards the equator means that regions of opposite polarity converge and cancel each other. On the other hand the followers, sent towards the poles, cancel and reverse the magnetic fields there (Fig. 2.7(*d*)). So after an 11-year period, the polarity of the Sun's general dipole field is reversed and a new cycle begins which will end with the finish of the 22-year cycle and the final restoration of the *status quo*.

2.4 Radio Output from the Sun

Since 1942, it has been known that the Sun emits radiation at radio wavelengths. Below a wavelength of 0·5 metre, the radiation comes from the lower chromosphere; at longer wavelengths it comes from the corona.

The term **quiet Sun** is used to describe the Sun during sunspot minimum, while the term **active Sun** implies that sunspots and flares are present in great numbers. The Sun's radio output is quite different at these times. The quiet Sun's output strength is near that expected from the black body temperature of the corona. When the Sun is active, irregular outbursts of radio noise occur, particularly intense radio storms being observed when large solar flares appear. It is thought that at such times, the proton and other ion streams ejected by the flares disturb the coronal gas, giving rise to the radio storms. At the speeds these ion jets travel, they reach the Earth about 24 hours later, their arrival producing a magnetic storm. The questions of such solar–terrestrial relationships will be dealt with in the sections devoted to the Earth.

2.5 The Sun as an Emitter of X-Rays and Cosmic Rays

As well as photons of infra-red, visible light, ultra-violet and radio waves, the Sun emits X-rays and cosmic rays. In recent years, high resolution X-ray pictures of the Sun have been made using suitable telescopes mounted in rockets and artificial satellites and it is found that outbursts of X-rays occur in localized regions of the Sun's surface, mostly in the vicinity of flares (see Fig. 2.8).

Solar cosmic rays consist of protons, electrons and atomic nuclei. These particles leave the Sun with high velocities and are thought to be associated with solar flares. The protons form most of the particles observed, alpha particles (helium nuclei) coming second in abundance. The electrons give up much of their energy in producing radio bursts in the solar corona and so are ejected with far lower speeds.

The origin of solar cosmic rays is still in some doubt; certainly solar magnetic fields are involved but no completely acceptable theory has emerged.

2.6 The Solar Neutrino Problem

We shall see later (Chapter 9) that at the centres of stars like the Sun, nuclear processes take place, producing the energy required to keep the stars shining. An additional product of such nuclear reactions are **neutrinos**, atomic particles that appear to have neither mass nor electromagnetic properties but carry energy and momentum. The mean free path of a neutrino is extremely large and the chances of one being absorbed *en route* through the Sun's bulk is about 1 in 10^{10}. However, the magnitude of the mean free path poses problems for the detection of the neutrino at the Earth, but such an exercise would provide direct evidence of the processes occurring at the Sun's centre. The first experimental arrangement to investigate the arrival of solar neutrinos used the interaction of the particle with chlorine atoms, the "apparatus" comprising 400 000 litres of cleaning fluid (perchloroethylene) in a tank in a gold mine at a depth of 1·5 km. The results so far indicate that something is wrong with our understanding of the processes at work in the solar centre, for the number of neutrinos detected is almost a factor of 10 down on the predicted figure. The solution of the neutrino problem will undoubtedly enlarge our knowledge of energy generation not only in the Sun but also in other stars.

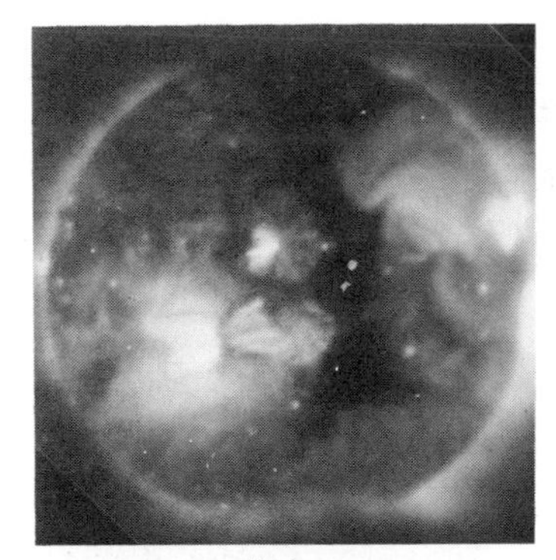
(a)

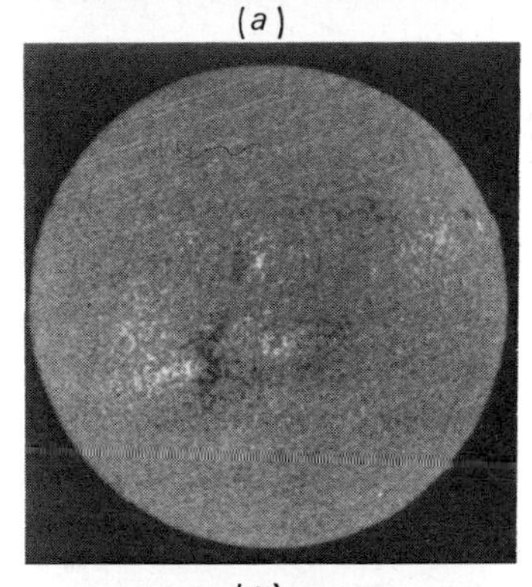
(b)

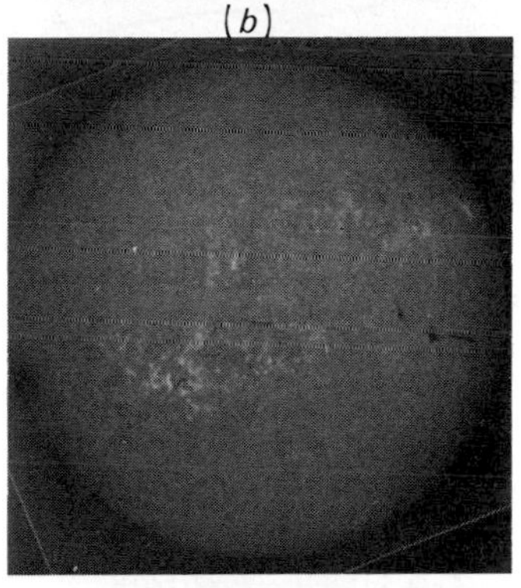
(c)

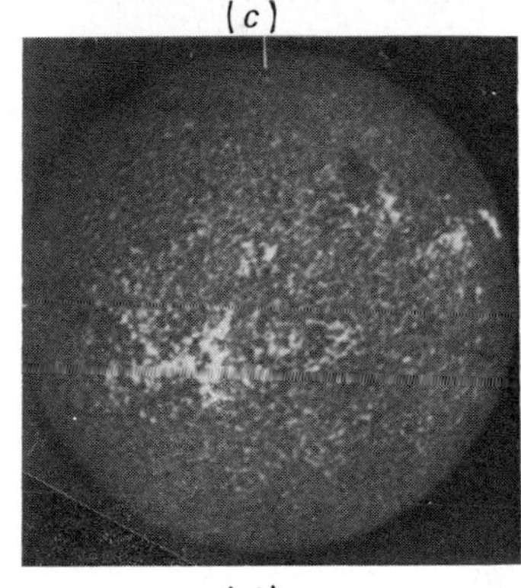
(d)

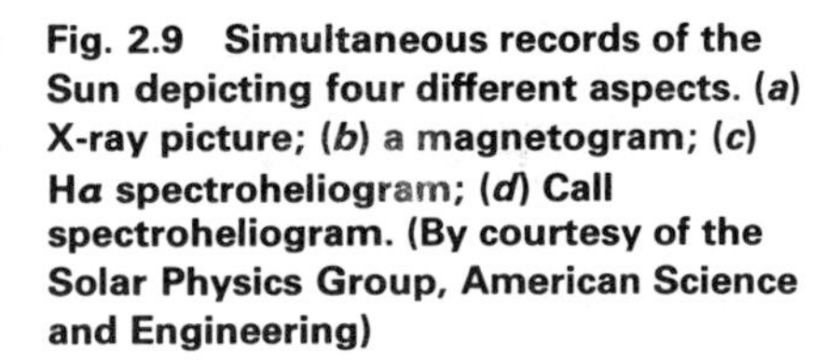
Fig. 2.9 Simultaneous records of the Sun depicting four different aspects. (*a*) X-ray picture; (*b*) a magnetogram; (*c*) Hα spectroheliogram; (*d*) Call spectroheliogram. (By courtesy of the Solar Physics Group, American Science and Engineering)

2.7 The Solar Wind

Interplanetary space is not empty. As well as dust and other meteoric particles, it contains a tenuous magnetized plasma originating from the Sun. This material is not stationary but is streaming away radially from the Sun at high velocity, of the order of 400 km s^{-1} at a distance of 1 AU. It is known as the **solar wind**. It has been suggested that the distance from the Sun at which the solar plasma becomes indistinguishable from the interstellar medium is about 50 AU. Earth satellites, and lunar and interplanetary probes have been used in recent years to investigate the properties of the solar wind. Instruments left operating on the Moon's surface also yield information about its nature.

Such measurements demonstrate a continuous flow of positive ions, principally protons, coming from the Sun. Magnetometer readings indicate an interplanetary magnetic field strength of the order of 10^{-5} gauss. Because of the rotation of the Sun and the hosepipe-like spraying outwards of the solar plasma, the interplanetary magnetic field lines of force spiral in the manner indicated in Fig. 2.10; the higher the energy of the wind, the less curved are the lines of force.

2.8 Solar Disk Phenomena: Practical Exercises

Occasionally, when there happens to be a large sunspot on the Sun's disk and the Sun is seen through a thick haze, it is possible to see the spot with the unaided eye. However, for a more serious study of the solar surface phenomena, optical equipment is required and in the first instance a basic telescope is essential. Useful work may be done with telescopes of 50 mm diameter or larger. Both a refractor and a reflector may be used although it is generally accepted that thermal distortions of the optics when the instrument is directed to the Sun are less severe for a refractor which gives better images than a reflector.

2.8.1 Visual Observations

BECAUSE OF THE POSSIBLE DANGER OF PERMANENT DAMAGE TO EYESIGHT, IT CANNOT BE EMPHASIZED TOO STRONGLY THAT THE SUN SHOULD **NEVER** BE VIEWED THROUGH OPTICAL INSTRUMENTS UNLESS SPECIAL EQUIPMENT IS USED.

Ordinary telescope eyepieces sometimes carry a screw thread to allow absorbing filters to be attached to them. This system is not suitable for use with the Sun as either the filter may not be sufficiently absorbing or the solar heat may cause it to splinter suddenly.

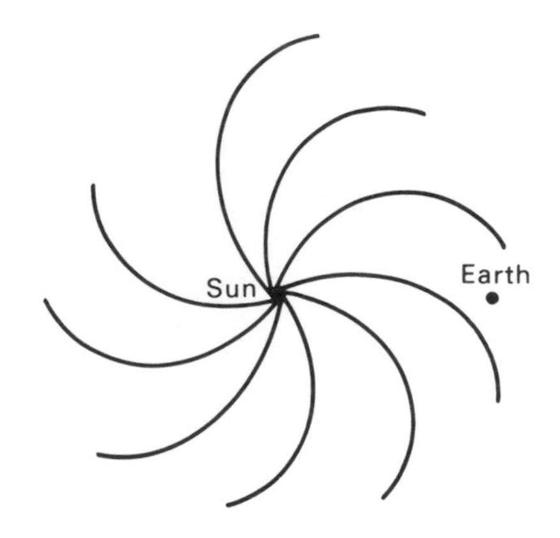

Low energy solar wind

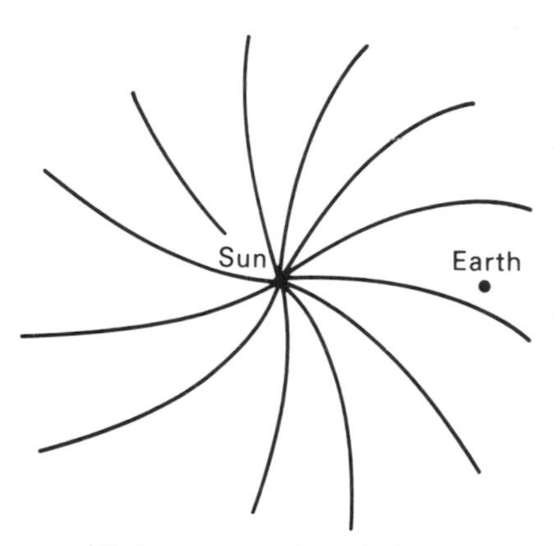

High energy solar wind

Fig. 2.10 The flow of the solar wind (after N. Ness, *Introduction to space science*, p. 355, ed. W. N. Ness & E. D. Mead. Gordon & Breach, New York (1968))

Direct telescopic views of the Sun should only be attempted if a special eyepiece is available. An example of such a device is the solar diagonal whereby the converging beam from the telescope is split into two very unequal components. Only about 10% of the collected light is used for viewing through an eyepiece, which also requires the addition of an absorbing filter; the remaining 90% which is in the second beam is passed out of the system.

2.8.2 Recording Sunspots by Drawing A more convenient way of looking at the features on the solar disk is by the projection method. If smooth white paper is held at a distance behind the eyepiece of a telescope which is directed to the Sun, a large image may be focused on to the paper, this being very suitable for viewing by groups of people. The best eyepiece to use is one of low power, preferably a Huyghens type in which the component lenses are not cemented together. (Cemented lenses are prone to damage by the heat from the Sun.)

By making a frame which can be attached to the telescope so that the paper may be clipped to a board, it is then an easy matter to draw the positions and details of the solar markings. The contrast of the projected solar image may be improved by also fitting a shadow board around the telescope tube. A schematic projection frame is depicted in Fig. 2.11.

The projection technique allows considerable detail to be seen and drawn. For example, it should be noticed that the solar disk is not uniformly bright and that it fades towards its edge (limb darkening). Sunspots and sunspot groups should be seen with structure—the spots showing a strong core (umbra) surrounded by a less black region (penumbra) which may show structure such as filaments radiating outwards. Small bright patches (faculae) may be evident particularly near to the spots or towards the edge of the Sun where the limb darkening effect produces a background with lower illumination. The detection of faculae may be made easier by gently shaking the frame which holds the paper.

When drawing the solar markings by the projection method, it is important to provide axes to which specific points may be referred and the north–south, east–west lines may be generated for this purpose. As convention, the north, south, east and west limbs of the Sun are defined according to the directions on the sky; to the unaided eye in the northern hemisphere, north is to the top and east to the left.

Keeping the telescope fixed, allow the solar image to drift across the projection board and rotate the paper on the board until the upper edge of the Sun moves along or parallel to the upper edge of the paper. This edge of the paper now corresponds to the east–west direction on the sky. By noting the direction in which the image moves when the telescope is kept fixed, the image drifting by the diurnal motion of the Sun across the sky, the west direction may be noted (with astronomical eyepieces, the left side of the paper, when viewed from the telescope, is the west side of the Sun). A line at right angles to the upper edge of the drawing paper corresponds to the north–south axis and the north direction may be marked easily by seeing which way the image moves when the telescope's altitude angle is raised slightly. (If the image moves downwards, north is to the top.) Re-centre the image on the projection board and mark the position of the Sun's limb. After ensuring that the image is kept to the markings corresponding to the limb, lightly draw over the details of the structures of the sunspots and mark any faculae.

If a suitable focusing camera is available, very reasonable pictures of the solar markings may be taken without too much distortion if the projected image is photographed from a position very close to the telescope tube, looking as directly as possible at the image.

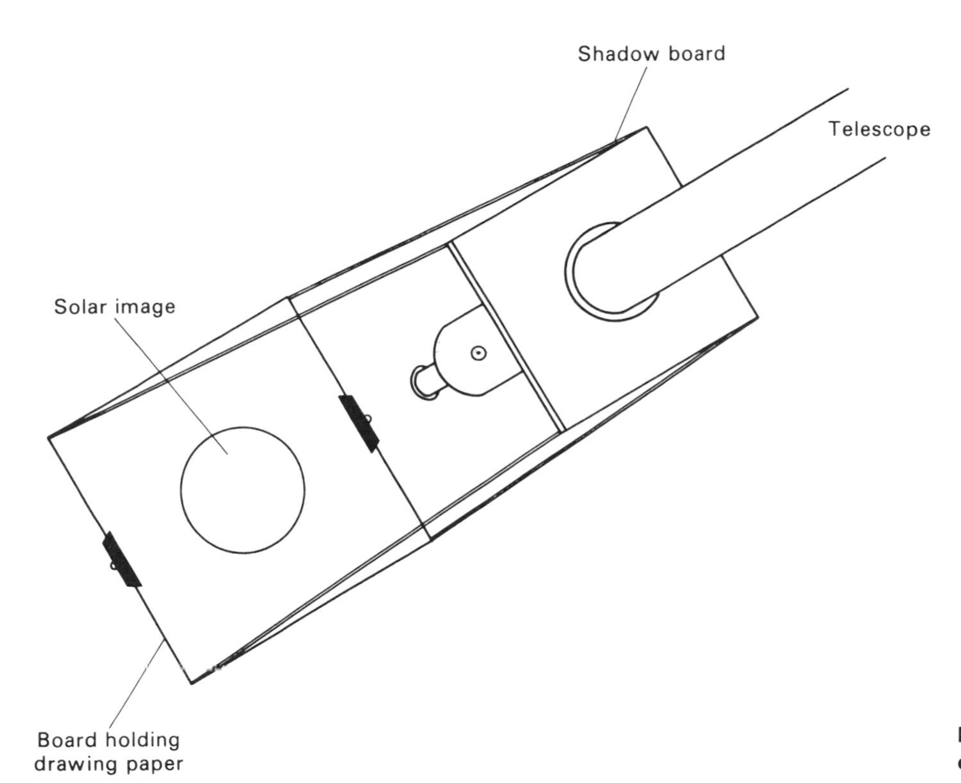

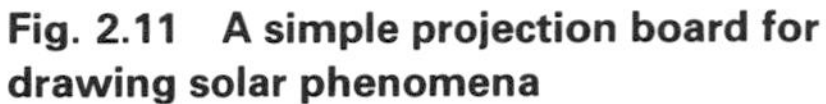

Fig. 2.11 A simple projection board for drawing solar phenomena

2.8.3 Determination of the Solar Rotation Period Drawings made daily of the Sun show that the features undergo slow evolutionary changes. In addition it will be noted that there is a general drift of the sunspots from east to west as a result of the Sun's rotation. Determination of the angular movement of a feature over a measured period of time obviously allows the Sun's rotation period to be determined. Close examination of the movements of sunspots at different latitudes shows that the Sun does not rotate as a solid body but that different latitudes rotate at different rates. Before the motion of spots across the image can be converted to a rotation period, allowance must be made for foreshortening effects of seeing the spots on a curved surface. In order that the progress of a spot across the solar disk may be noted in terms of a shift in apparent longitude, it is important to know the relationship between the marked reference of the north–south, east–west axes of the sky and heliographic (solar) latitude and longitude.

The Sun's rotational axis is at an angle of about 7°25 to the perpendicular of the plane of the ecliptic, this being the plane defined by the Earth's orbit. Consequently there will be times of the year when the north pole of the Sun will be slightly inclined towards the Earth and visible and other times when it is inclined away from the Earth on the other side of the Sun. In early December and June, the Earth is in a position so that the Sun's pole is at right angles to the line of sight, passing through the projected limb of the Sun. Because the Earth's pole is also inclined by about 23°5 to the ecliptic, the projection point of the Sun's north pole does not correspond to the sky's north–south axis but is inclined to it. However, at these times in December and June, the projection of latitude lines on the solar disk produces a series of straight lines parallel to the solar equator. In general the motion of a sunspot follows a path of constant latitude and so at these times of the year they may be seen to move across the Sun's disk in straight lines.

At other times of the year the motions of the spots across the disk follow elliptical paths, each path being the orthographic projection of a solar latitude as seen from the Earth. Since at these times the solar pole is inclined to the line of sight, it follows that the centre of the projected image of the Sun does not lie on the Sun's equator. In determining the Sun's period of rotation from the movement of sunspots, allowance must be made for this effect on the projection of the image. It is necessary to determine the heliographic co-ordinates of any spot from the drawings.

The heliographic latitude, B_0, of the centre of the image is tabulated; it may be found for each day in *The Astronomical Almanac* (*AA*)† or for every fourth day in the appropriate year of *Handbook of the British Astronomical Association*. Also tabulated is the position angle P which the projection of the solar axis makes with the north–south direction and it is measured positively from the north point of the solar image towards east. For any particular day the values of B_0, P may require interpolation.

The formulae from spherical trigonometry which allow the heliographic latitude, B, and apparent longitude, L, of a spot to be determined are:

$$\sin B = \sin B_0 \cos \rho + \cos B_0 \sin \rho \cos (P-\theta)$$

$$\sin L = \sin \rho \sin (P-\theta) \sec B,$$

wherc θ is the position angle of the sunspot on the disk, again measured from the north point of the disk towards the east, and ρ is the angle formed by the direction of the sunspot, the centre of the Sun and the direction of the Earth. The angle ρ is given by

$$\sin (\rho + \rho_1) = r/r_0$$

where r is the distance of the spot from the centre of the solar image and r_0 is the radius of the solar image; the angle ρ_1 in turn is given by

$$\rho_1 = Dr/2r_0,$$

where D is the angular diameter of the Sun, the values also being tabulated in the *AA* or the *B.A.A. Handbook*.

For a determined latitude B, the time taken for a spot to move from one longitude to another may be used to calculate the time necessary for the spot to complete one rotation over the solar disk.

As an example of a heliographic co-ordinate determination, consider the drawing made for 1973, Jan 2nd, 12–10 UT (see Fig. 2.12). From the drawing:

$$r_0 = 75 \text{ mm}$$

$$r = 50 \text{ mm}$$

$$\theta = 79^\circ.$$

From the *AE*†, we have interpolated values for B_0, P and $D/2$ as follows:

$$B_0 = -3^\circ\!.23$$

$$P = +1^\circ\!.35$$

$$D/2 = 16' \, 17''\!.5.$$

† Between 1960 and 1980 the *Astronomical Almanac* was called the *Astronomical Ephemeris*.

Hence

$$P-\theta = -77^\circ\!.65$$

$$\rho_1 = (16'\ 17''\!.5)(50/75)$$

$$= 0^\circ\!.18,$$

$$\sin(\rho + 0^\circ\!.18) = 50/75$$

$$\rho + 0^\circ\!.18 = 41^\circ\!.81$$

$$\therefore\quad \rho = \underline{41^\circ\!.63.}$$

$$\sin B = \sin(-3^\circ\!.23)\cos(41^\circ\!.63) + \cos(-3^\circ\!.23)\sin(41^\circ\!.63)\cos(-77^\circ\!.65)$$

$$\therefore\quad B = \underline{5^\circ\!.72.}$$

$$\sin L = \sin(41^\circ\!.63)\sin(-77^\circ\!.65)\sec(5^\circ\!.72)$$

$$\therefore\quad L = \underline{-40^\circ\!.70.}$$

On the following day, this same sunspot was determined to have an apparent longitude of $-26^\circ\!.91$. Thus in one day its apparent change of longitude is $13^\circ\!.79$. In order to have an apparent change of 360°, the time required is $360/13^\circ\!.79$ days, i.e. 26·10 days.

By following a sunspot over a longer period, a more accurate value for the Sun's rotation will ensue and if spots are measured at different solar latitudes, it should be possible to determine the differential rotation effect.

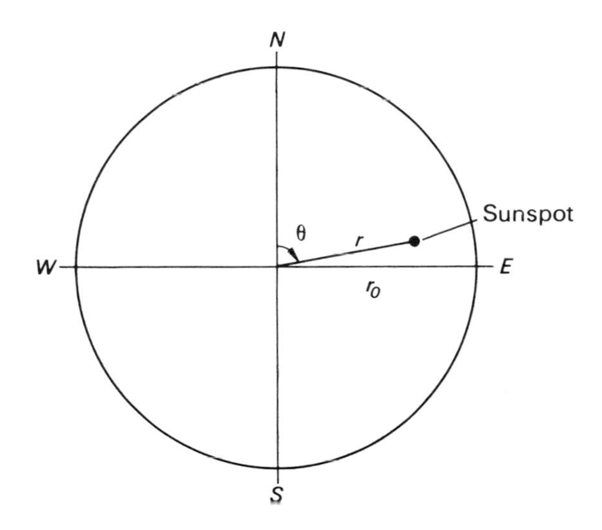

Fig. 2.12 **The co-ordinates of a sunspot**

2.8.4 Use of a Pinhole Camera The most simple means of taking photographs of the Sun is with a pinhole camera. In this system a tiny hole acts as a lens and the image that it produces can be recorded on film. A very convenient arrangement has the pinhole at one end of a long tube, blackened on its inside, with the body of a single lens reflex camera, without its lens, at the other end.

For the pinhole camera to give reasonable images, the focal ratio must be $f/1000$ or higher. Thus a 1 mm hole attached to a 1 metre tube is sufficient. Since the Sun subtends about $\frac{1}{2}^\circ$, its image with a 1 metre tube is equal to

$$\frac{1000}{2} \times \frac{\pi}{180}\ \text{mm} = 8{\cdot}7\ \text{mm}.$$

Even with such a focal ratio, the illumination of the Sun's image is high and a typical exposure of 1/1000th of a second is required using Pan X film. The demand for an extremely high shutter speed may be relaxed by using either a higher focal ratio, or filters over the pinhole, or a slower film, or even a photosensitive photographic paper rather than celluloid film.

A pinhole camera under good conditions is capable of allowing pictures of sunspots to be taken and is useful for recording the shape of the Sun, say at the time of a partial eclipse, or when the Sun is just rising or setting.

When a celestial object is on the horizon, refraction alters its true altitude by over half a degree. For an extended object such as the Sun, refraction has a differential effect so that the lower limb is more refracted than the upper limb. Thus when the Sun is close to the horizon, it has an elliptical shape with the lower and upper limbs closer together than the east and west limbs. The distortion is easily seen with the unaided eye at a location where there is an uninterrupted view of the horizon. It may be photographed using a pinhole camera. (An ordinary camera is generally insufficient because of its short focal length.)

3
The Earth and its Moon

3.1 The Earth as a Planet

We are concerned in this chapter with the Earth and its only natural satellite, the Moon. The Earth's orbit, lying between the orbits of Venus and Mars, is an ellipse of small eccentricity. The elements of this orbit, like those of the other planetary orbits, suffer perturbations. These changes are measured with respect to some fixed reference plane and direction such as the position of the ecliptic and vernal equinox at a given epoch. They are not only caused by the gravitational attractions of the other planets but also by the Moon, because of its proximity.

It has already been shown how astronomers found it convenient to use data connected with the Earth's orbit and the Sun as their units of time, distance and mass and how the **solar parallax** and **astronomical unit** were defined and measured.

Fig. 3.1 A view of the Earth taken from Apollo 8

3.2 The Earth's Form

The Earth's shape is roughly that of an oblate spheroid, its equatorial diameter being 12 756·2 km. About 70% of its surface is covered with water and it possesses an atmosphere reaching to a height of about 1000 km. Even taking into consideration the mountain ranges and ocean depths, the Earth is remarkably smooth—to about one part in 600. In other words, if the Earth was a ball about 1 metre in diameter, the highest mountain and deepest ocean would depart less than one millimetre from mean sea level.

Although Clairaut and others had worked out in broad outline the theory of the Earth's figure by the 18th century, it is only within the last 100 years, and especially since the advent of artificial Earth satellites, that most of our knowledge of our planet has been gathered. The figure of the Earth itself has been found by geodetic measurements, by measuring the separation of places of known latitude and longitude. These triangulation measurements were referred to suitable spheroids of reference such as the Hayford spheroid. Artificial satellites, tracked from the Earth's surface, have augmented and refined our knowledge of the Earth's form (see Roy & Clarke, *op. cit.*, Section 13.6.2).

3.3 The Earth's Interior

Information about the interior of the Earth is obtained indirectly from the motions of Earth satellites, the study of earthquake waves and the physics and chemistry of matter under high temperatures and pressures.

The interior of the Earth appears to be made up of a number of regions (see Fig. 3.2). There is an increase of density towards the Earth's centre. The refraction and reflection of earthquake waves show the presence of a liquid **core**, with a diameter of more than 6400 km. Its density is from ten to twelve times that of water. There seems no doubt that the core is fluid, though the presence of a small solid inner core is possible.

Above the core lies a shell called the **mantle**. Its mean density is about four times that of water. The mantle, possibly composed of heavy basic rocks, has on

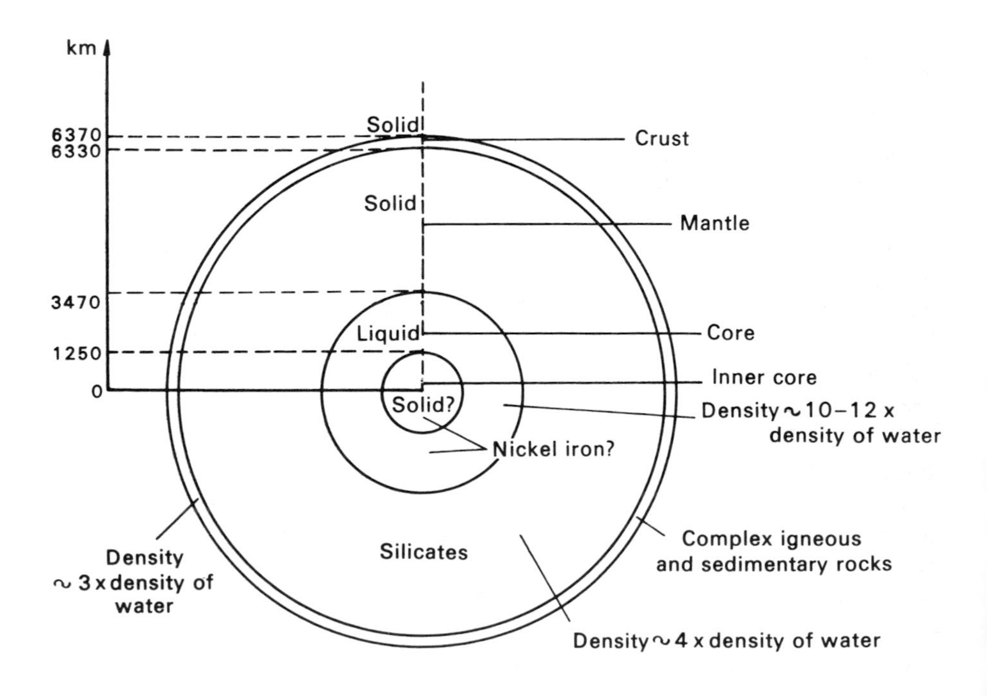

Fig. 3.2 The interior of the Earth

top of it the very thin layer of the **crust**, less than 80 km thick. It is made up of igneous rocks such as granite and basalt, with sedimentary rocks such as sandstone and limestone. The mean density of the crust is of the order of three times the density of water. In landward areas the crust is much thicker than in places under the oceans.

Where the constitution of the Earth's deep interior and the temperatures and pressures that exist there are concerned, we are on more speculative ground. The core may be predominantly nickel–iron, but Ramsey has shown that this older view contains serious difficulties and has given theoretical grounds for the alternative view that the dense core may come about partly because the pressure at such depths is high enough to collapse the molecules.

3.4 The Earth's Magnetic Field

To a first approximation, the Earth's magnetic field simulates that of a simple dipole embedded within and near the centre of the Earth at an angle of about $11^{\circ}\!.4$ to the Earth's axis of rotation. In fact, the line connecting the two geomagnetic poles misses the centre by some hundreds of kilometres. The vertical field-strength at the geomagnetic poles is 0·63 gauss; at the equator it is 0·31 gauss.

More accurately, it is found that the field departs from a simple dipole field at various places due to the presence of magnetic materials in the Earth's crust. In addition, as a result of solar outbursts, fluctuations of short period also occur. At any particular place on the Earth's surface the magnetic field changes slowly, such a change being called the **secular variation**. Much information about the extent and strength of the field out to distances where it merges with the interplanetary magnetic field has been gathered in recent years by putting magnetometers in artificial satellites that in some cases move in highly-elliptic orbits, taking them out beyond the Moon's orbit.

The source of the Earth's magnetic field almost certainly lies in the Earth's core, possibly in a self-acting dynamo action set up by motions in the fluid electrically-conducting core. Thermal convection provides a satisfactory mechanism for such motions.

3.5 The Earth's Atmosphere

The Earth's atmosphere is the thin fluid skin separating and protecting terrestrial life from the vacuum of space and damaging solar radiations. In Fig. 3.3 the main regions making up the structure of the Earth's atmosphere are depicted.

The regions **troposphere**, **stratosphere**, **mesosphere, thermosphere** and **exosphere** are classified on a thermal basis; the layers dividing them are named by substituting the suffix "pause" for the suffix "sphere". If the classification is by chemical composition, the main regions are the **homosphere** and **heterosphere.** The structure of the atmosphere can be classified from a number of other viewpoints such as its degree of ionization.

In Fig. 3.3 the variation of temperature with height is also sketched. The boundaries of the ionosphere are indicated.

In recent years, work with rockets, satellites and other instruments of atmospheric research has increased enormously our knowledge of the constitution and extent of the atmosphere up to the 700 km level, where it finally merges into the interplanetary medium.

Up to a height of 70 km the composition is unchanging. By volume, the principal constituents are molecular nitrogen (78%), molecular oxygen (21%), with argon, water vapour and carbon dioxide taking up most of the remaining 1%. In addition, other permanent gases such as neon are present in very small quantities. Ozone (O_3) appears in a layer some 25 km up as a result of the dissociation of molecular oxygen by ultra-violet radiation, the atomic oxygen then combining with oxygen molecules.

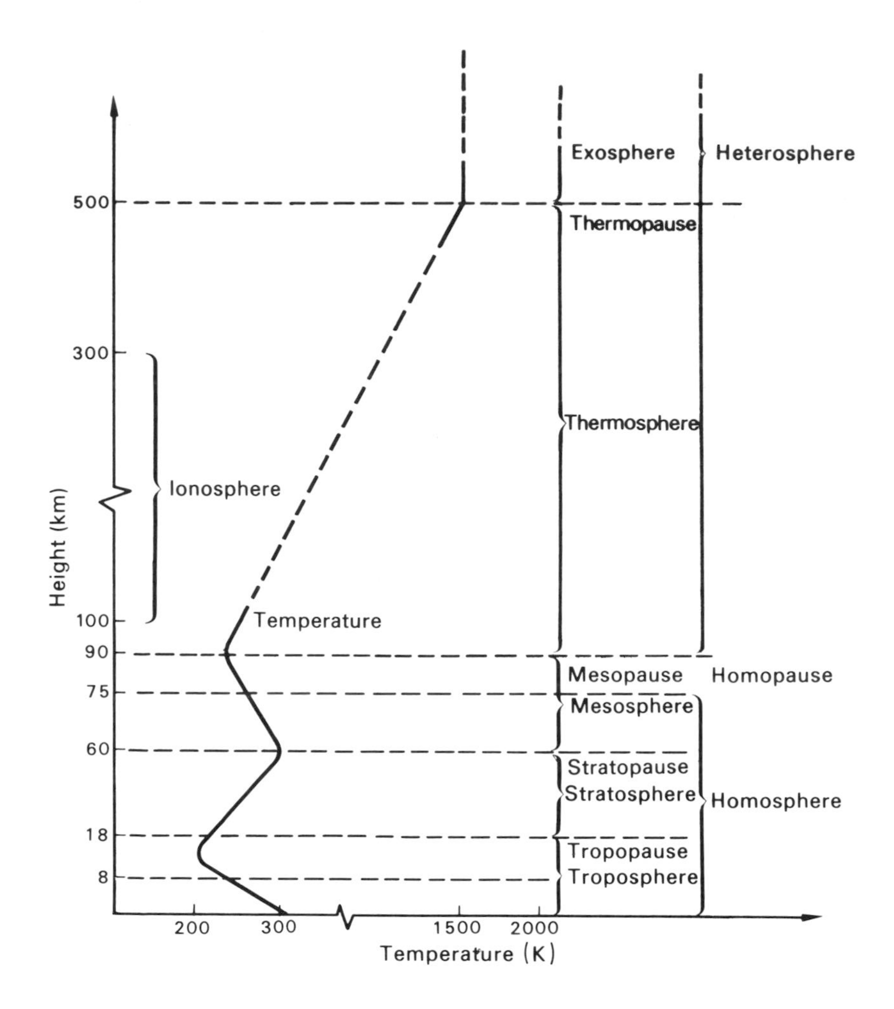

Fig. 3.3 The layers in the atmosphere

At the homopause, the composition begins to change and within the heterosphere a number of processes such as diffusion, mixing, photodissociation, and so forth, are at work changing the make-up of this tenuous region.

The **ionosphere** is a region of ions and electrons created by the Sun's short-wave radiation and by cosmic rays. This region is usually divided into several layers called the D, E, F_1 and F_2 layers in order of ascending height. The ionosphere is extremely variable, the number of electrified particles depending on sunspot numbers, season, latitude and the change from day to night.

In attempts to gain insight into the relations between pressure, density and temperature throughout the atmosphere, model atmospheres have been constructed mathematically. Their predictions have then been compared with data derived from vertical rocket flights and observations of atmospheric drag on artificial satellites.

Density and air pressure fall off rapidly with increase of height. Half the total mass of air lies below a height of 6 km. By a height of 160 km, the ratio of air density to sea-level air density is down to 10^{-9}, the air density being

10^{-9} kg m^{-3} at that level. By the time a height of 400 km is reached, the respective figures for the ratio are 4×10^{-12} and 5×10^{-12}.

3.6 Solar–Terrestrial Relationships

The correlation of such terrestrial events as auroral displays and magnetic storms with solar activity reveals an intimate relationship between the output of electromagnetic and corpuscular radiation from the Sun and changes in the Earth's atmospheric density, magnetic field and atmospheric electrical activity. In addition to the fluctuations in air density caused by solar radiation, streams of charged particles (especially at times of solar flare outbursts) impinge on the atmosphere causing violent magnetic storms, changes in air density and auroral displays. These streams also contribute to the numbers of charged particles trapped in the Earth's van Allen radiation belts.

A **magnetic storm** is a sudden change in the strength of the Earth's magnetic field caused by corpuscular radiation from the Sun. Such a storm can interfere seriously with radio communications. Other less violent and longer-lasting storms are periodic, the period being the time it takes the Sun to make one rotation with respect to the Earth. This period of 27 days suggests that the area of activity on the Sun's surface causing the storm is hosepiping material out continuously.

The **aurorae** (aurora borealis and aurora australis in northern and southern hemispheres of the Earth respectively) are luminous phenomena in the upper atmosphere of the Earth between 100 km and 1000 km caused by solar charged particles that excite the atoms of air. Aurorae are observed mainly in polar regions, a consequence of the effect of the Earth's magnetic field that deflects the particles towards these regions. The colours that predominate in the auroral curtain-like streamers are green, red and blue, corresponding to emission lines from oxygen and nitrogen.

The Earth's van Allen radiation belts, discovered in 1958 by Geiger counters in the artificial satellites Explorer 1 and Pioneer 3, owe their existence to solar activity and the Earth's magnetic field. Streams of charged particles, protons and electrons, are trapped in the Earth's magnetic field and spiral along the lines of magnetic force. The pitch of the spiral becomes smaller as the particle approaches the Earth's surface until it reverses its direction and roughly retraces its path. There is also a drift in longitude so that an injection of charged particles at a point above the atmosphere quickly results in a spread about the Earth. The radiation zones and the process are sketched in Figs. 3.4 and 3.5.

There are two doughnut-shaped belts or regions of maximum concentration of such particles, one about 4000 km above the Earth's surface, the other about 16 000 km up. Regions of maximum intensity are shown cross-hatched in Fig. 3.4.

The orbits of these particles are quasi-stable in that irregularities in the Earth's magnetic field and collisions with air molecules ultimately reduce the numbers in the belts, but solar outbursts supply fresh particles to "top-up" the numbers.

A further ring-current of electrons at a distance of some 60 000 km circles the Earth. Finally the solar wind (protons and electrons ejected by the Sun) pushes in the Earth's magnetic field on the sunward side of the planet and stretches it out on the opposite side. The term **magnetosphere** has been given to the resulting tear-drop-shaped region about the Earth in which the Earth's magnetic field is dominant.

In Fig. 3.6, we summarize the effects of solar flares and give their schedule of occurrence. Also included are the instruments by which some of these effects are detected.

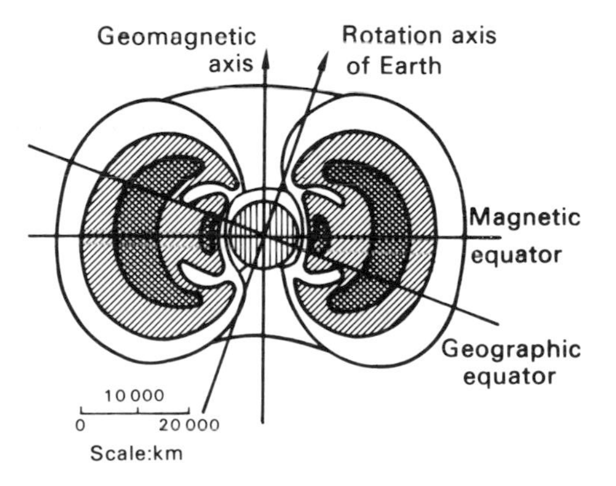

Fig. 3.4 The van Allen radiation belts

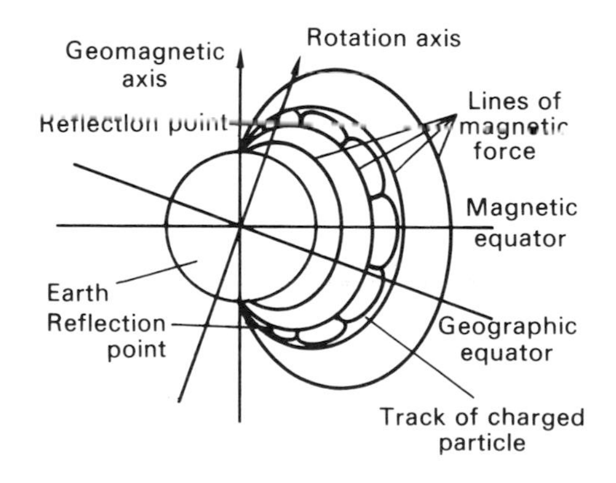

Fig. 3.5 Movement of the charged particles in the van Allen belts

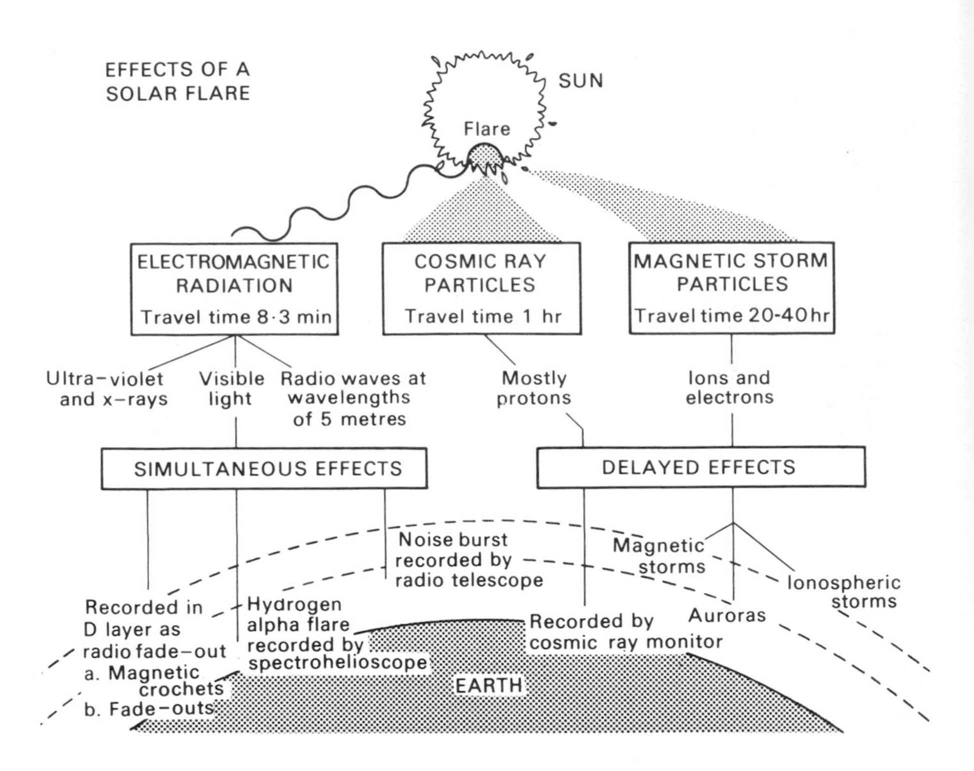

Fig. 3.6 The effects of a solar flare

3.7 The Moon's Physical Nature

3.7.1 Basic Parameters The classical parallactic method of measuring the Moon's distance has now been superseded by radar measurements and by tracking artificial lunar satellites. The placing of corner reflectors on its surface to reflect laser beams also gives a measure of its distance and of its diameter. A value of 3476 km for the latter is obtained.

The centre of mass of the Earth–Moon system moves in an elliptic orbit about the Sun, both Earth and Moon moving in orbits about this centre with a period of one sidereal month. This orbital movement of the Earth produces small monthly oscillations in the directions of external bodies such as the Sun and planets. From measurements of such oscillations a value for the Moon's mass of 1/81·3 that of the Earth is deduced, so that the centre of mass of the Earth–Moon system is actually inside the Earth. Much more accurate values of the Moon's mass have been obtained more recently by the careful tracking of lunar artificial satellites.

Knowing the mass and linear diameter of the Moon, its mean density may be calculated immediately. It is found to be about 3·33 times that of water, very close to that of the basic rocks under the thin surface crust of the Earth.

3.7.2 The Moon's Surface Features From the earliest days of telescopic observations, it was clear that the Moon had a solid surface. The main features on the large scale are the **maria**, the **mountains**, the **craters** and the **rays**.

The maria are the most conspicuous. These dark, almost level plains, at one time thought to be liquid oceans, are depressions filled with lava or volcanic ash. The largest, *Mare Imbrium*, is almost 1200 km across. There are some thirty of them, covering almost half the Moon's near side, the other side being singularly devoid of such areas. These "seas" are possibly the result of collisions of comets, asteroids or large meteors with the Moon. The object that created the *Mare Imbrium* may have been 60 km in radius. Coming in at a flat angle, it caused vaporization at the impact point—"ground zero"—with a region of melting around that. The collision possibly raised a wavelike structure surrounding this region, fractured bedrock, created mountain ranges radiating from the area and sprayed an enormous quantity of material in all directions, helping to produce secondary incidents. The other maria are probably the result of similar collisions.

Fig. 3.7 The full Moon taken by the 100-inch Mt. Wilson telescope. (By courtesy of the Mt. Wilson and Palomar Observatories)

Perturbations in the orbits of artificial lunar satellites show that there are large, ***mas***sive ***con***centrations of material (**mascons**) buried beneath a number of the maria. It is likely that the lava basaltic rock forming the maria is now overlain with 10–20 metres of rubble, the accumulation of bedrock pulverized and thrown out by large meteors since the maria were formed.

The chains of mountains found on many parts of the Moon's surface rival the largest and most extensive terrestrial mountains. For example, the Doerfel range has summits over 6000 metres in height. Many are named after familiar mountain ranges on Earth, such as the Apennines and the Alps. Their heights may be measured from the lengths of their shadows and a knowledge of the Sun's altitude in the lunar sky. Photography from artificial lunar satellites has provided the means of preparing accurate maps of the lunar topography, a study known as **selenography**.

Even a pair of binoculars is sufficient aid in viewing the largest lunar craters, these being over 200 km in diameter. The *Mare Orientale*, only partially seen from the Earth, is perhaps the Moon's most spectacular feature. It is 1000 km across and consists of three concentric rings of mountains enclosing a relatively featureless plain. Seen from directly overhead it resembles a gigantic bullseye. The three crater rings are possibly frozen shock-waves created by the object that struck there. Indeed on the Moon's far side there are a number of double-walled craters of large size, similar in overall features to the *Mare Orientale* super-crater.

Many of the Moon's craters have central peaks; many craters are superimposed on presumably older ones. They occur randomly all over the Moon, except in the maria regions where less than 10% exist. Estimates of as many as 1 000 000 craters over 100 metres in diameter have been made. It is almost impossible to estimate the number of smaller ones that have now been revealed by the cameras of space craft. Although the ring-walls of the largest craters may be as high as 3000 metres, their diameters are such that an observer placed at the crater centre could find them out of sight, below his horizon (see Fig. 3.8).

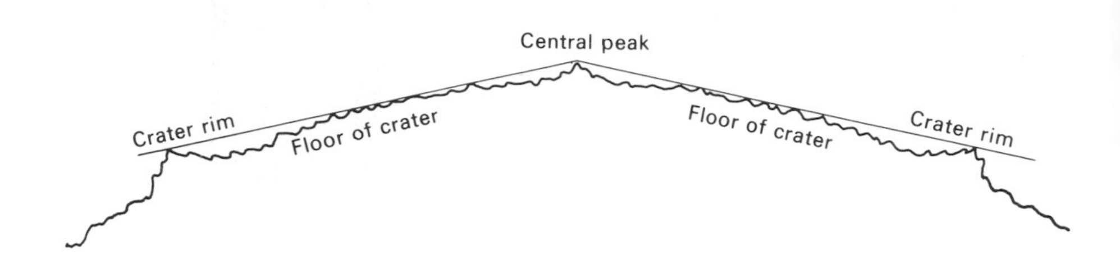

Fig. 3.8 A schematic diagram of the profile of a lunar crater

The lunar rays resemble gigantic splash-marks up to 2000 km in length, running outwards from some of the main craters. Lighter in colour than the surface they cross, they do not cast shadows. They are probably what they appear to be—material ejected from the craters at the time of their formation.

As well as these main features, the Moon's surface contains clefts or rilles, up to three kilometres in width and sometimes twisting for 500 km in length. Their depth is uncertain.

No weathering by water or air occurs on the Moon. Without an appreciable atmosphere, changes in its physical features are slow to take place or occur suddenly, as when a meteor strikes the surface and disturbs the overlay of rubble. Armstrong's footprints could well remain crisp for millions of years.

Alternate heating by the Sun and cooling by radiation produce an effect called **insolation** whereby a slow crumbling of rock under differential expansion and contraction takes place. It may be mentioned that previous ground-based brightness and polarization studies had suggested that the lunar surface structure was porous.

Summing up, the large- and small-scale features of the Moon's surface suggest a gigantic battlefield disrupted by continual bombardment in the Moon's past on such a scale that its material has been vaporized, melted, shattered, ejected, fused, churned up, cooled under vacuum and under pressure. The samples of rocks brought back by astronauts and unmanned craft to be examined in terrestrial laboratories confirm this picture. As expected, they contain many of the chemical elements such as oxygen, iron, aluminium, titanium and so on, found in terrestrial rocks. The glossy black fragments and spheres are basaltic glasses. Other fragments are of whitish anorthosite glass. There are also breccias.

The lunar basalts are similar in character to the terrestrial basalts and predominate in the maria. They are formed by the cooling of surface lava flows. Anorthosite rocks, on the other hand, form in deeply buried masses of magma (or molten rock). Such rocks are common in the lunar highlands. The breccias are rocks composed of angular fragments cemented together. Minerals found

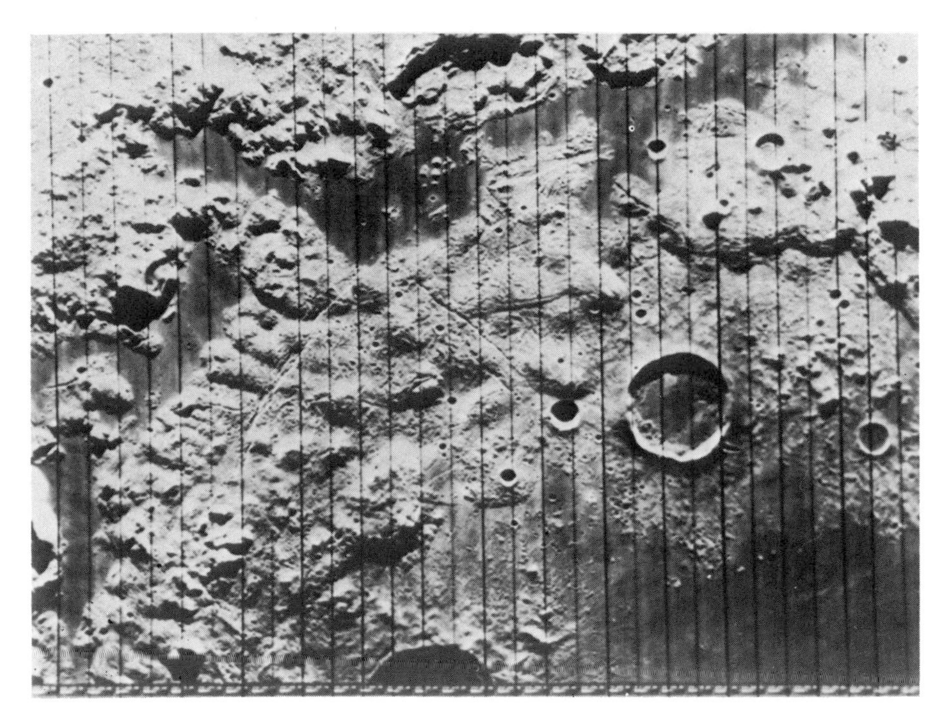

Fig. 3.9 *Mare Orientale* and the Rook Mountains photographed from Orbiter IV. (By courtesy of the Royal Astronomical Society)

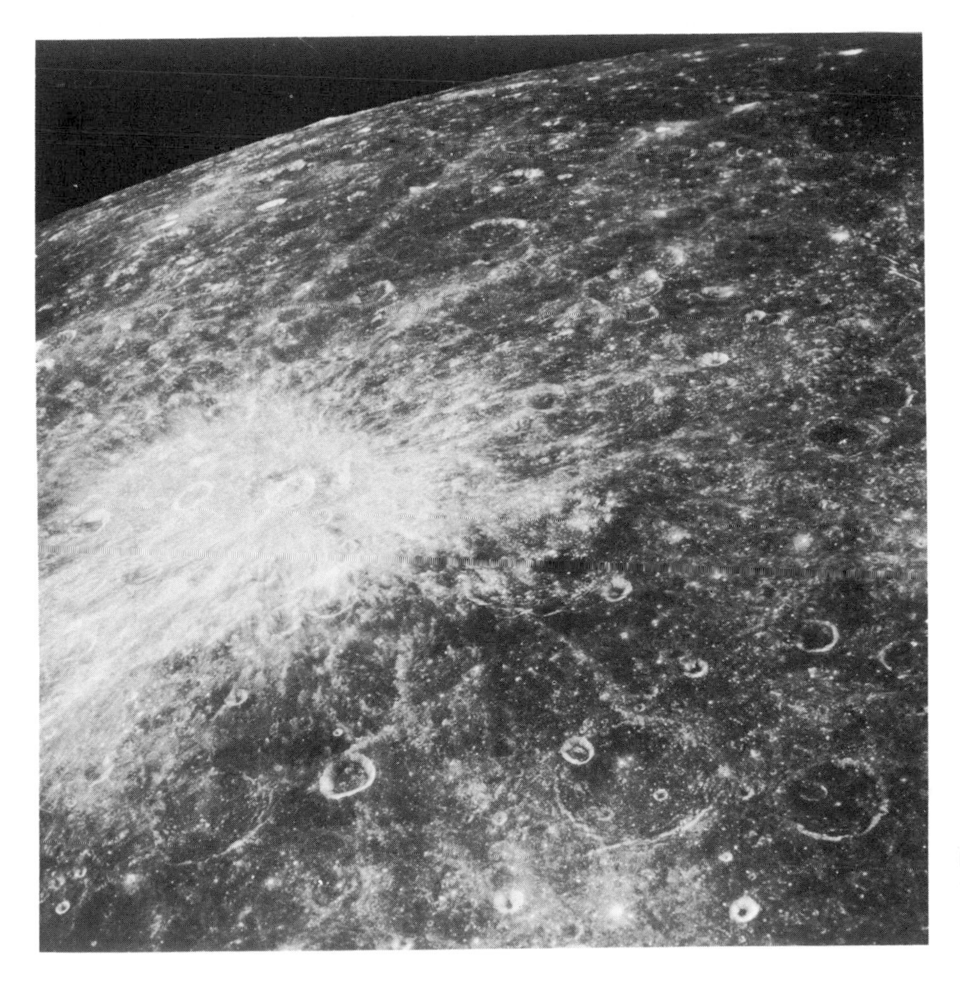

Fig. 3.10 The hidden side of the Moon photographed from Apollo 8. (By courtesy of the Royal Astronomical Society)

in terrestrial igneous rocks are also found in lunar rocks—pyroxene, plagioclase and ilmenite. The continual bombardment of the lunar surface by meteors has enriched the overlay of rubble with all these fragments of evidence reporting the Moon's violent past.

It may also be mentioned that radioactive dating of the samples puts the Moon's age at 3×10^9 years at least.

Fig. 3.11 Lunar surface experiments deployed by Apollo 12 mission

The first geological surveys of the Moon have now been completed, though much remains to be done. Moonquakes initiated by impacting spent lunar modules and rocket stages have durations supporting the theory that the lunar interior cannot have a liquid core.

The seismometers left on the Moon by the successive Apollo missions show that the Moon is much quieter seismically than the Earth. Natural moonquakes occur most frequently when the satellite is near perigee, a fact supporting the belief that tidal stresses produced periodically in the Moon's structure are at their maximum at such times.

The ages of rock samples from different parts of the Moon's surface lead to the conclusion that the main surface features produced by cosmic bombardment were formed over a few hundred million year period some 4×10^9 years or more ago, possibly not long after the creation of the Solar System. The extraordinary resemblance of the surface of the planet Mercury to that of the Moon is supporting evidence that the birth of the Solar System was not a peaceful event. Pictures of the satellites of Jupiter and Saturn sent back by the Voyager spacecraft confirm this opinion. On our own planet and its twin Venus, such tell-tale signs of early cosmic trauma must have been almost completely removed by weathering processes thousands of millions of years ago; Mars, on the other hand, still retains many traces of such events as well as evidence of more recent vulcanism.

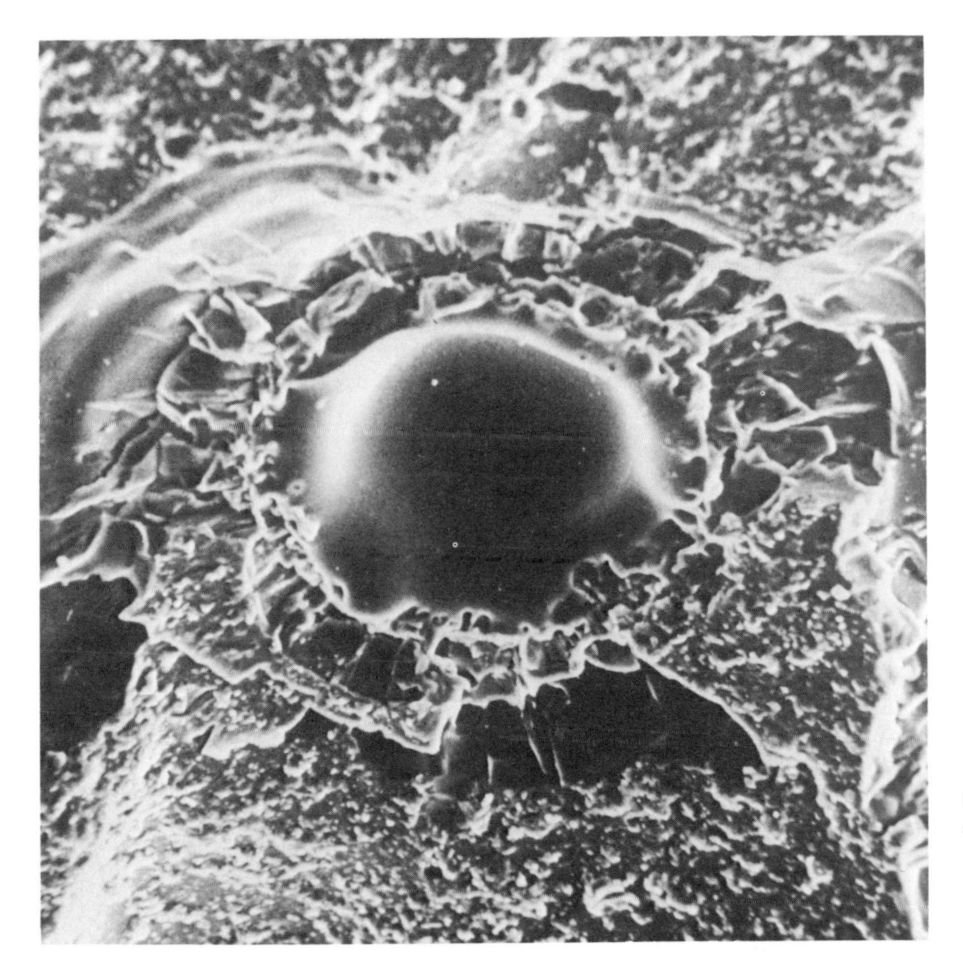

Fig. 3.12 A micro-crater in a lunar rock sample returned by Apollo 11; the crater has been magnified about 600 times by a scanning electron microscope. (By courtesy of the National Aeronautics and Space Administration)

The complete unravelling of the tangled skein of lunar geological evidence to give us an unambiguous history of the Moon and probably of the Solar System is only beginning. As yet it is only safe to say that at least three distinct aspects of lunar history are tentatively recognized:

(i) the creation of the Moon by accretion;
(ii) the main crater formation era by bombardment;
(iii) the flooding of the maria sites with basaltic lavas from the interior of the Moon caused by heat generated by the radioactive decay of uranium and thorium.

3.7.3 The Lunar Atmosphere Knowledge of the mass of the Moon and its radius allows us to estimate the lunar surface gravity. The value obtained is approximately one-sixth that of the Earth's surface gravity. As a consequence, it would not be expected that the Moon retains an atmosphere of any significance. This is confirmed by observations made by ground-based astronomers and by astronauts on the Moon's surface.

By the absence of refraction and absorption of the light from a star at the time of an occultation, an upper limit of a density of one two-thousandth that of the terrestrial atmosphere can be placed on the lunar atmosphere. The inability to detect light which would be scattered by any lunar atmosphere reduces this upper limit. Such scattered light would be detected more easily by polarization measurements rather than by ordinary brightness measurements and the most sensitive technique puts the upper limit at 10^{-9} times that of the density of the Earth's atmosphere. Radio source occultation experiments reduce the density limit even further to a figure of 10^{-14} times that of the terrestrial atmosphere.

These figures do not mean that the Moon has no atmosphere whatsoever, as there must always be a few gaseous molecules in the lunar environment either being swept up from the solar wind or set free from the lunar interior; the number of molecules, however, would provide an atmosphere barely distinguishable from a hard vacuum. Indeed it has been reported that the lunar surface occasionally exhibits localized transient phenomena which appear to be small eruptions, when gaseous and particulate material is ejected. The continuous meteoritic bombardment must also release gases trapped in the lunar surface material and some gases will be produced by the vaporization of particles on impact.

4
The Earth–Moon System

4.1 The Phases of the Moon

The Earth's one natural satellite, the Moon, must early in Man's history have excited interest and speculation. Casual observations over a number of nights would reveal that not only did the Moon change its shape but also that its position with respect to the fixed stars altered. More precise observations by the Greeks were required to detect that its apparent size changed as well.

It was early recognized that the Moon's shape depended upon its **age**, that is, the number of days that had elapsed since the previous new moon, when the satellite was in conjunction with the Sun. In Fig. 4.1, the inner circle shows the Moon's orbit, assumed circular, the Earth being at its centre. The Sun's direction is indicated and since the Sun's distance is some 400 times the Moon's, we can also assume that the Sun's direction as seen from the Moon is always parallel to its geocentric direction. The Moon's illumination being provided by the Sun, the day and night sides of the Moon at different parts of its orbit will be as shown.

The outer circle of figures then shows the appearance of the Moon as seen from the Earth, in other words the phases of the Moon. At A the Moon is **new**; at B a crescent is seen. **First quarter** occurs at C; between C and E more than half the Moon's illuminated face is visible, a condition known as **gibbous**. At D it is **full moon**; at E the position is called **third quarter**.

Two other terms are used. In general the phase may be drawn by using a semi-circle and a semi-ellipse or a straight line. The points of intersection of

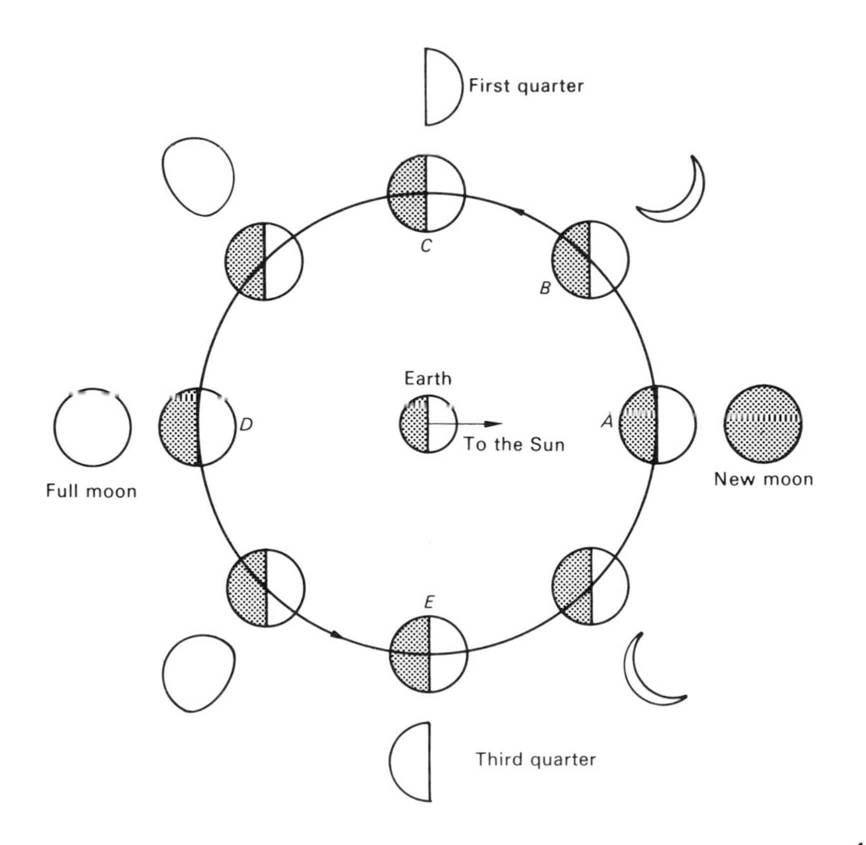

Fig. 4.1 The phases of the Moon

these curves are the **cusps**; the line separating the illuminated part of the Moon's surface from the dark part is the **terminator**.

In passing it may be mentioned that when the Moon is a thin crescent, the illuminated side of the Earth reflects sunlight to the dark part of the Moon. This **Earthlight** causes the Moon to appear as a bright crescent with a greyish light over the rest of the globe.

We can now define the Moon's **synodic period**, **lunation** or **lunar month**. Although the Moon's orbit suffers changes, an average value can be given for this period, defined to be the time interval between successive new moons. Its mean value is 29·53059 days.

The Moon's **sidereal period** or **sidereal month** is the time interval required by the Moon to make one complete revolution of the Earth with respect to the stellar background. Again an average value can be found, of 27·32166 days.

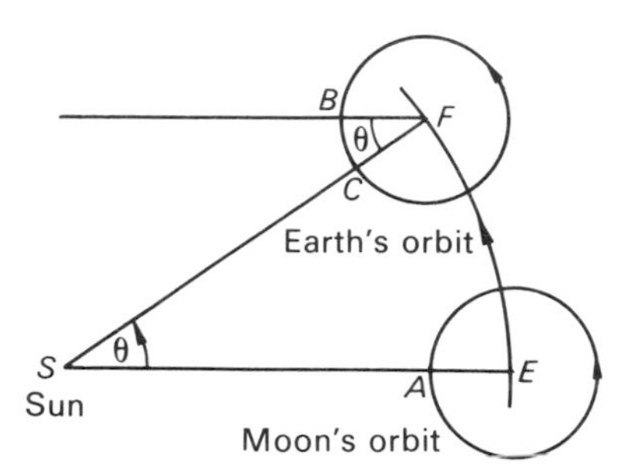

Fig. 4.2 The movement of the Earth between consecutive new moons

The difference between these two periods of time is due to the fact that the Moon has to travel a little further in its orbit to catch up with the Sun which, from the geocentric point of view, is also revolving round the Earth. The three quantities, namely the sidereal periods of revolution of the Moon about the Earth and the Earth about the Sun, and the Moon's synodic period, must then be related. This relation is easily found.

Let S, M and T denote the Moon's synodic period, its sidereal period and the Earth's sidereal period respectively.

In Fig. 4.2 let E and F be the positions of the Earth in its orbit at the times of two consecutive new moons. If S is the Sun, the Moon at those times will be at A and C respectively. The Moon will have arrived at B in its orbit one lunar sidereal period after it left point A where FB is parallel to EA.

Now $\angle$ESF is the angle travelled through by the Earth in time S, the Moon's synodic period. If θ is the value of this angle, then

$$\theta = \frac{360}{T} \times S. \tag{4.1}$$

During this time the Moon has revolved through an angle $(360 + \theta)$. But it revolves 360° in a time M, its sidereal period, so that we may write

$$360 + \theta = \frac{360}{M} \times S. \tag{4.2}$$

Eliminating θ between Equations (4.1) and (4.2), we obtain

$$360 + 360\frac{S}{T} = 360\frac{S}{M},$$

giving, after a little reduction, the required relation

$$\frac{1}{S} + \frac{1}{T} = \frac{1}{M}. \tag{4.3}$$

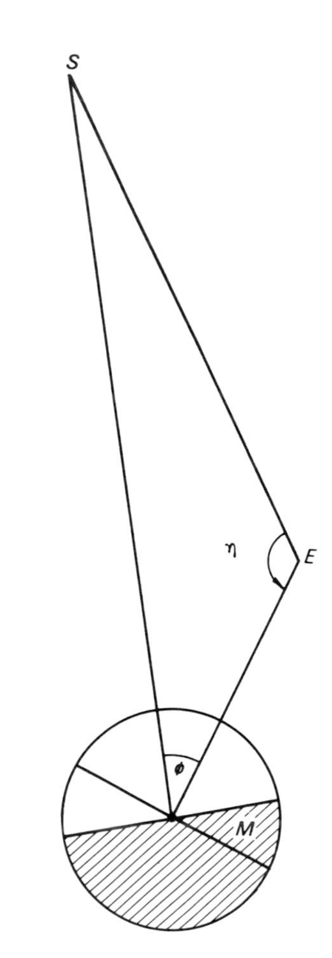

Fig. 4.3 The phase of the Moon

The value of the phase may be found in terms of the age of the Moon, A, and its synodic period S. The expression for the phase P of a planet was given by

$$P = \tfrac{1}{2}(1 + \cos\phi),$$

where ϕ was the angle between the planetocentric directions of Sun and Earth. If we now substitute the Moon M for a planet, as in Fig. 4.3, then we can make use of the fact that EM/ES $\sim 1/400$. That being so, $\angle$MSE is never more than a few minutes of arc and we can take MS to be parallel to ES without appreciable loss of accuracy.

Then if $\angle$SEM, the Moon's elongation, has value η, we may write

$$\phi = 180 - \eta,$$

giving

$$P = \tfrac{1}{2}(1 - \cos \eta).$$

But

$$\eta = 360(A/S),$$

so that the phase can be calculated, knowing A and S.

Finally in this section, the Metonic cycle, discovered by Meton in 433 B.C. is of interest. He discovered that the number of days in 19 years is very nearly a multiple of the Moon's synodic period. Thus

$$235 \times 29{\cdot}53059 = 6939{\cdot}689 \text{ days}$$

while

$$19 \times 365{\cdot}25 = 6939{\cdot}75 \text{ days},$$

taking the year to contain 365·25 days.

The small difference between these two figures ensures that if a full moon, or new moon, occurs on a particular date, full moon, or new moon, will occur on the same date 19 years later.

4.2 The Retardation of the Moon's Transit

Because the Moon's eastward movement against the stellar background is some 13° per day, the time interval between successive transits of the Moon is about $24^h\,50^m$ of mean solar time. Accordingly the Moon rises and sets about 50^m later each day *on average*. It will be remembered, however, that the length of the apparent solar day, while about 4^m longer than the sidereal day *on average*, varies throughout the year because the Sun moves in the ecliptic at a non-uniform rate. Similarly, because the Moon's orbit is inclined to the equator and the satellite's angular velocity is not constant, the daily retardation of the Moon's transit and moonrise vary from the average figures.

Let us assume for the moment that the Moon moves in the plane of the ecliptic. Then in $27\tfrac{1}{3}$ days its declination will change from about $+23\tfrac{1}{2}°$ to $-23\tfrac{1}{2}°$ and back again. Now it was seen that for the Sun the number of hours it is above the horizon on a particular day in latitude $\phi°$ N is given by $2H$, where

$$\cos H = -\tan \phi \tan \delta,$$

δ being the Sun's declination (see Roy & Clarke, *op. cit.*, Equation 7.6).

If δ is increasing, H is increasing, that is the day gets longer, sunrise being earlier each day. Similarly if δ decreases, H decreases and the day shortens, the Sun rising later each day. We can apply this reasoning to the Moon. While its declination is increasing, moonrise takes place earlier each night than it would have done if its declination had been constant; while its declination decreases, moonrise takes place later each night than it would have done if the normal average delay in the time of transit were added to the previous night's time of moonrise.

A case of particular interest is the full moon that occurs nearest the time of the autumnal equinox, around September 21st. The Sun is then on the equator so that the full moon must be near the vernal equinox and therefore crossing the equator from south to north. In Fig. 4.4 we see that at the autumnal equinox the ecliptic is least inclined to the horizon and so on a number of successive nights, the full or nearly-full moon rises shortly after sunset. The use farmers traditionally made of having bright moonlight in the early evenings led to this full moon being called the **Harvest Moon**.

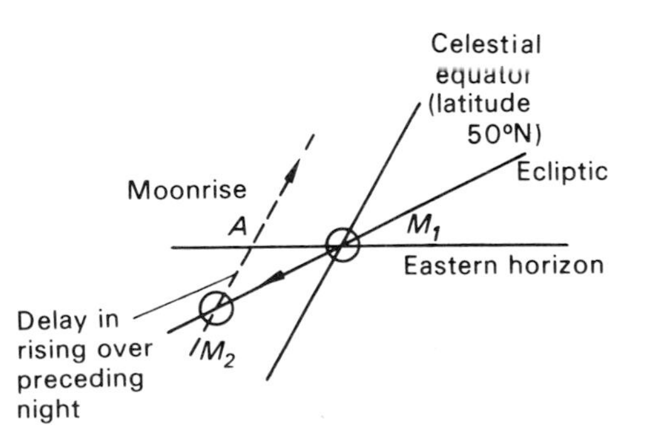

Fig. 4.4 Delay in moonrise on consecutive nights

Even the full moon in October following the harvest moon exhibits a similar effect, though not so pronounced. It is called the **Hunter's Moon**.

The above phenomena also take place for anyone living in the southern hemisphere, but with a six months shift in calendar dates.

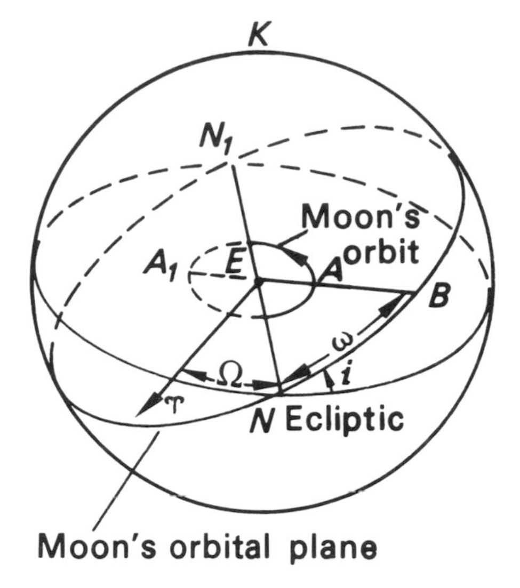

Fig. 4.5 Elements of the lunar orbit defining its orientation in space

4.3 The Moon's Orbit

In the previous section, we assumed the Moon to move in the plane of the ecliptic. In fact the Moon's orbital plane lies at an angle of inclination to the former plane. The orbit is elliptic. The mean values of the semi-major axis, a, the eccentricity, e, and the inclination, i, are given below:

$$a = 384\,400 \text{ km} = 238\,850 \text{ miles}$$

$$e = 0{\cdot}05490$$

$$i = 5^\circ\, 09'.$$

Because of solar perturbations, all three elements are subject to periodic variations about these values. In particular the eccentricity varies from 0·044 to 0·067 while the inclination oscillates between 4° 58′ and 5° 19′. The planets themselves, and the departures of the Earth and the Moon from perfect spheres, also affect these three elements of the Moon's orbit, but to a much lesser extent. Their effects are also periodic in character.

The other three elements of the Moon's orbit are the **longitude of the ascending node**, Ω, (measured in Fig. 4.5 from ♈ along the ecliptic to the ascending node N), the **argument of perigee**, ω, (measured from N along the orbital plane to the projection of perigee A on the celestial sphere at B), and the **time of perigee passage** τ, which, as its name implies, is simply a Greenwich Date when the Moon was at perigee.

Because of the Sun's gravitational disturbance of the Moon's orbit about the Earth, all three elements suffer not only periodic changes but also secular ones as well. The **line of apsides** (the line joining perigee and apogee) advances while the line joining the ascending node N to the descending node N_1 regresses, making the circuit of the ecliptic in 6798·3 days or almost $18\frac{2}{3}$ years. Again, the planets' attractions and the shapes of Earth and Moon contribute in a minor degree to these disturbances.

Various periods of revolution of the Moon in its orbit may therefore be defined, namely the **sidereal** (the time required for the Moon to move through 360°), the **synodic** (the time between successive new moons), the **nodical** (the time between successive passages through the ascending node), the **anomalistic** (the time between successive passages through perigee), and the **tropical** (the time between successive conjunctions with the vernal equinox). The mean values of these *months* are given in Table 4.1 below.

Although in any revolution of the Moon in its orbit these months may differ by a few hours from the mean values given in the table, the mean values remain steady over many centuries to within one second.

Table 4.1 The lengths of lunar months

Length of month	Days	d.	h.	m.	s.
Synodic	29·53059	29	12	44	03
Sidereal	27·32166	27	07	43	12
Anomalistic	27·55455	27	13	18	33
Nodical	27·21222	27	05	05	36
Tropical	27·32158	27	07	43	05

The construction of a complete lunar theory, an analytical description of the Moon's motion that includes the effects of Earth, Sun, planets and the figures of Earth and Moon, is one of the most difficult in astronomy. Newton, Euler, Clairaut, Hansen, Delaunay, Hill and Brown are some who have worked on this problem using many different approaches. Brown's lunar theory and his *Tables of the Motion of the Moon* (1919) are the most exhaustive treatment of the problem. His theory contains 1500 separate terms and is still used in preparing the lunar ephemeris. For example the first few terms in the expression for the Moon's longitude, λ, are given approximately by

$$\lambda = L + 377 \sin l + 13 \sin 2l + 74 \sin (2D - l) + 40 \sin 2D - 11 \sin l' + \cdots,$$

where the numbers are expressed in minutes of arc, L is the Moon's mean longitude, l is the angular distance of a fictitious mean moon from the mean perigee, D is its distance from the mean sun and l' is the mean sun's distance from the perigee point of the Sun's apparent orbit about the Earth.

Essentially similar series give the Moon's latitude and parallax. The terms in l and $2l$ are ordinary elliptic two-body terms. The term in $(2D - l)$ is the **evection** and is due to the variation in the eccentricity of the orbit caused by the Sun's gravitational pull. Its period is 31·8 days. The term in $2D$ is the **variation**, a disturbance in the Moon's motion due to a variation in the magnitude of the solar perturbing force during a synodic month. The other main inequality, the **annual equation**, given by the term in l', has a period of one anomalistic year and is due to the annual variation of the Earth's distance from the Sun.

4.4 The Moon's Librations

The Moon's sidereal rotation period is equal to its sidereal period of revolution about the Earth. Because of this the Moon always presents the same hemisphere towards the Earth. This locking of the Moon is a dynamical consequence of the Moon's shape. The Moon is a triaxial ellipsoid, the largest axis always pointing more or less towards the Earth's centre. If the axis were displaced slightly, the Earth's gravitational attraction would tend to restore the former situation. A rough analogy is that of a pendulum. If slightly displaced from the vertical, gravity brings it back to the vertical.

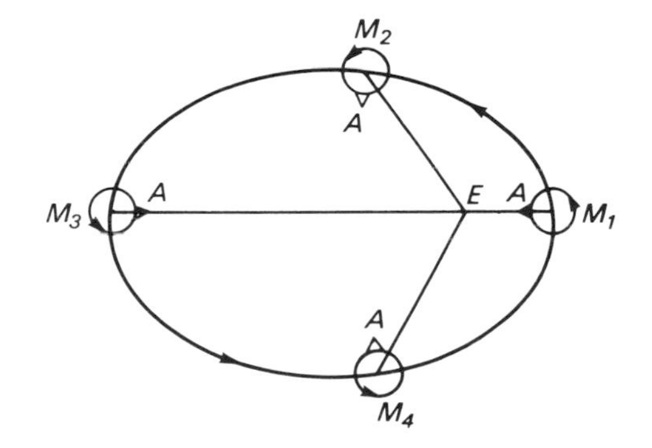

Fig. 4.6 The phenomenon of libration in longitude

At the same time, although the Moon rotates with constant angular velocity, the radius vector joining its centre to that of the Earth does not, but sweeps out area in accordance with Kepler's second law. In Fig. 4.6 the effect of this on the position of a mountain A seen in the centre of the Moon's visible disk at perigee (M_1) is considered.

One quarter of the period of revolution after the time of perigee passage, the Moon is at M_2, having *rotated* through 90° but having *revolved* in its orbit through about 97°. The mountain A therefore appears to have moved about 7° in longitude from the central point in the disk as measured from the Moon's centre. By the time half the period has elapsed, the Moon is at M_3 and the mountain is back at the centre of the disk. In the other half of the orbit, the mountain shifts about 7° in the opposite direction before returning to its original position. This libration is called **the libration in longitude**.

The Moon's axis of rotation is inclined at an angle of about $6\frac{1}{2}°$ to the pole of its orbit about the Earth. For half the sidereal period, therefore, the Moon's north polar regions are tilted towards the Earth; during the other half, the south polar regions are presented, as illustrated in Fig. 4.7. This is called **the libration in latitude**.

The third libration, **the diurnal libration**, results from observing the Moon from a finite-sized, rotating Earth. We have already made use of this fact in

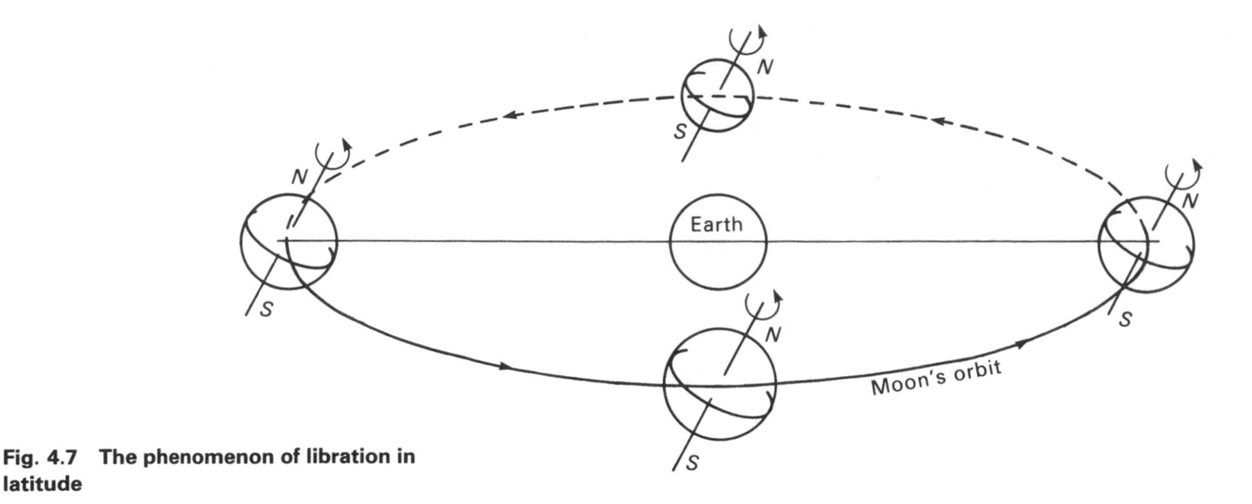

Fig. 4.7 The phenomenon of libration in latitude

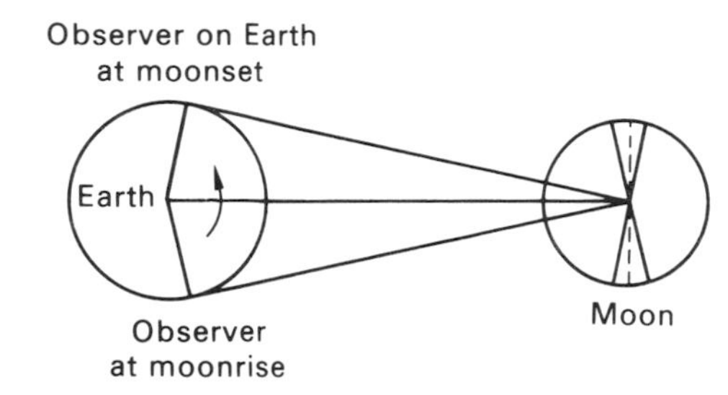

Fig. 4.8 The phenomenon of diurnal libration

measuring the Moon's distance. The observer's shifting position due to the diurnal rotation of the Earth (of diameter about one-thirtieth of the lunar distance) enables him to see farther over the western limb at moonrise and the eastern limb at moonset (see Fig. 4.8, where the effect is grossly exaggerated).

In the days before the advent of lunar probes and artificial lunar satellites, from which the entire surface of the Moon has been photographed and mapped, these librations—noted by Galileo and correctly interpreted by him—allowed examination of rather more than 59% of the Moon's surface.

4.5 The Tides

An important terrestrial phenomenon caused by the combined gravitational fields of Sun and Moon is the rise and fall of the ocean twice a day. Newton was the first to give an adequate explanation, although the connection between the tides and the phase of the Moon had been noted from antiquity.

The main tidal phenomena are that the times of successive high waters or **high tides** are separated by about $12^h 25^m$, as are the times of successive low waters or **low tides**, low tides occurring half-way between high tides. In addition, particularly high tides—the **spring** tides—occur at new and full moon. Half-way between these spring tides, when the Moon is at first or third quarter, the day's high tides are at their lowest. Such weak high tides are called **neap** tides. The behaviour of the tides at a particular place during the first half of a lunation is illustrated in Fig. 4.9.

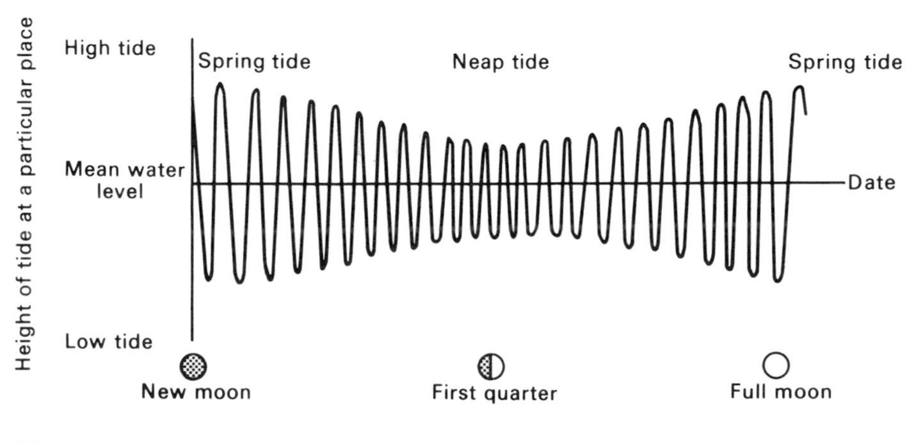

Fig. 4.9 The variations of heights of tides according to the lunar age

We can best understand the above tidal phenomena by imagining the Earth to be non-rotating and covered with water. Let the Moon's gravitational pull act alone. At the Earth's centre, centrifugal force due to the Earth being in orbit about the centre of mass of Earth and Moon is balanced by their gravitational attraction. On the side of the Earth nearest the Moon, however, the Moon's gravitational attraction is greater than centrifugal force while on the opposite side the centrifugal force is greater. In both cases the resultant force acts to produce a tidal bulge in the oceans. So two tidal bulges are produced that follow the Moon in its orbit round the Earth. An island on the non-rotating Earth would experience a high tide *once every 13.65 days* or half the Moon's sidereal period of revolution.

The Earth is rotating, however, once every 24^h in the direction of the Moon's movement. The combination of these two periods produces two high tides every $24^h\ 50^m$, in other words at the same interval we found in discussing the retardation of the Moon's transit.

Again the picture is distorted by the fact that the Earth's axis of rotation is not at right angles to the Moon's orbital plane. The complicated configuration of the land masses and the varying depths of the oceans further modify the picture.

If we now introduce the Sun, it will be seen that when the Moon is new or full, the Sun's effect will reinforce the Moon's, producing particularly high tides. At first and third quarters the Sun's tide-raising force, less than half as effective as the Moon's will tend to cancel the lunar tide-raising force as indicated in Fig. 4.10.

The tides produce important long-term effects in the Earth–Moon system.

The Earth rotates faster than the Moon moves in its orbit. Because the tides are linked to the more slowly-moving Moon, they act by friction as a brake on

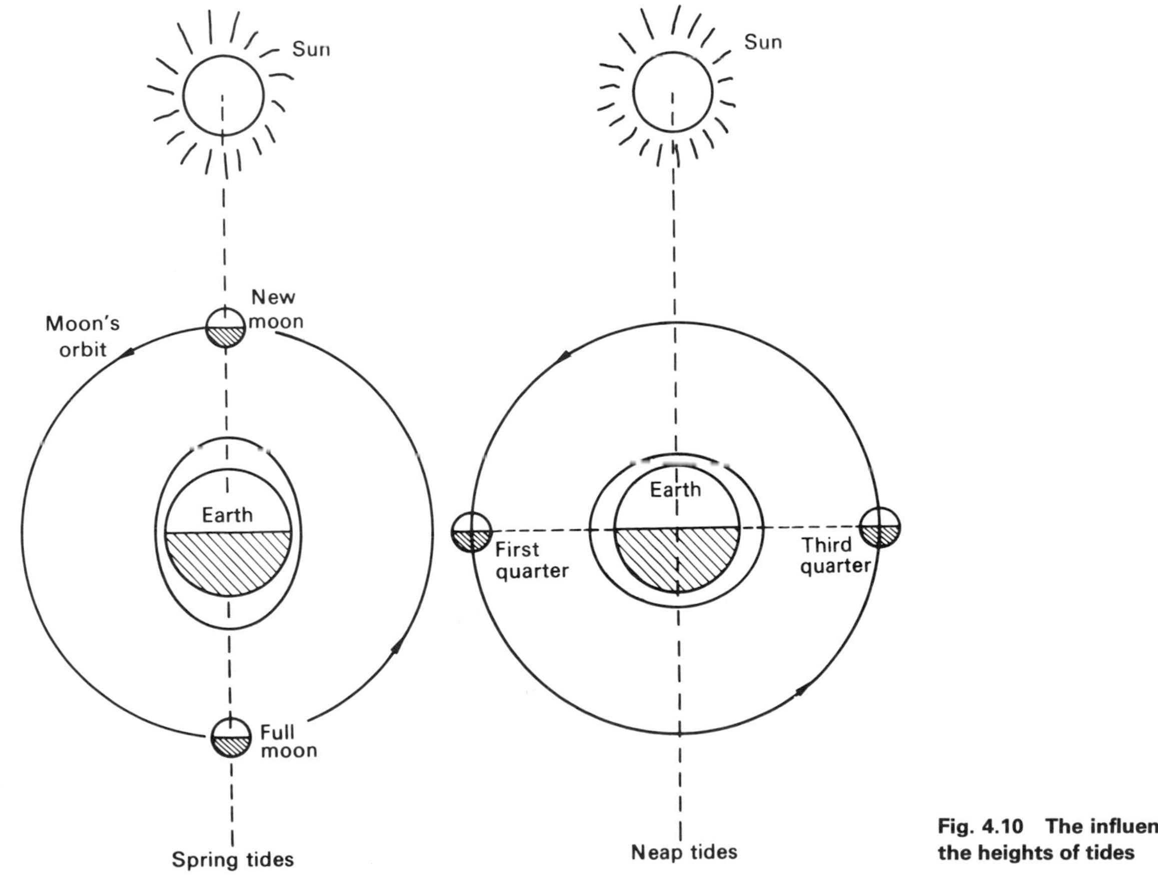

Fig. 4.10 The influence of the Sun on the heights of tides

the Earth's rotation, gradually slowing it down. The rate of increase in the rotational period of the Earth is found to be $0^{\mathrm{s}}.0016$ per century, a figure arrived at by a study of ancient eclipse records.

The angular momentum lost by the rotating Earth in this process is transferred to the Moon's angular momentum. It is accelerated in its orbit, causing it to spiral slowly outwards. The day and the month are thereby lengthening at different rates and the state of affairs both in the remote past and the remote future has been calculated. Traced backwards to a time when the Moon was only about 16 000 km from the Earth, the month was approximately seven mean solar days (present value) long; the day was only a few hours in length. In the future, thousands of millions of years hence, the day and month will be equal, about 47 of our present days long, and the Earth will always turn the same face towards the Moon.

It may be shown that this state of affairs will not persist for ever; solar tides will continue to act. Gradually the Earth and Moon will be unlocked from their confrontation and the satellite will retrace its steps until it once more approaches the Earth.

4.6 The Moon's Orbit: Practical Projects

A simple exercise, allowing the eccentricity of the Moon's orbit to be determined, is available from *Sky Publishing Corporation.* The material for the data consists of a series of photographs of the Moon, depicting how the apparent size of the Moon changes during its orbit. These changes are a reflection of the variations in the distance of the Moon from the Earth as the orbit is described. Measurements made on the photographs allow the production of a scaled drawing for the lunar orbit.

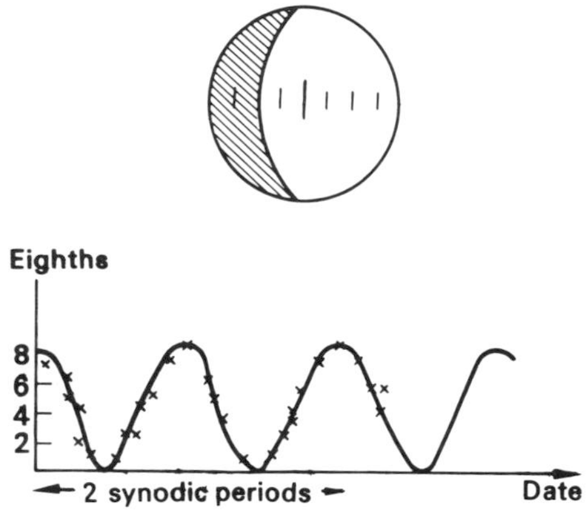

It is possible, without having accurate astronomical instruments, to take observations over a few months that enable values of the Moon's orbital elements to be deduced. Observations are made of the Moon's phase, its sidereal position and its angular size.

The phase measurements are the easiest to make. If a diameter of the Moon's disk is divided into eighths, then it is possible with the naked eye to obtain fairly accurately the number of eighths of the diameter at right angles to the line joining the cusps that are illuminated. For example, in Fig. 4.11 the phase would be noted as 6. The date and time of the observation is also noted. If observations are made in this fashion over some months and graphed against time, an accurate value of the synodic period S can be found.

Fig. 4.11 Estimation of the fraction of the lunar disk that is illuminated and the variation of this fraction with time

By using a sky map such as *Norton's Star Atlas*, the Moon's sidereal positions on those nights when stars and Moon are visible may be plotted on the map. The Moon's right ascension and declination (read off from the star charts) and the time and date are recorded. As the months pass, it will be seen that not only does the Moon make circuits of the heavens, crossing from one side of the ecliptic to the other, but the ascending and descending nodes regress. It will be remembered that the nodes are the points on the ecliptic at which the Moon is seen to cross that plane. After six months, the nodes will have regressed by almost 10 degrees. From a track on the map, or from a graph of right ascension against date and time, a value for the Moon's sidereal period T may be found.

Changes in the Moon's angular size are more difficult to detect by the naked eye. The Moon's orbital eccentricity has a value of the order of 1/20, so that the ratio of minimum to maximum angular semi-diameter is about 9/10.

The type of equipment required is a long, graduated rod (AB in Fig. 4.12). Two metre sticks nailed end to end are suitable. A small piece of wood that can be slid along the rod is also required. A pencil stub attached to the metre sticks with an elastic band is good enough.

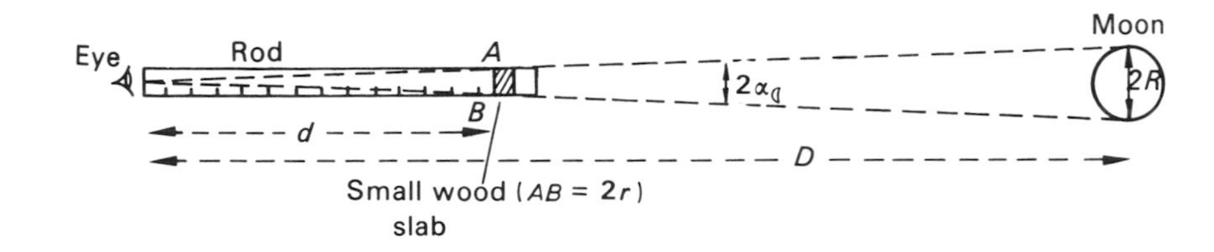

Fig. 4.12 A simple device for measuring the angular diameter of the Moon

The rod is pointed in the direction of the Moon. By trial and error, a position is found for the piece of wood where its angular size equals that of the Moon. Let this position be distance d from the eye.

Then if $\alpha_{☾}$ and R are the semi-diameter and radius of the Moon respectively, while $2r$ is the length of the piece of wood as shown in Fig. 4.12

$$\alpha_{☾} = r/d = R/D, \tag{4.4}$$

where D is the Moon's distance from the observer.

Let D_P, D_A be the Moon's distances at perigee and apogee. Then

$$D_P = a(1-e); \qquad D_A = a(1+e),$$

so that

$$\frac{D_P}{D_A} = \frac{1-e}{1+e},$$

giving

$$e = \frac{D_A - D_P}{D_A + D_P}. \tag{4.5}$$

But by Equation (4.4) d is proportional to D, so that we may rewrite Equation (4.5) as

$$e = \frac{d_A - d_P}{d_A + d_P}. \tag{4.6}$$

A large number of readings over many months are taken. A graph of d against time is desired in order to obtain the maximum (d_A) and the minimum (d_P) values. In practice, however, it is found that the intrinsic errors in this crude method make it difficult to draw a suitable curve among the points as plotted (see Fig. 4.13(*a*)). Fortunately we know that the variation in distance is periodic and we can use this to enable us to abstract the required information.

The period in question is the anomalistic period, the time it takes the Moon to go from perigee to perigee. The anomalistic month is not all that different in value from the length of the sidereal month T, a value of which should have been obtained by this time.

If the x-axis or time-axis is then divided into such periods, and the readings within each period lifted and re-plotted into the first sidereal period, the trend of the graph is usually obvious, as shown in Fig. 4.13(*b*).

From the graph (see Fig. 4.13(*b*)), values of d_A and d_P can be found, enabling a value of the eccentricity of the Moon's orbit to be calculated from Equation (4.6).

Having obtained the eccentricity e, we can find a date for the time of perigee passage from the same graph. It is the date at which the Moon's distance D was a minimum, and therefore the date τ in Fig. 4.13.

From the track of the Moon's orbit plotted on the star atlas, it is easy to find a value for the longitude ☊ of the ascending node N of the orbit on the ecliptic by estimating the angular distance along the ecliptic from ♈ to N.

The inclination i is best found by estimating the angular distance between the ecliptic and the orbit, 90° from the nodes (see Fig. 4.14).

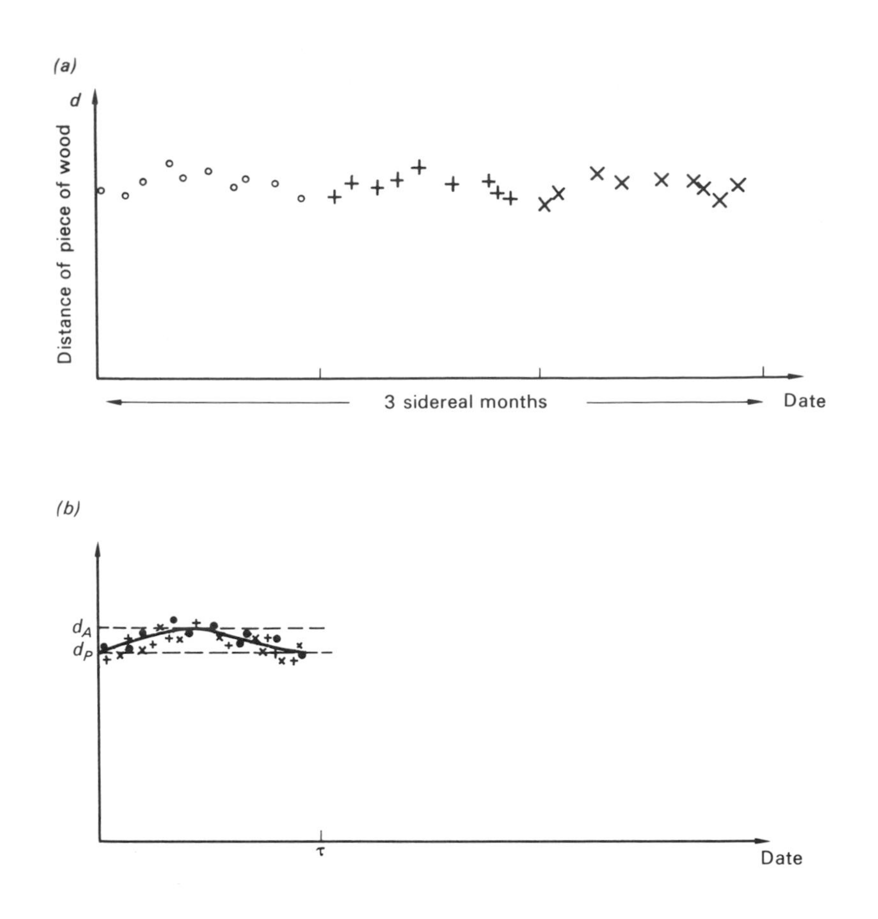

Fig. 4.13 The variation of lunar distance with time: (*a*) over three months, (*b*) by superimposing the results into one sidereal month

The argument of perigee, ω, that is the angular distance between the direction of perigee and the ascending node N, is found by noting on the star chart the position of perigee. This can be done from our knowledge of the time of perigee passage and from the observations of the Moon's sidereal position at known times of observation.

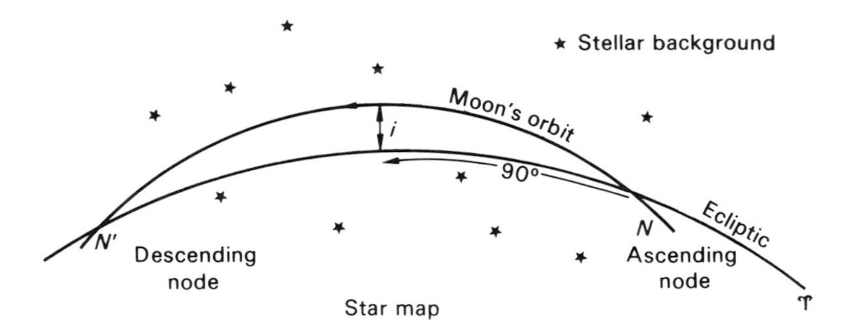

Fig. 4.14 Estimation of the inclination to the ecliptic of the lunar orbit

The five elements thus obtained, namely e, i, Ω, ω, τ, give a scale model of the Moon's orbit. To obtain the scale we require a knowledge of a, the semi-major axis. It is possible to obtain this also from simple observations carried out by one person with a star atlas.

4.6.1 Measuring the Moon's Distance Suppose that at full moon, an observer notes carefully the position of the Moon on a star atlas every hour, say from four hours before midnight to four hours after midnight. The difficulty of seeing stars relatively close to the bright Moon can be solved by holding a small disk at arm's length over the Moon to eclipse most of its light. Then the Moon's observed shift in sidereal position will be caused by (i) the Moon's motion in its orbit, (ii) the movement of the observer due to the Earth's rotation.

(*a*) *The parallactic shift* In Fig. 4.15 the observer in latitude ϕ° N is moved from O_1 to O_2 during the night's observations, of duration t hours. Forgetting for the moment the Moon's orbital movement, let it be at M while observations

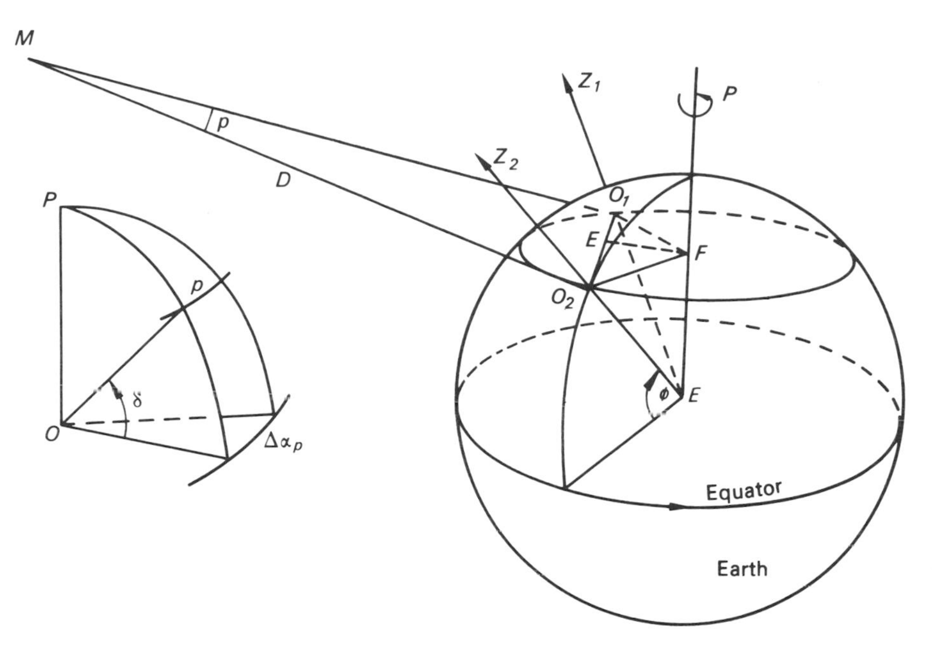

Fig. 4.15 Determination of the Moon's distance by measuring the effect of diurnal parallax on the Moon's sidereal position

are being made. Then, if δ is the Moon's declination, and p is the parallactic angle moved through by the Moon during t hours, we may write

$$p = \Delta\alpha_p \cos \delta, \tag{4.7}$$

where $\Delta\alpha_p$ is the parallactic displacement in right ascension. It may be noted that as the observer moves along a small circle parallel to the equator there is no corresponding parallactic shift in declination.

Let the Moon be on the meridian at a time halfway through the period of observation. Then, since p is a small angle,

$$p = O_1O_2/D, \tag{4.8}$$

D being the Moon's distance from the observer.

By Equations (4.7) and (4.8),

$$D = \frac{O_1O_2}{\Delta\alpha_p \cos \delta}. \tag{4.9}$$

Now in t hours, the angle O_1FO_2 rotated through by the observer is θ, where

$$\theta = \frac{t}{24} \times 360°.$$

If $R_\oplus$ is the radius of the Earth,

$$O_2F = O_2E \cos \phi = R_\oplus \cos \phi.$$

Also

$$O_1O_2 = 2 \times O_2E = 2 \times O_2F \times \sin(\theta/2).$$

Hence

$$O_1O_2 = 2R_\oplus \cos\phi \sin\psi, \tag{4.10}$$

where $\psi = 7{\cdot}5\, t^{\mathrm{h}}$ and is measured in degrees.

Substituting in Equation (4.9), we obtain

$$\frac{D}{R_\oplus} = \frac{2\cos\phi \sin\psi}{\Delta\alpha_{\mathrm{p}} \cos\delta} \tag{4.11}$$

giving the Moon's distance in Earth radii.

If, for example, observations were made over a six-hour period, the value of ψ is 45° and by Equation (4.10), $O_1O_2 \approx R_\oplus$. Since the Moon's distance is about 60 Earth radii, the parallactic angle is approximately 1 degree and, for an observer near 60° north latitude, $\Delta\alpha_{\mathrm{p}}$ is almost two degrees.

(*b*) *The Moon's Orbital Motion* During the night the Moon will have moved in its orbit. Let us suppose we are hoping to measure the Moon's distance to an accuracy of about 10%. We remember that the Moon's mean motion is n, where $n = 360°/27\frac{1}{3}$ days $\approx \frac{1}{2}°$/hour. In six hours, therefore, the Moon's true anomaly v will increase by about 3 degrees. But by Kepler's second law, the angular velocity varies because of the eccentricity of the Moon's orbit. We must take this variation into account, at least to the first order.

It is known that to the first power of the eccentricity, the small increase Δv in the Moon's true anomaly, v, in a small time interval, Δt, is given by

$$\Delta v = n\Delta t(1 + 2e\cos v). \tag{4.12}$$

If we take e to be of the order of 0·055, then

$$\Delta v = n\Delta t(1 + 0{\cdot}11\cos v),$$

showing that the shift $n\Delta t$ in the Moon's true anomaly could be different from its mean value by an amount of the order of ±10% of its mean value, depending upon which part of its elliptical orbit it is in.

Equation (4.12), however, enables us to calculate Δv accurately enough. The eccentricity e will have been measured; n and $\Delta t(=t^{\mathrm{h}})$ are known and v, the angle between the direction of perigee and the Moon's direction, can be found from the observer's graphs.

It remains to compute what change in the Moon's right ascension, $\Delta\alpha_0$ say, will have been made because the Moon's orbital movement has led to an increase in its true anomaly of amount Δv.

In Fig. 4.16 the Moon's orbital plane, its orbit and the plane of the Earth's equator are shown.

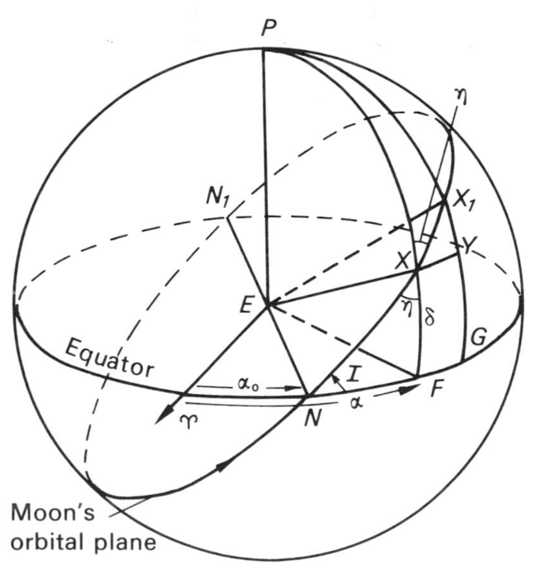

Fig. 4.16 The celestial sphere depicting the Moon's orbital plane

Now the angular shift Δv is small so that we may consider triangle XX_1Y to be plane, XY being the arc of a small circle parallel to the equator. If $\angle PXX_1$ is taken to be of size η, we can write

$$XY = XX_1 \sin\eta, \tag{4.13}$$

or

$$XY = \Delta v \sin\eta.$$

In $\triangle XNF$, $\angle XNF$ is the inclination I of the Moon's orbital plane *to the plane of the equator*, $\angle NXF = \eta$, $\angle XFN = 90°$.

If α and α_0 are the right ascensions of the Moon and the ascending node N of the Moon's orbit (again with respect to the equator),

$$NF = \alpha - \alpha_0.$$

Then by the sine formula in $\triangle NFX$,

$$\sin \eta = \frac{\sin(\alpha - \alpha_0) \sin I}{\sin \delta}. \tag{4.14}$$

Equation (4.13) becomes

$$XY = \frac{\Delta v \sin(\alpha - \alpha_0) \sin I}{\sin \delta}.$$

But $XY = \Delta\alpha_0 \cos \delta$, and so

$$\Delta\alpha_0 = \frac{\Delta v \sin(\alpha - \alpha_0) \sin I}{\sin \delta \cos \delta}, \tag{4.15}$$

from which $\Delta\alpha_0$, the shift in the Moon's right ascension due to its orbital motion can be found. It may be noted in passing that, as in finding i, the most accurate way of measuring I is to measure on the star chart the angular distance between the graph of the Moon's orbit and the equator at points 90° from the nodes (N and N_1 in Fig. 4.16).

Now the observed shift in right ascension $\Delta\alpha$ will be made up of the parallactic shift $\Delta\alpha_p$ and the orbital shift $\Delta\alpha_0$. Thus

$$\Delta\alpha = \Delta\alpha_p + \Delta\alpha_0,$$

giving

$$\Delta\alpha_p = \Delta\alpha - \Delta\alpha_0.$$

Care should be taken with regard to the algebra, since $\Delta\alpha_p$ is essentially a negative quantity while $\Delta\alpha_0$ is positive.

Having found $\Delta\alpha_p$, Equation (4.11) gives the Moon's distance in Earth radii. If use is made of the polar equation for the ellipse, namely,

$$r = \frac{a(1 - e^2)}{1 + e \cos v},$$

a value for a can then be calculated.

5 Eclipses

5.1 Introduction

In its orbital movement round the Earth, the Moon on occasion casts its shadow on the Earth's surface. Such an event is called a **solar eclipse** or **eclipse of the Sun**. On other occasions the Earth's shadow is intercepted by the Moon and these events are termed **lunar eclipses** or **eclipses of the Moon**. Because both Earth and Moon are smaller than the Sun, the shadows cast by these bodies are cone-shaped. The length EV of the Earth's shadow-cone (Fig. 5.1) is about 1 370 000 km while that of the Moon MW is of the order of 374 000 km.

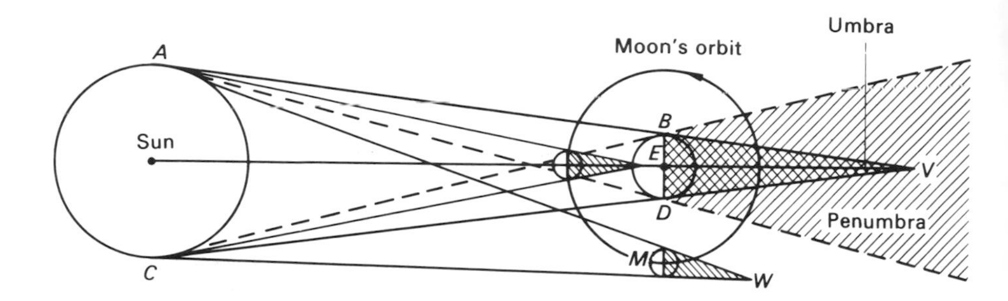

Fig. 5.1 The shadows cast by the Earth and the Moon

The Earth's shadow-cone is called the **umbra** and is formed by drawing the direct tangents (such as AB and CD) to the surfaces of Sun and Earth. The transverse tangents define another region, the **penumbra**, within which only part of the Sun's disk would be visible. It is therefore a region wherein objects such as the Moon or an artificial satellite shine with diminished brightness. On entering the umbra from the penumbra the artificial satellite, being a starlike object, is seen to wink out of visibility; in the case of an extended object, the edge of the Earth's shadow begins to cross its surface.

The Moon's umbra and penumbra are likewise defined by the direct and transverse tangents drawn to the surfaces of Sun and Moon.

It is obvious that a solar eclipse can take place only when the Moon is new. A lunar eclipse, on the other hand, can occur only around the time of full moon. The lengths of the two shadow-cones, taking into consideration the radius of the Moon's orbit (namely, 384 000 km), would suggest that the Moon should be eclipsed once every lunation but that it would be impossible for the Moon's umbral cone to reach the Earth's surface except on favourable occasions when the Moon was at perigee. However, a solar eclipse of a special kind, namely **annular**, would still take place at each new moon. At such a time the limb of the Sun is seen as a ring or annulus surrounding the black disk of the Moon.

However, lunar and solar eclipses do not occur each lunation. The Moon's orbital plane is inclined at an angle of about 5° (twenty semi-diameters of the Moon) to the plane of the ecliptic, so that it is only when the Moon is near one of its nodes at a time of new or full moon that a solar or lunar eclipse can take place. This further condition cuts down the number of possible eclipses drastically, the maximum number each year being seven. Of the seven, either four are solar and three lunar, or five are solar and two lunar. In some years there may be as few as two eclipses.

Thus in Fig. 5.2, no eclipse can take place at new or full moon at *A*, but a solar or lunar eclipse could take place at these times at *B* when the line of nodes is very near the line joining Sun and Moon.

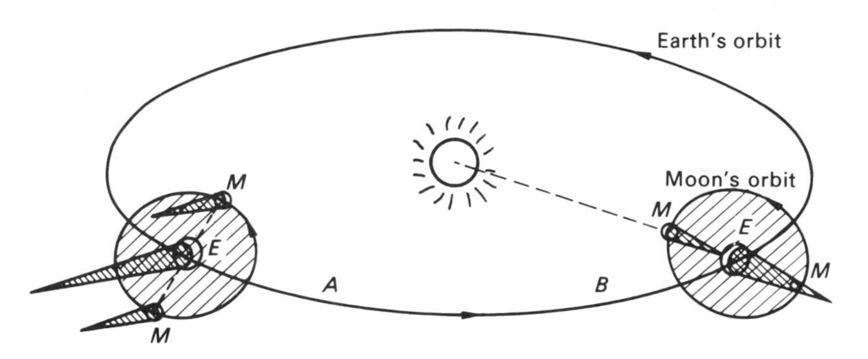

Fig. 5.2 The effect of the inclination of the Moon's orbit to the ecliptic on the positions of the shadow cones

5.2 Types of Eclipses

Lunar eclipses may be **total** or **partial**. A total lunar eclipse occurs if the entire disk of the Moon lies within the Earth's umbra or shadow-cone at some time during the eclipse. Otherwise, the eclipse is said to be a partial one.

Solar eclipses can be **total, partial** or **annular**. For a total eclipse to take place, the Moon's disk must cover the Sun's disk completely. This will happen if the Moon's semi-diameter is greater than the semi-diameter of the Sun. From the observer's point of view, the Moon's shadow-cone must intersect the Earth's surface and the observer must be within the circular area of intersection if he is to see a total solar eclipse. The situation is illustrated in Fig. 5.3(*a*).

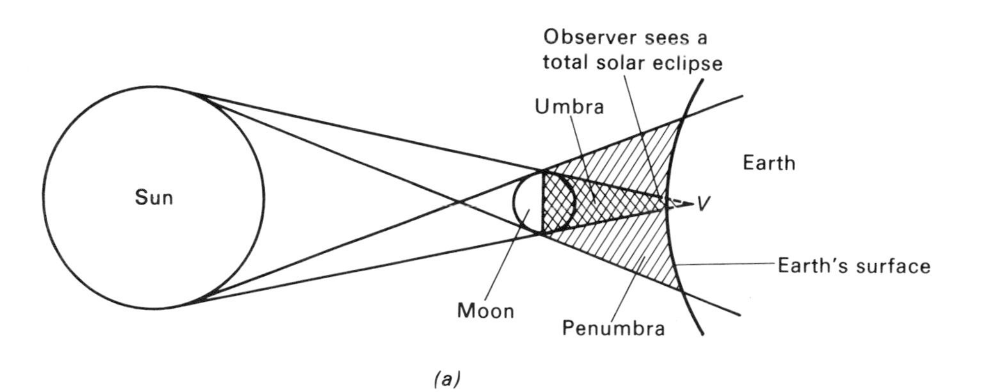

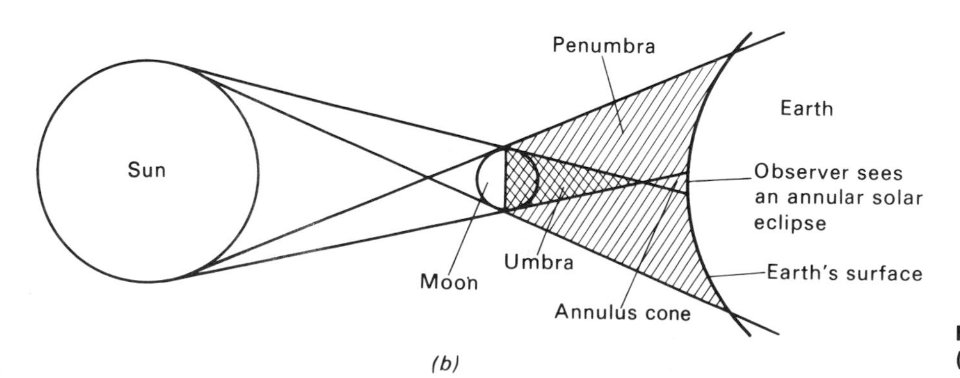

Fig. 5.3 Eclipses of the Sun: (*a*) total, (*b*) annular

Just prior to the beginning and end of totality, a bright broken arc of light can sometimes be seen around the lunar limb. The phenomenon is known as **Baily's beads**. It results from the fact that the Moon has an irregular surface and does not present a disk which is exactly circular. The "beads" of light correspond to depressions or craters that are seen in profile on the lunar limb.

If the Moon's semi-diameter is less than the Sun's, and the vertex V of the Moon's umbra falls short of the Earth's surface, the observer within the area bounded by the extension of the direct tangents to Sun and Moon will see an annular solar eclipse.

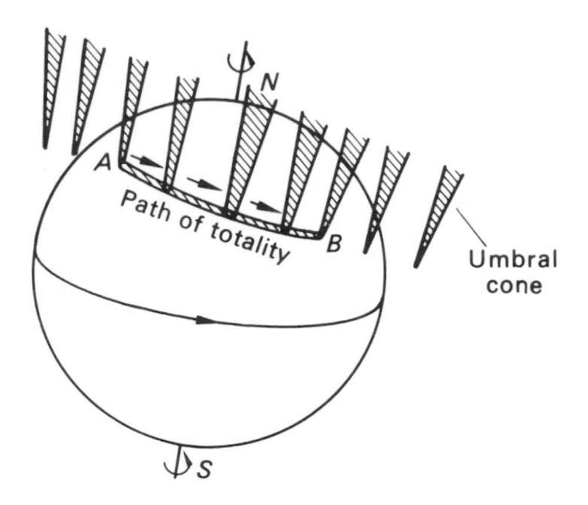

Fig. 5.4 The path of totality

The limit on the Earth's surface within which a solar eclipse is visible is always less than 300 km. Surrounding this spot there is the region intersecting the Moon's penumbra. This zone may be 5000 km across. Anyone within it will see part of the Sun's disk blotted out by the Moon and so will be able to observe a partial solar eclipse.

The Moon not only revolves in an elliptic orbit inclined to the ecliptic, but the Earth rotates about an axis which is not perpendicular to that orbital plane. The small circle of intersection of the Moon's umbral cone or the annulus cone with the Earth's surface therefore traces out a path—the **path of totality**—that depends upon the exact geometry of the Earth–Moon–Sun system for the particular eclipse.

In Fig. 5.4 is sketched the path of totality AB resulting from the intersection of the Moon's umbral cone with the Earth's surface. The speed with which the shadow moves across the surface varies from place to place. It can be as low as 1800 km per hour or as high as 8000 km per hour. For a particular place, therefore, totality never lasts more than about $7\frac{1}{2}$ minutes, although in recent years fast planes such as Concorde have been used to fly along the path of totality and so increase for an observer on board the duration of total solar eclipse to well over an hour.

It might be thought that with such a complicated and variable geometry, every eclipse would be entirely different from every other one. We shall see, however, that this is by no means the case. Firstly, we consider the behaviour of the Moon's nodes in more detail.

5.3 The Eclipse Year

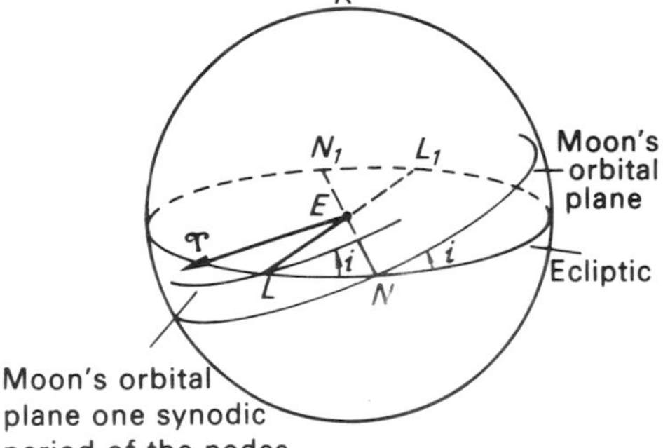

Fig. 5.5 Progression of the Moon's node over one synodic period

It has been seen that the Moon's orbital plane precesses backwards, the line of nodes NN_1 in Fig. 5.5 making one circuit of the ecliptic in a period Q of length $18\frac{2}{3}$ years. We can define the synodic period of the Moon's ascending node to be the time S between successive conjunctions of the node with the Sun. Let T be the length of the sidereal year.

Let the Sun and the ascending node N have the same longitude at some epoch. The Sun will move in the direction of increasing longitude at a rate of about 1° per day. The node N will precess in the opposite direction and will move through an angle θ, equal to NL, before it meets the Sun again at L. This period S is necessarily shorter than one year. Then

$$\theta = 360 \times S/Q.$$

Also, in this time, the Sun increases its longitude by an amount $(360-\theta)$. Hence

$$360-\theta = 360 \times S/T.$$

Eliminating θ, we obtain

$$360-360\frac{S}{Q} = 360\frac{S}{T},$$

giving

$$\frac{1}{S}=\frac{1}{T}+\frac{1}{Q}.$$

Putting values of 365·25 mean solar days for T and 6798·3 mean solar days for Q, we find that S is 346·62 days.

This synodic period of the Moon's node is also called the **eclipse year** because of its importance with respect to eclipses. It should be noted that the Sun is in line with the line of nodes every 173·31 days, or every half eclipse year.

5.4 The Saros

It was discovered by the Babylonians from their eclipse records extending over many centuries that a particular time interval is useful for predicting not only the occurrence of eclipses but also their circumstances. This period is known as the saros (the word in fact means "repetition").

Now 19 synodic periods of the Moon's node are equal to 6585·78 days, while 223 lunations equal 6585·32 days. In addition, it is also found that 239 anomalistic months are equal to 6585·54 days. The saros period of 18^{yr} 10 or 11^{d}, depending upon the number of leap years in the interval, is therefore a time interval at the end of which the geometrical configuration of Sun, Moon and Earth is almost exactly repeated.

To illustrate this, let the line of nodes NN_1 of the Moon's orbit in Fig. 5.6 coincide with the direction of the Sun S as seen from the Earth E at a time when it is full moon, the Moon being at perigee P. The line of apsides PA also lies in this direction. Then a total lunar eclipse will take place visible to all observers on the night side of the Earth. One saros later, there will have occurred by that time 19 eclipse years, 223 lunations and 239 anomalistic months, so that the configuration will be repeated and another total lunar eclipse will take place. Because the agreement in the three times is not exact, the second eclipse is not exactly a copy of the first in all its circumstances, but closely resembles it. After yet another saros has elapsed, a third eclipse, rather less similar to the first but very like the second, occurs, and so on. It is obvious that there must exist a saros series lasting many hundreds of years, containing these lunar eclipses, beginning with a grazing partial eclipse, progressing saros by saros to total eclipses before becoming more and more partial and finally concluding with another grazing partial eclipse. Any particular lunar eclipse must belong, therefore, to a particular saros cycle.

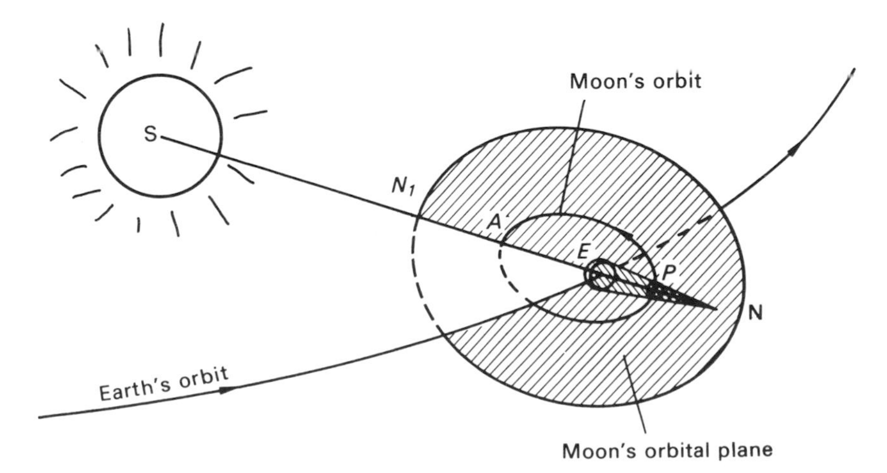

Fig. 5.6 The line of the nodes N_1N coincides with the direction of the Sun and the apse line at full Moon, giving a total lunar eclipse: One saros later, the configuration is almost exactly repeated

The saros also governs the occurrence of solar eclipses and they too belong to particular saros series. The picture is complicated, however, by the fact that the Earth is rotating once per day. Neighbouring solar eclipses in a saros cycle will therefore closely resemble each other in such matters as their type (partial, total or annular), but will be seen from different parts of the Earth's surface. For example the total solar eclipse of June 30th, 1973, was visible from Guiana, Central Africa and the Indian Ocean. The next in that series, dated July 11th, 1991, will be seen from the Marshall Islands, Central Mexico and Brazil.

It is of interest that the saros is also significant in the dynamics of the Earth–Moon–Sun system. The main disturber of the Moon's orbit about the Earth is the Sun. If the Sun was itself at perigee when the configuration in Fig. 5.5 occurred, it may be shown that the dynamical behaviour of Moon, Earth and Sun, considered as an isolated system of point-masses, will mirror after that time the behaviour prior to that time. Solar perturbations that had been changing the shape of the lunar orbit are reversed in their effect. If the saros was exact and the configuration was repeated in every detail, the changes would once more be reversed. The saros is not quite exact in this respect, but is so close that any residual dynamical changes are of very long period.

5.5 Duration of a Total Lunar Eclipse

Although the calculation of the circumstances of a particular eclipse is complicated, a simple example can be treated. We assume the orbits of Sun and Moon about the Earth to be circular and coplanar and calculate the duration of the resulting total lunar eclipse.

The calculation is split into two parts. Firstly we find the angular radius of the Earth's umbra at the Moon's distance, then we obtain the time it takes the Moon to cross it.

(i) Let P and P_1 be the horizontal parallaxes of Sun and Moon respectively. Let S be the Sun's semi-diameter. In Fig. 5.7, the vertex V of the Earth's

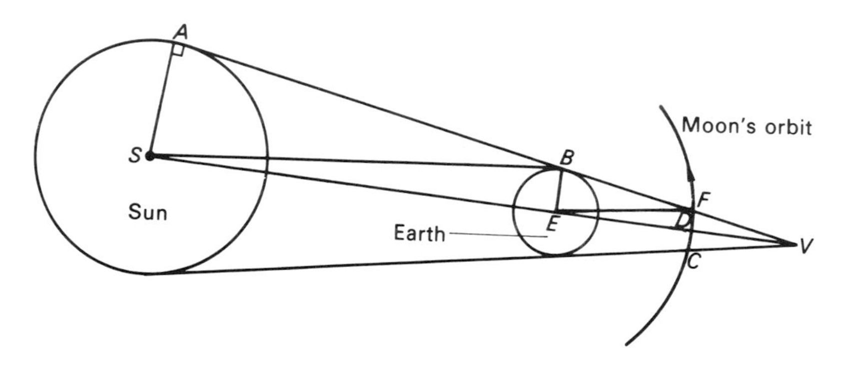

Fig. 5.7 The angular radius *FED* of the Earth's umbral shadow at the Moon's distance

shadow-cone lies beyond the Moon's orbit CF and the angular radius of the cone at the Moon's distance will therefore be given by $\angle FED = s$, say.

Consider $\triangle FEV$.

$$\angle FEV + \angle FVE = \text{external angle } BFE.$$

But $\angle BFE = P_1$ and $\angle FEV = \angle FED = s$, so that

$$s = P_1 - \angle FVE,$$

or

$$s = P_1 - \angle BVS. \tag{5.1}$$

Join SB. In $\triangle SBV$ we have

$$\angle BSV + \angle BVS = \text{exterior angle } ABS. \tag{5.2}$$

But $\angle BSV = \angle BSE \approx$ Sun's horizontal parallax P. Also $\angle ABS =$ Sun's semi-diameter S. Hence by Equation (5.2),

$$P + \angle BVS = S.$$

Substituting for $\angle BVS$ in Equation (5.1), we obtain

$$s = P + P_1 - S. \tag{5.3}$$

(ii) In Fig. 5.8, the circular cross-section of the umbra is seen. The Moon begins to enter it at C when the Moon's centre is at M. Totality begins when the Moon's centre reaches L and ends when it arrives at K. The angular distance moved through by the Moon's centre during totality is therefore $LK = FC - 2S_1$, where S_1 is the Moon's semi-diameter.

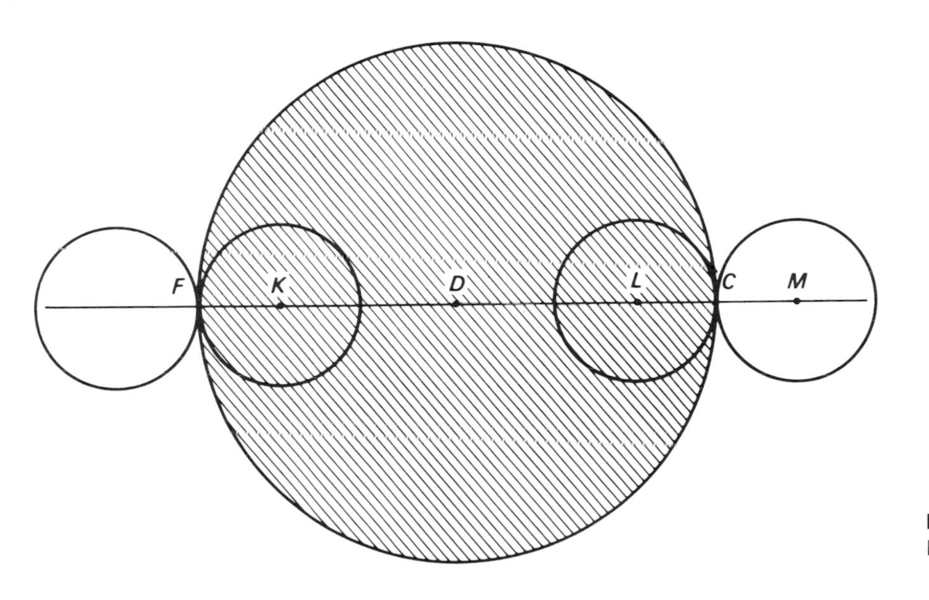

Fig. 5.8 The Moon's disk and the Earth's penumbra

Now $FC = 2s$, where s is given by Equation (5.3).
Hence

$$LK = 2(P + P_1 - S - S_1). \tag{5.4}$$

The average values of P, P_1, S, S_1 are 9″, 57′, 16′ and 16′ respectively. If they are inserted in Equation (5.4), it is found that $LK \approx 50'$.

Now it must be remembered that the Earth's shadow-cone swings round the Earth at the same angular rate at which the Sun moves in its apparent geocentric orbit. Hence the Moon's synodic period of 29·53 days is relevant here and not its sidereal period. The time of totality is the time it takes the Moon to advance 50′ on the moving shadow and is therefore T, where

$$T = \frac{50'}{360 \times 60} \times 29{\cdot}53 \times 24 \text{ hours}$$

that is,

$$T = 1{\cdot}64 \text{ hours}.$$

We have already said that a total lunar eclipse is visible from one hemisphere of the Earth. In fact, because of the Earth's rotation and the duration of totality, the eclipse is visible from rather more than one half of the Earth's surface.

Thus in Fig. 5.9, a person who is at A when totality begins (the Moon being at L) will have been carried round to B by the time the Moon has progressed to K.

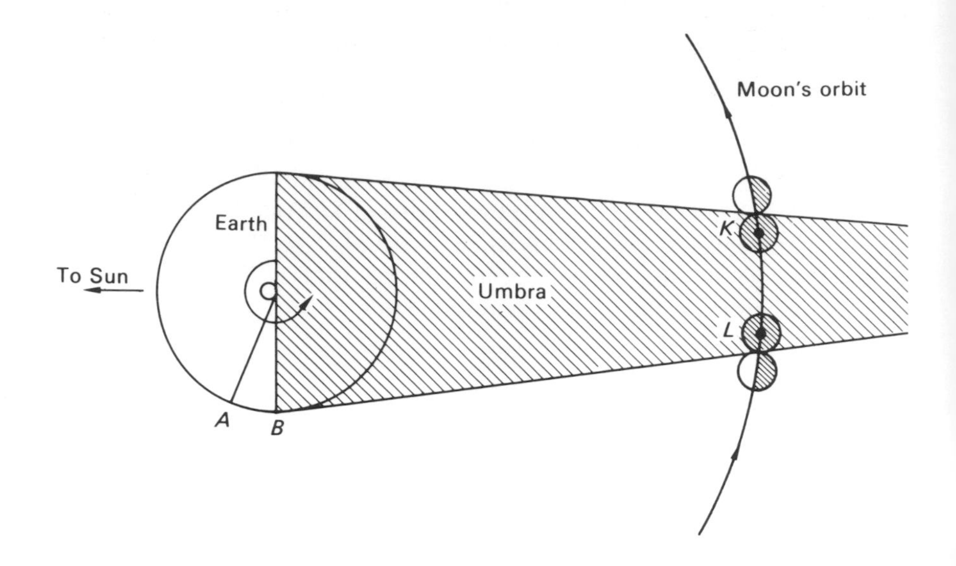

Fig. 5.9 **Effect of the Earth's rotation on the visibility of the total eclipse**

This angle AOB is obviously given by $(360/24) \times 1{\cdot}64$. The additional fraction of the equator is therefore

$$\frac{360 \times 1{\cdot}64}{24 \times 360} \approx \frac{1}{15}.$$

5.6 The Ecliptic Limits

We have seen that because the Moon's orbital plane is in fact inclined to the plane of the ecliptic, both Sun and Moon have to be near the line of nodes if an eclipse is to take place.

The **lunar ecliptic limit** is the greatest angular distance of the Sun (or the centre of the Earth's shadow) from the node when the Moon arrives at the node, if a lunar eclipse is to take place. The actual value of the limit is a function of the distances of Moon and Sun from the Earth, and therefore of the angular sizes of the Moon and the radius of the shadow-cone at the Moon's distance. It is also a function of the inclination of the Moon's orbit which is subject to periodic changes.

Suppose that in Fig. 5.10, the Moon suffers a momentary total eclipse at C when the shadow centre has reached the point B on the ecliptic. Then if the shadow centre was at A when the Moon passed through the node N, the arc NA

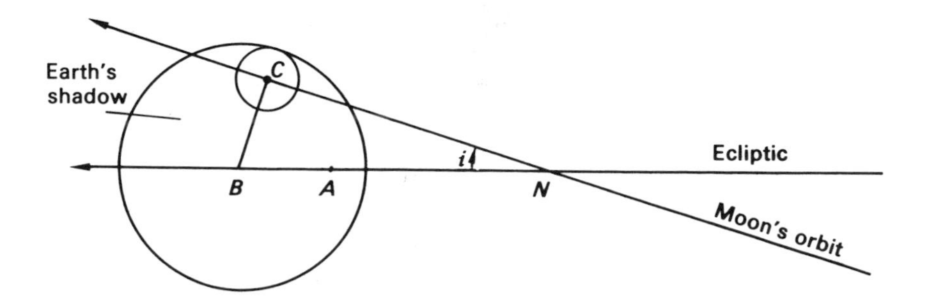

Fig. 5.10 The lunar ecliptic limit

is the lunar ecliptic limit. It should be remembered that the Moon's mean motion is much faster than that of the Sun and therefore of the shadow. A similar argument applies to finding the ecliptic limit if a partial lunar eclipse is to take place. The greatest and least possible values of AN in that case are found to be 12°.1 and 9°.5 and are referred to as the **superior** and **inferior ecliptic limits**. If AN is less than 9°.5, a partial eclipse must occur. No eclipse can take place if AN is greater than 12°.1. For values of AN between these limits, an eclipse may or may not occur.

In the case of a solar eclipse we can likewise define ecliptic limits. The **solar ecliptic limit** is the angular distance of the Sun from the node when the Moon arrives there if a partial solar eclipse is to take place. Thus in Fig. 5.11, the Sun's disk was at A when the Moon's disk arrived at the node N, the value of AN being such that by the time the Sun had reached B, the Moon's disk had progressed to C and was able to make grazing contact with the Sun's disk.

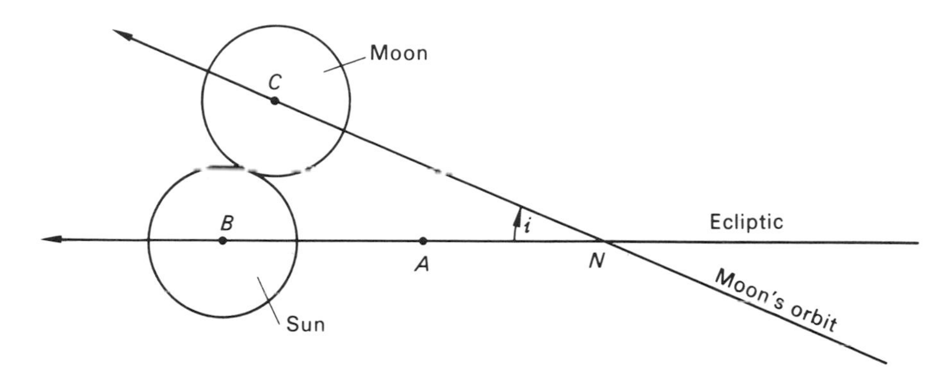

Fig. 5.11 The solar ecliptic limit

Again, the value of the limit is a function of the geometry of the Earth–Moon–Sun system. The **superior ecliptic limit** is the maximum value of AN if a partial solar eclipse is just possible; the **inferior ecliptic limit** is the minimum value. Their sizes are 18°.5 and 15°.4 respectively. We can then say that if the Sun's distance from a node exceeds 18°.5 when the Moon arrives there an eclipse cannot take place. For distances less than 15°.4 a partial eclipse will take place.

5.7 Megalithic Man's Lunar Observatories

The remarkable megalithic civilization that flourished in western Europe, in particular in the British Isles, during the second millennium B.C., revealed its interest in the Moon, as well as the Sun, by the stone alignments it set up. We know that megalithic man used the Sun as a calendar. By noting the azimuths of sunrise and sunset at the solstices and equinoxes, he was able to divide the year into four parts. These crucial azimuth directions at midwinter, midsummer, spring and autumn were singled out by alignments of stones with natural foresights such as mountain peaks or notches on the horizon.

Recent work by Thom shows that a large number of alignments still exist which must have been lunar in purpose. On studying these, it is difficult to resist the conclusion that megalithic man was able to predict eclipses. The method apparently adopted by megalithic man is outlined below.

If the Moon's orbit lay in the ecliptic, the Moon's declination would shift from $-23\frac{1}{2}^\circ$ to $+23\frac{1}{2}^\circ$ and back again in a period of one sidereal month.

Because of its orbital inclination of just over 5°, the Moon's declination can vary per month from $-28\frac{1}{2}^\circ$ to $+28\frac{1}{2}^\circ$, at one extreme, or from $-18\frac{1}{2}^\circ$ to $+18\frac{1}{2}^\circ$ at the other extreme, depending upon the position of the Moon's nodes, as shown in Fig. 5.12. The period of oscillation of the declination from maximum northerly to maximum southerly and back again is still one sidereal month. On top of this, however, we see that there will be imposed a period of 18·6 years, being the period of revolution of the Moon's nodes. This period will be the time it takes the maximum northerly and southerly declinations in one month to change from $28\frac{1}{2}°$ *in magnitude* to $18\frac{1}{2}°$ *in magnitude* and back again.

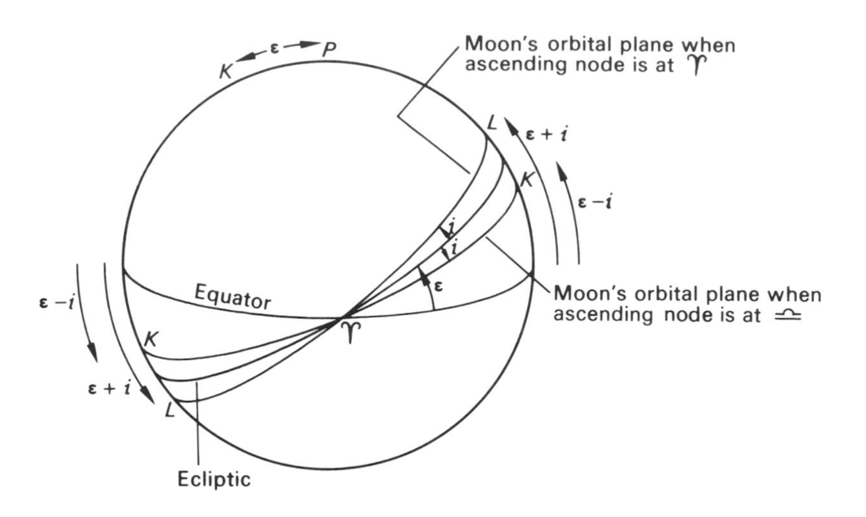

Fig. 5.12 The extremes of lunar declination

Thus in Fig. 5.13(*a*) we see the changes in the Moon's declination throughout June, July and August in 1950 A.D. while in Fig. 5.13(*b*), the effect of the 18·6 year period on the maximum northerly and southerly declinations is shown.

It is certain that many of the megalithic sites still surviving have alignments that single out the four turning points in azimuth of moonset dictated by the four turning points in declination, namely $-28\frac{1}{2}°$, $-18\frac{1}{2}°$, $+18\frac{1}{2}°$ and $+28\frac{1}{2}°$. In fact the values of declination found from the azimuths, altitudes and latitudes of the various sites are slightly different from the above figures, providing a value for the obliquity of the ecliptic of 23°91. Calculations show that this is the value it had about 1800 B.C. The implication that observations were carried out over many scores of years in order to fix these alignments with the precision found is interesting enough; but there is a great deal of evidence at such sites supporting the view that these megalithic astronomers had also discovered the third oscillation in the Moon's declination.

The third oscillation is caused by the Sun's disturbing effect on the inclination of the Moon's orbit to the ecliptic. The inclination oscillates in value from 5° 18′ to 5° 00′ and back in a period of 173·31 days, or one-half the eclipse year. This declination change is graphed in Fig. 5.13(*c*) over the course of a few years.

It is the third oscillation that enables predictions to be made about eclipses. Every half eclipse year, the Sun is at one of the nodes of the lunar orbit. At such times, the third oscillation is at a maximum. It is also only at such times that eclipses can take place. If the Moon is new, a solar eclipse will occur; a full moon, on the other hand, means that a lunar eclipse will take place.

Again, this 9′ oscillation in declination would show up as an oscillation in azimuth of the setting Moon over the period of 173·31 days. Many of the sites that megalithic man chose had horizons whose slopes made very small angles with the path of the setting Moon. Small changes were therefore magnified, increasing the sensitivity of observing phenomena of this kind. If the observers could keep track of the 173·31 day variation for long enough, and observe its correlation with eclipses, they would be able to predict future eclipses with a high degree of success.

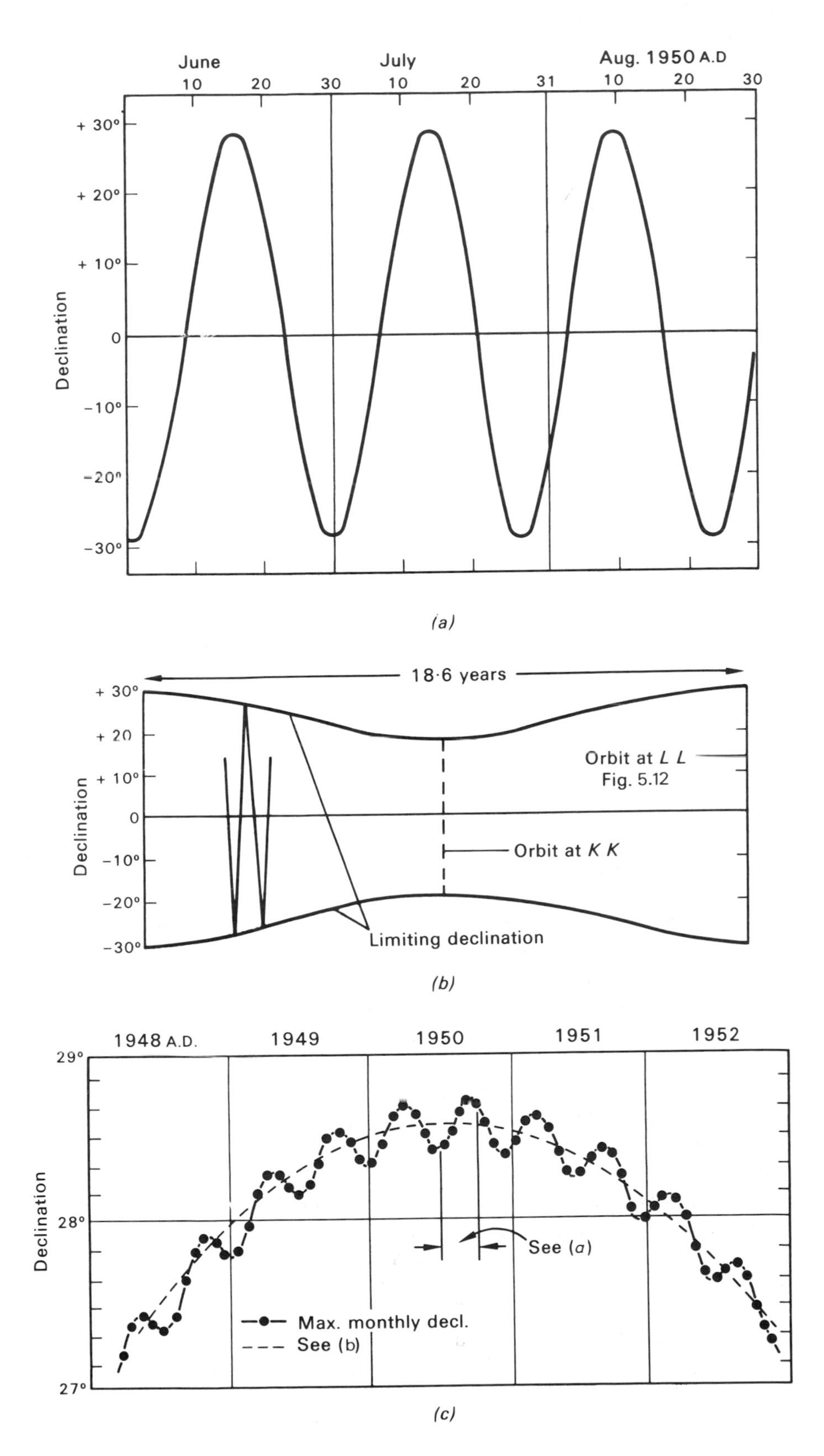

Fig. 5.13 **The variation of lunar declination**

Fig. 5.14 The solar corona as seen at the eclipse of 1966, Nov 12th, Pulacayo, Bolivia. Structural detail has been enhanced by using a radially-symmetric neutral filter and fluorododge printing; Venus is also shown (Photo by G. A. Newkirk, Jr. by courtesy of the Royal Astronomical Society)

5.8 The Value of Eclipses

Historical records often contain accounts of particular eclipses, especially those that occurred at or around a significant event such as a battle or the death of a king. Modern astronomical methods enable the exact circumstances of any eclipse, solar or lunar, to be determined—in particular the date. By this method, long stretches of history that contain eclipse records can be dated unambiguously, for example in Assyrian chronology. The chronicler's accuracy of detail in describing the eclipse can also be checked. In the Bury St. Edmunds Chronicle one solar and eleven lunar eclipses, with dates of occurrence, are described which happened between 1258 and 1297 A.D. The agreement in date and accuracy of description is excellent, thus increasing the historian's confidence in the veracity of the scribe.

Scientifically, total eclipses of the Sun are still of importance. With the Moon blotting out the brilliant photosphere, it is possible to make observations of the chromosphere and the corona (see Fig. 5.14). This is not quite so important now that methods exist of masking out the Sun's disk at any time (see Roy & Clarke, *op. cit.*, Section 17.6).

Checks on the agreement of predicted positions of Sun and Moon with actual positions can be made at the beginning and end of totality, so suddenly do they occur.

A total solar eclipse can also be used to test one of the predictions of Einstein's theory of relativity. Rays of light passing through a strong gravitational field should be slightly deflected. By measuring the positions of stars close to the Sun's limb during totality and comparing with their normal positions, such a test can be made. The total solar eclipse of 1919 was the first occasion when this was done.

Lunar eclipses have been also of value scientifically. In the days before packages of instruments were landed on the Moon, it was possible to acquire knowledge about the structure of the Moon's surface by measuring with a thermocouple the rapid rate of cooling that followed the sudden immersion of any region of the surface within the Earth's shadow.

6
Members of the Solar System

6.1 Mercury

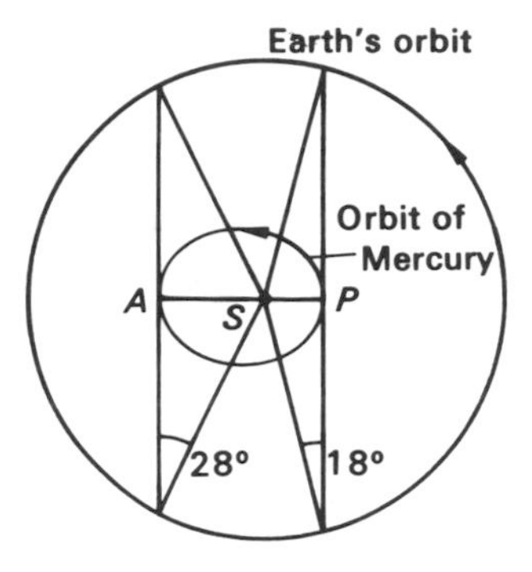

Fig. 6.1 The variation in maximum elongation of Mercury

The smallest of the nine principal planets, Mercury's radius is about two-fifths that of the Earth. Of the two planets that have orbits within the Earth's, it is the nearer to the Sun. Its orbit is markedly elliptic with eccentricity about 0·21, which means that the observed maximum elongation of the planet can vary substantially depending upon whether Mercury is viewed near perihelion or aphelion (Fig. 6.1).

The planet is always seen as a morning or evening "star" and is a difficult

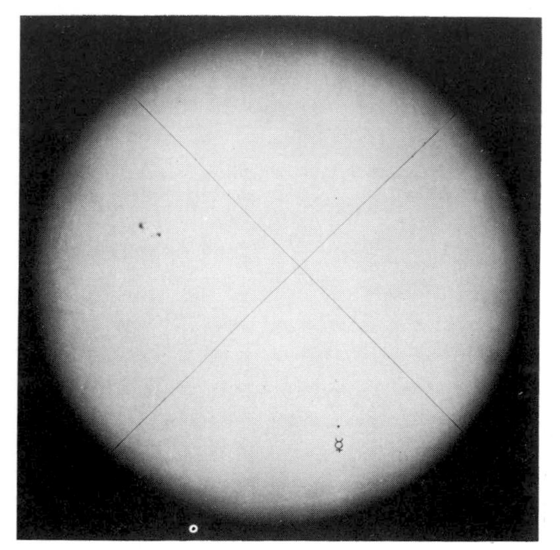

Fig. 6.2 The transit of Mercury recorded on 1914, Nov 7th. (By courtesy of the Royal Astronomical Society)

Fig. 6.3 A Mariner 10 photomosaic of Mercury from 210 000 km. A large circular basin, about 1300 km in diameter, is emerging from the day–night terminator at left centre. Bright rayed craters are prominent; one ray seems to join in both east–west and north–south directions. (By courtesy of the National Aeronautics and Space Administration)

object to observe because of its close proximity to the Sun. **Transits** of Mercury (see Fig. 6.2) can occur at certain inferior conjunctions, in that at such times the projected solar image is seen to be crossed by a small black spot.

In a telescope the planet shows phases in accordance with the Copernican theory of the Solar System. Visual observers sometimes note permanent markings on the planet resembling those seen on the Moon with the unaided eye. By a study of a series of observations a rotation period equal to the period of revolution of the planet in its orbit was deduced, implying that one face was always kept toward the Sun, apart from librations due to the orbital eccentricity.

Radar studies demonstrate, however, that the period of rotation is about 59 days, in contrast to the 88-day period of revolution about the Sun. Because this rotation period is so near two-thirds of 88 days and because most observations were made at a particular phase angle and certain time intervals, the behaviour of the markings is consistent with both periods. Preconceived notions that the planet like our Moon always kept the same face towards its primary led observers to adopt the 88-day period without considering the possibility of an alternative one.

Fig. 6.4 The largest structural feature on Mercury discovered by Mariner 10. A ring basin 1300 km in diameter, it is bounded by mountains 2 km high. The floor is intensely disrupted with many fractures and ridges. Named *Caloris,* it is similar in size to the *Mare Imbrium* on Earth's moon and was almost certainly caused by the impact of a body tens of kilometres in diameter. (By courtesy of the National Aeronautics and Space Administration)

Fig. 6.5 The heavily-cratered south polar region on Mercury (north is towards the top). Scarps and ray systems are prominent. (By courtesy of the National Aeronautics and Space Administration)

Because of the planet's slow rotation, the Sun is above the horizon of any particular place on Mercury for many months at a time. The surface temperature may rise to about 350 °C. From polarization studies, the presence of a tenuous atmosphere on the planet is suspected, but still not definitely substantiated. Indeed, with Mercury's low surface gravity and close proximity to the Sun's heat, it is difficult to see how the planet could have avoided losing its atmosphere long ago.

If the Moon, almost the same size and mass as Mercury, were placed in Mercury's orbit, it would resemble it in appearance. Even before the successful flypast of Mariner 10, this led a number of astronomers to guess that Mercury's surface would be found to be of the same nature as that of the Moon. The pictures of the planet sent back by Mariner 10 confirmed this (see Figs 6.3–6.6). Craters, scarps, basins and plains exist on the planet in all sizes. Many of the craters have central peaks; at least some show ray systems. The distributions of crater sizes and shapes (from sharply-rimmed craters with well-developed ejecta surrounding them to flattened, worn, bombarded features) closely resemble those on the Moon. One huge basin, about 1300 km in diameter, is similar in a number of respects to the Moon's *Mare Imbrium*. Some of the scarps are hundreds of kilometres in length, traversing large craters.

Fig. 6.6 Cratered area near Mercury's south pole, pictured from a distance of 54 000 km. The picture covers an area 460 by 350 km. North is at the top. (By courtesy of the National Aeronautics and Space Administration)

There is no evidence of atmospheric erosion, confirming the belief that, at most, only an extremely tenuous atmosphere has ever existed.

6.2 Venus

In size and mass Venus is almost a twin of the Earth, implying a similar constitution. Possessing no natural satellite, its mass, prior to the flyby of Mariner 2, was measured indirectly by its gravitational effect on other bodies. Prior to 1967, measurements of its radius during solar transits and during an occultation of Regulus gave a value of 6150 ± 25 km. Pettengill and his fellow workers, using radar, have revised the figure to 6056 ± 1 km. The artificial satellite orbiting Venus has further improved the accuracy of this figure.

Because its orbit lies inside that of the Earth, the planet at its closest approach (inferior conjunction) presents its night side to us. This, together with its dense opaque atmosphere, removes the observational advantage afforded by Venus approaching Earth more closely than any of the other sizeable planets.

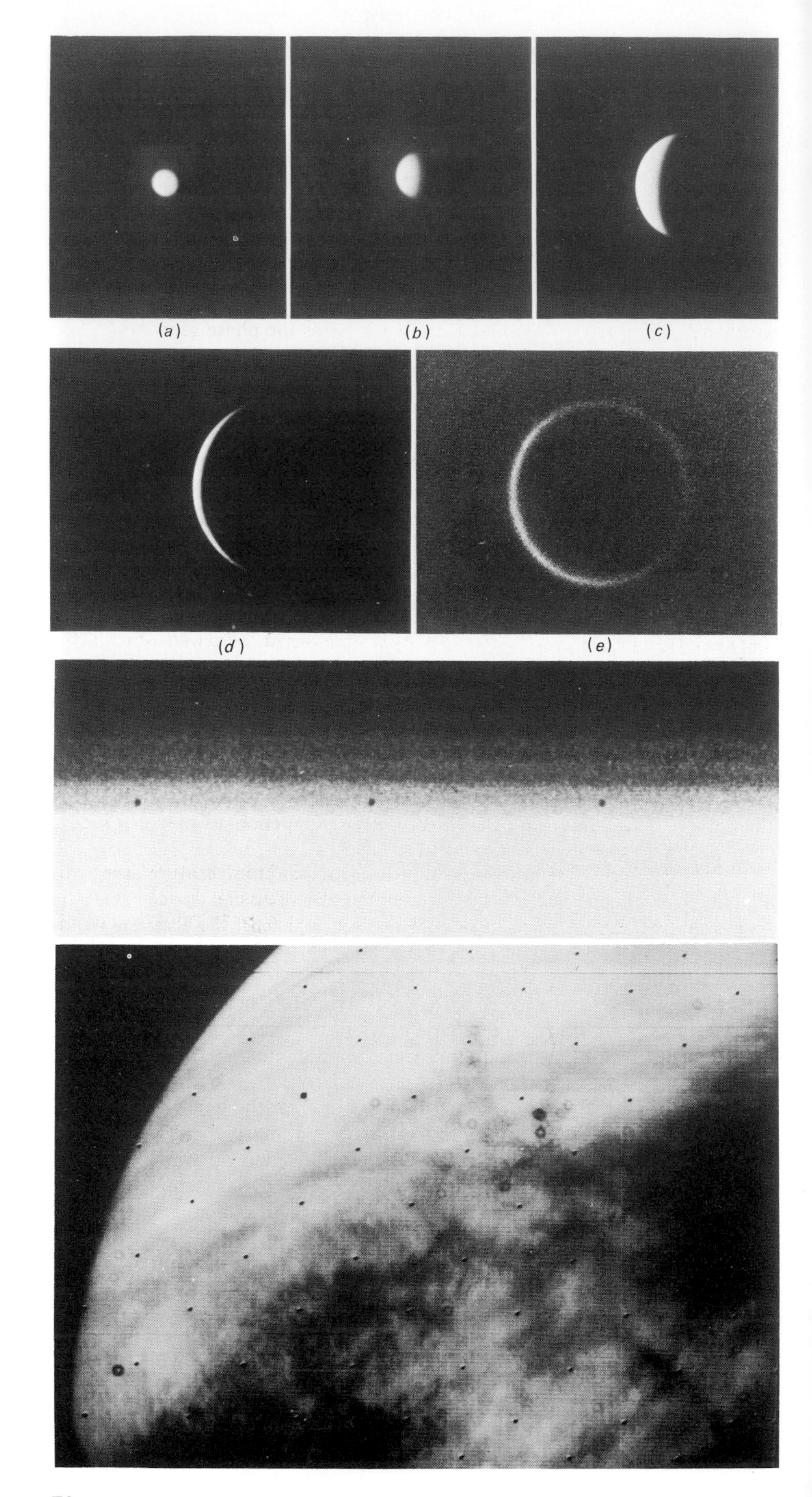

Fig. 6.7 Five aspects of the phase and apparent size of Venus. Photographs taken on: (*a*) September 27th, 1910; (*b*) June 10th, 1910; (*c*) October 24th, 1927; (*d*) September 25th, 1919; (*e*) June 19th, 1964

Fig. 6.8 Haze layers on the limb of Venus photographed in orange light by Mariner 10. The thickness of the haze above the visible clouds is about 6 km and appears to cover the entire planet. (By courtesy of the National Aeronautics and Space Administration)

Fig. 6.9 Part of an ultra-violet filter photomosaic of Venus by Mariner 10. The dark belt at the bottom straddles the planet's equator. Rising and descending air currents typical of convection in the Earth's atmosphere can be seen within the belt. (By courtesy of the National Aeronautics and Space Administration)

Like Mercury, Venus is seen as a morning or evening object. Its almost circular orbit ensures that its maximum elongation does not change appreciably from a mean value of 48° no matter what its heliocentric longitude is when that elongation is reached. After superior conjunction, Venus comes out from behind the Sun, appearing to the east of the Sun and setting later in the evening after sunset. By maximum elongation, it is the brightest star-like object in the sky, being visible even before the Sun has set and remaining above the horizon for hours after sunset. At its maximum brightness, Venus can actually cast shadows on the Earth when the Sun has set. As its elongation diminishes and it approaches inferior conjunction, it becomes fainter, the smaller phase more than offsetting its closer approach to Earth. As a morning object it lies westward of the Sun and increases in brightness as the phase grows.

The planet transits less frequently than Mercury, such transits occurring in pairs at present with a separation of 8 years between the members of a pair. The next pair will be on June 8th, 2004 and June 5th, 2012. Once used as a method of obtaining the solar parallax, transits of the planet are of little astronomical importance now.

The Venusian atmosphere is dense and opaque, effectively preventing optical observation of the planet's surface, though recent Russian probes have sent back pictures from the surface, revealing a rough, stony landscape (see Fig. 6.11). Radar observations by the Pioneer radar altimeter indicate that there are wide-spread elevated regions on the planet. The planet possesses flat, rolling plains, some large shallow basins and valleys and two major elevated plateaus, *Ishtar Terra* and *Aphrodite Terra*. *Maxwell Montis*, the highest feature in *Ishtar Terra* is almost 11 km above the mean planetary radius (see Fig. 6.12). Other radar studies have enabled the planet's period of rotation to be measured. It appears to be 243 days *retrograde*, with the rotation axis almost perpendicular to the orbital plane. A slow rotation of the planet is consistent with its closer proximity to the Sun than Earth, resulting in solar tides on Venus being about four times greater than on the Earth. But the reason why Venus rotates in the opposite direction to that in which it revolves round the Sun remains unknown.

The composition of the atmosphere, its density and temperature at various heights, have been measured by American and Russian spacecraft. The American Mariner spacecraft made flybys (see Fig. 6.8); the Russian Venus probes entered the atmosphere and descended by parachute, their instruments sampling and measuring during the descent.

Carbon dioxide is a major constituent of the atmosphere, amounting to as much as 90%, while the remaining gases are chiefly nitrogen and the rare gases. The Venus 4 probe measurements gave an upper limit of 1·4% for water vapour in the lower atmosphere.

Mariner 10 pictures in ultra-violet light showed for the first time details of the complicated circulation currents in the planet's atmosphere (see Fig. 6.10). North and south temperate and polar latitudes showed formations resembling cirrus clouds while the tropical regions showed regions suggesting rising and descending convection cells (see Fig. 6.9). The overall picture resembles the general circulation in the Earth's atmosphere.

The temperature at the surface of Venus has been measured to be at least 325 °C, the atmosphere producing a marked **greenhouse** effect. The atmospheric surface pressure has been measured by barometer to be around 60 atmospheres, but the actual pressure may be much higher.

If Venus has a simple magnetic dipole field similar to that of the Earth, then the strength of the magnetic dipole moment indicated by Mariner 2's magnetometer is less than 5% of Earth's.

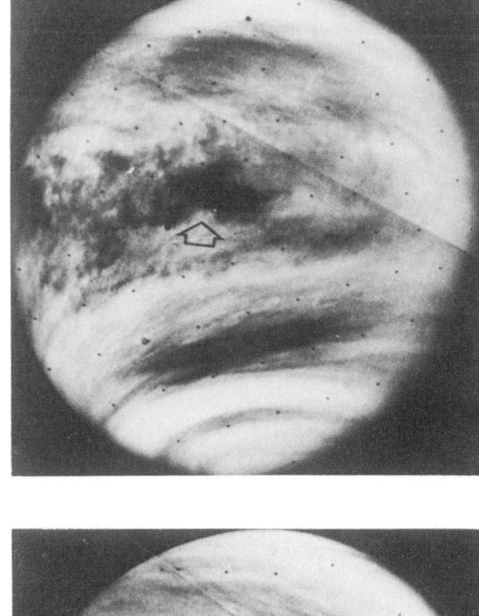

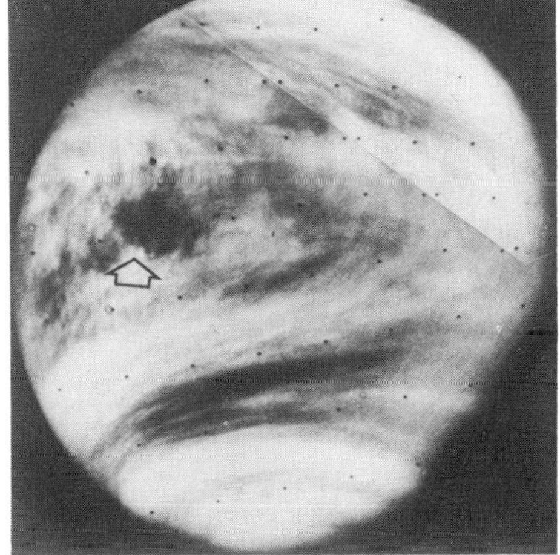

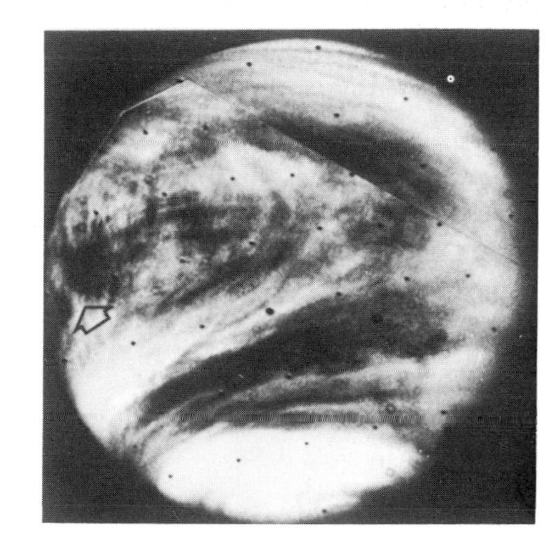

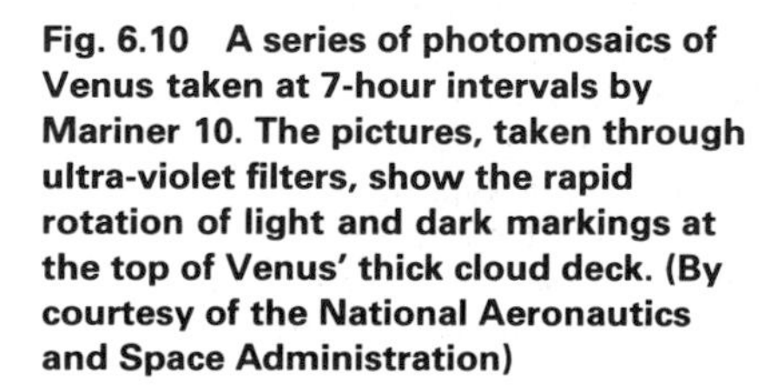

Fig. 6.10 A series of photomosaics of Venus taken at 7-hour intervals by Mariner 10. The pictures, taken through ultra-violet filters, show the rapid rotation of light and dark markings at the top of Venus' thick cloud deck. (By courtesy of the National Aeronautics and Space Administration)

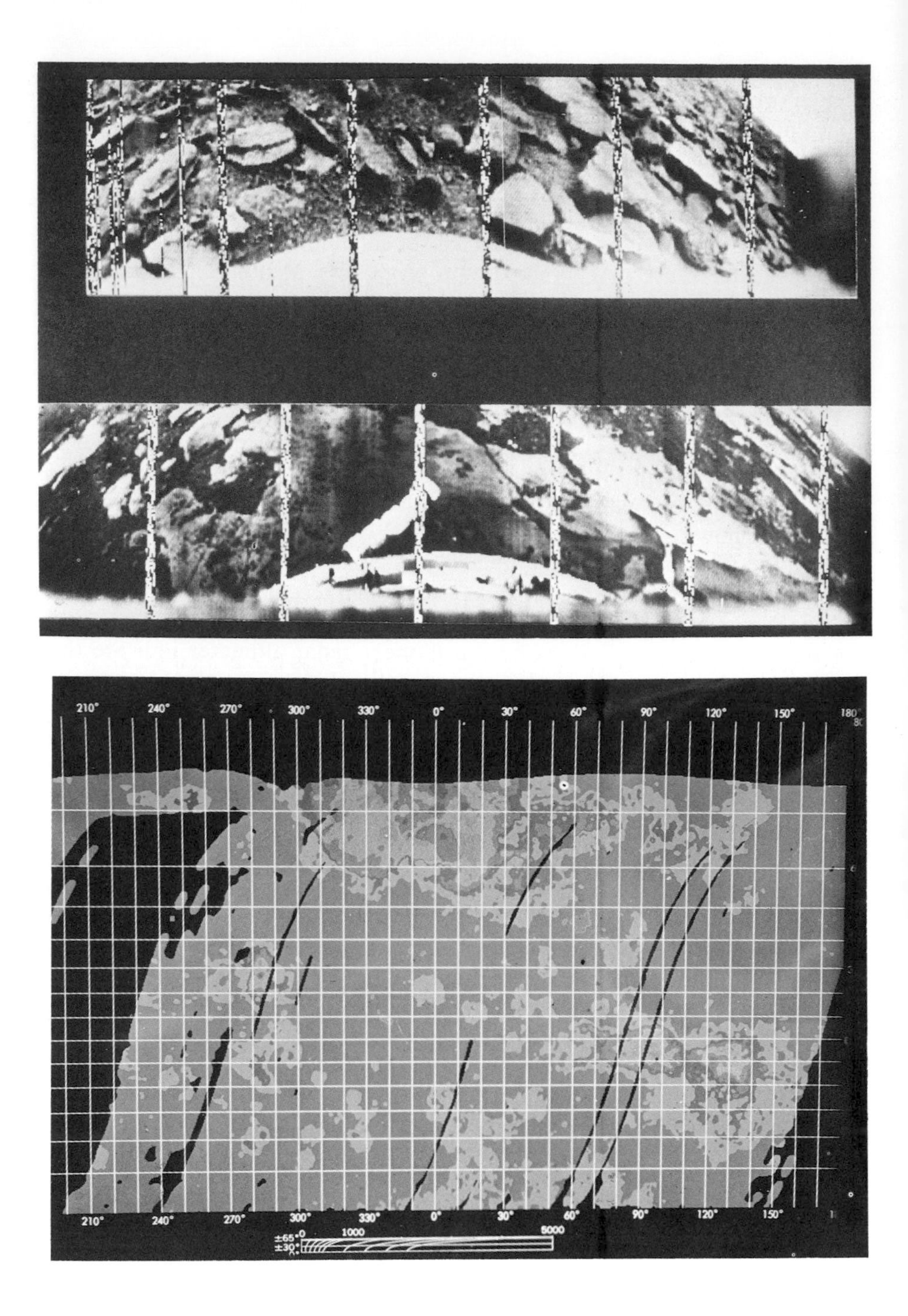

Fig. 6.11 Surface areas of Venus taken from Venera 9 (above) and Venera 10 (below)

Fig. 6.12 Map of Venus based on radar altimetric data from the Pioneer–Venus orbiting spacecraft. The black curved gaps are missing data. The map shows two high continent-sized areas now named *Ishtar Terra* and *Aphrodite Terra*

6.3 The Earth

The next planet in order of distance from the Sun is the Earth. The Earth and its satellite have already been discussed in Chapters 3, 4 and 5, and no further comment is made here.

6.4 Mars

6.4.1 The Planet Mars appears in the night sky as a bright reddish object. From the form of its apparent path, exhibiting a slow loop at about the time of opposition, it is immediately evident by Copernican theory that Mars lies outside the Earth's orbit; The sizes of the annual loops tell us that Mars is the nearest of the outer planets. Its mean distance from the Sun is 1·52 AU and it has an orbital period of 1·88 years. The Martian orbit is elliptical with an eccentricity of 0·09, this being relatively high. Opposition distances from the Earth differ markedly according to the position of Mars in its orbit at the time of opposition. For example, if opposition occurs when Mars is close to aphelion, its distance from the Earth is of the order of 100×10^6 km but if the opposition is close to Mars being at perihelion, this distance is approximately halved.

Between the extreme aspects of Mars being at conjunction and being at a close approach (favourable) opposition, the apparent angular size varies from 3·5 arc sec to 25·1 arc sec. This sizeable disk allows accurate measurements to be made of its diameter. Its equatorial radius is 3400 km, being a little more than half that of the Earth. A slight amount of flattening, a little larger than the Earth's, has also been measured.

In 1877, two small Martian satellites were discovered by Hall. They were given the names of Phobos and Deimos. Observations of their orbits allowed the mass of Mars to be known accurately. Indeed, perturbations to their orbits, caused by the Martian equatorial bulge, gave a figure for flattening which is in good agreement with that obtained more directly. The mass of Mars is approximately one-tenth that of the Earth. By relating this to its volume, we find that the Martian mean density is about 0·7 that of the Earth's. It seems likely that in comparison with the Earth, Mars is less condensed at its centre and may not have a liquid core. The mass and figure of Mars have both been determined much more accurately in recent years by tracking artificial Martian satellites such as Mariner 9.

The Martian surface has a poor albedo, similar to that of the Moon. Its value is strongly wavelength-dependent, being approximately twice as great at the red end of the spectrum than at the blue; the difference is sufficiently large to make Mars appear red to the naked eye. Over the range of phase angle which the orbit of Mars provides, the polarization–phase curves are remarkably similar to those of the Moon. Thus, from albedo and polarization measurements, it might have been suspected that the Martian surface structure resembles that of the Moon and indeed this was confirmed directly from photographs taken by Mariner 4 in 1965. These photographs revealed that Mars was covered by crater formations, similar to those on the Moon. In the flybys of Mariners 6 and 7 in 1969, many thousands of craters were photographed with much greater resolution (see Figs. 6.13 and 6.14). Study of the pictures sent back by Mariner 9 and the Viking orbiters, however, greatly modified the "lunar terrain" interpretation but discussion of this will be left until later in this section.

As would be expected from Mars being an outer planet, crescent phase appearances are never seen by telescope. However, the phase is distinctly gibbous near the times of quadrature.

Telescope views of Mars reveal permanent markings on the surface, some of them showing slight colour. The features that are seen and photographed have been charted and given names. For example, a large dark feature known as Syrtis Major, looking in shape like the continent of Africa, may sometimes be seen. Timed observations of the positions of the markings have allowed the Martian rotation period to be measured. Observations made by Hooke and Huygens in the 17th century have provided a long temporal base-line by which the rotation period is known very accurately. The Martian day is just over half an hour longer than our own.

Other prominent telescopic surface features are those of the polar "ice" caps. These white markings are found at the poles of the planet. Their sizes change markedly with the Martian seasons and each cap may even disappear during the Martian summer in that hemisphere. The speed at which they

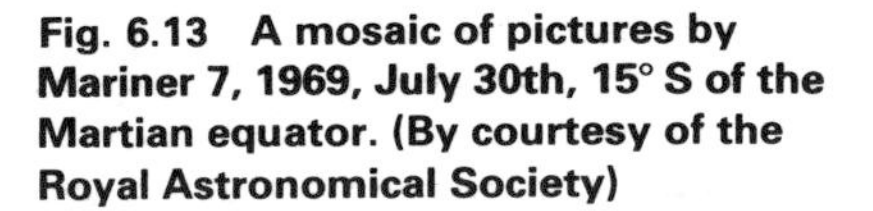

Fig. 6.13 A mosaic of pictures by Mariner 7, 1969, July 30th, 15° S of the Martian equator. (By courtesy of the Royal Astronomical Society)

change their appearance shows that the ice layer must be extremely thin. From infra-red measurements made on the south polar cap during the Mariner 7 mission, a temperature of −120 °C was recorded. This temperature agrees closely with the freezing point of carbon dioxide under a reduced pressure corresponding to the Martian atmosphere. On the other hand, measurements from Viking II suggest that the north polar cap is predominantly water ice.

Fig. 6.14 A Mariner 7 picture from 5300 km above the south polar cap of Mars (1969, August 4th). (By courtesy of the Royal Astronomical Society)

It is worth remarking that water vapour has been detected in the clouds over the south polar cap by Mariner 9. In addition, as we will see, certain remarkable surface features are difficult to interpret other than by the action of surface water in large quantities. Perhaps the most reasonable view, therefore, is that at the present time, the polar caps are composed of frozen carbon dioxide with some percentage made up of water ice.

The similarity of the Martian seasons to our own is perhaps forecast by the planet's orbital and rotational parameters. The orbit of Mars is very nearly in the same plane as the Earth's orbit, being inclined at 1° 51′ to the ecliptic. The inclination of the Martian equator to its orbit is 23° 59′, a value very close to that for the Earth. When Mars is near perihelion, its south pole is inclined towards the Sun and, in a similar way to the Earth, the southern hemisphere has a hotter but shorter summer than the northern hemisphere.

Temperature measurements have been made of different parts of the Martian surface by Earth-based telescopes using sensitive thermocouple detectors and by thermal radio-wave observations. These have now been supplemented by measurements taken by spacecraft in orbit about the planet or from the Viking landers on the Martian surface. An average equatorial mid-day temperature is about 10 °C but this drops markedly during the night to below −75 °C. Such sharp changes in temperature illustrate that the Martian atmosphere has little blanketing effect.

That Mars has an atmosphere is evident from visual and photographic studies made by telescopes. Disturbances are sometimes seen to obscure the ground surface details. These may take the form of clouds, which in some cases are seen to project beyond the limb of Mars. They appear at different heights within the atmosphere and have distinctive colours—their appearance on photographs varies according to the spectral sensitivity of the photographic plates used. In the lower atmosphere, yellow clouds may be seen, rising to an altitude of about 5 km. White clouds, probably consisting of ice crystals, may be seen in the upper atmosphere up to heights of 25 km; blue clouds may be seen at a lower level than this.

The Mariner 9 photographic mission itself was endangered by a particularly severe dust storm that obscured the entire surface of the planet for some weeks. When the storm abated, the surface debris of sand-dune patterns provided much information about prevailing wind systems. The Viking packages have now enabled hour-to-hour studies of the surface meteorology to be made.

The low velocity of escape for Mars means that any atmosphere must be thinner than the Earth's. Light elements such as hydrogen and helium would escape fairly easily and it would be expected that only the heavier gases such as oxygen, carbon dioxide, etc., might be present. Positive identification of the Martian atmospheric constituents by Earth-based spectrometry is made very difficult, however, by the light having to pass through the Earth's atmosphere before entering the analysing equipment. Our own atmosphere imposes strong absorption features, and additional effects produced by the gases in the Martian atmosphere are small by comparison. Spectrometers carried by Mariner spacecraft have revealed the presence of hydrogen, oxygen, carbon monoxide and carbon dioxide in the upper atmosphere. They also indicate the

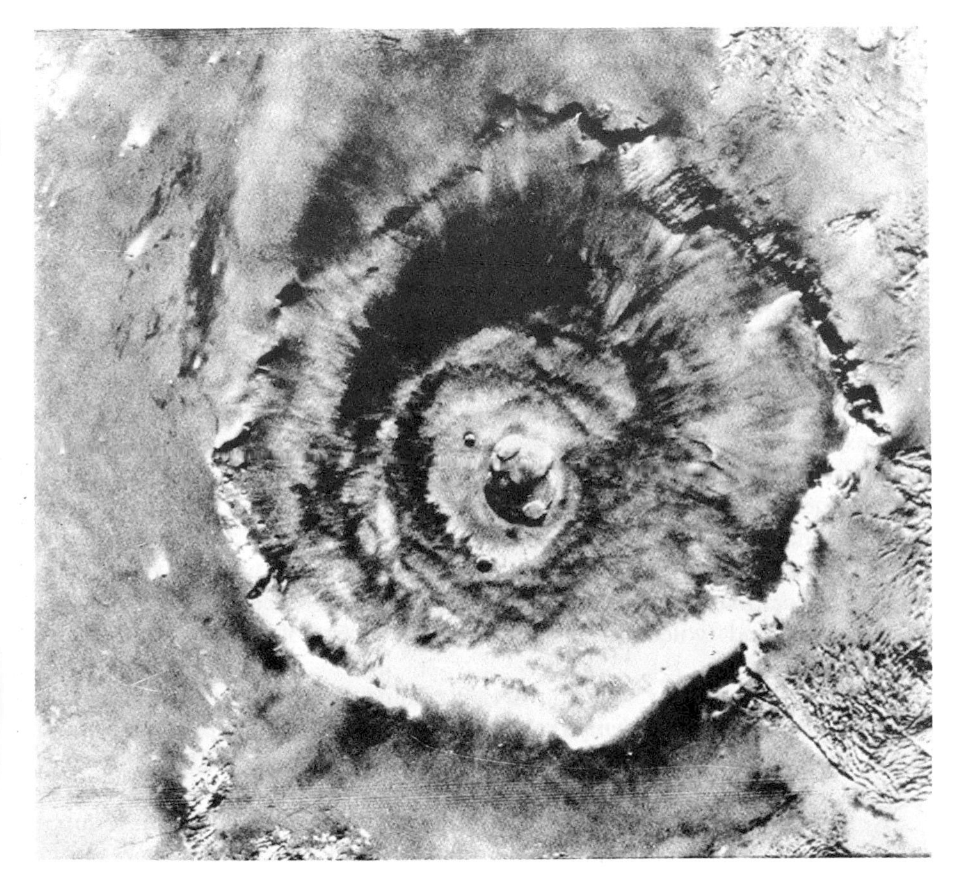

Fig. 6.15 ***Olympus Mons*****, a gigantic volcanic mountain on Mars in the** ***Tharsis*** **region, is 27 km high with a base of diameter 500 km surrounded by a steep cliff. Extensive lava flows beyond the cliff enlarge the volcano to about 700 km across. At the summit is a crater of diameter 65 km formed by a complex multiple volcano vent. (By courtesy of the National Aeronautics and Space Administration)**

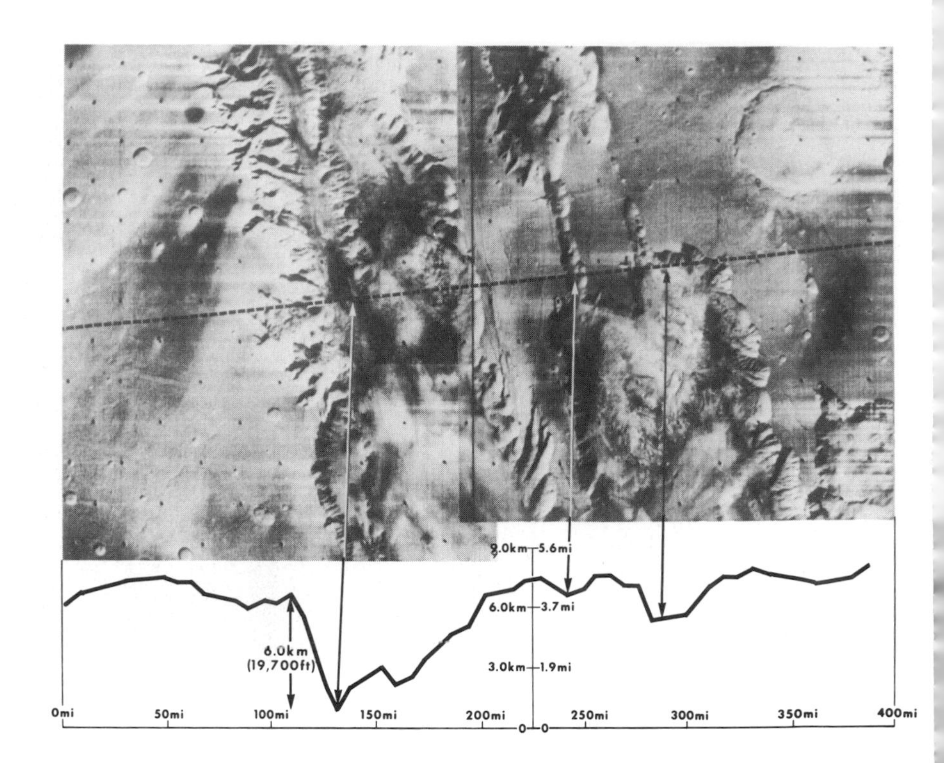

Fig. 6.18 Two photographs of the *Tithonius Lacus* region on Mars, revealing a canyon nearly four times as deep as the Grand Canyon in Arizona. Width of the Martian canyon is 120 km, depth is 6 km (compare with Earth's Grand Canyon of width 21 km and depth 1·6 km). The graph gives a cross section profile along the dotted line. (By courtesy of the National Aeronautics and Space Administration)

some sections this great feature is almost 100 km wide and 6 km deep; tributary valleys enter it in profusion and it is difficult to explain them other than as erosion valleys cut by water.

The interpretation of the planet's complicated history from the accummulating evidence is only beginning. It may be that as Sagan has suggested, orbital and precessional changes over long periods conspire at times, changing surface conditions so that atmospheric pressure is raised to the extent that liquid water can exist, in relatively large quantities. At present, however, the pressure is so low that water ice sublimes straight into water vapour. The exploration of the planet by instrument packages landed on its surface will doubtless answer some

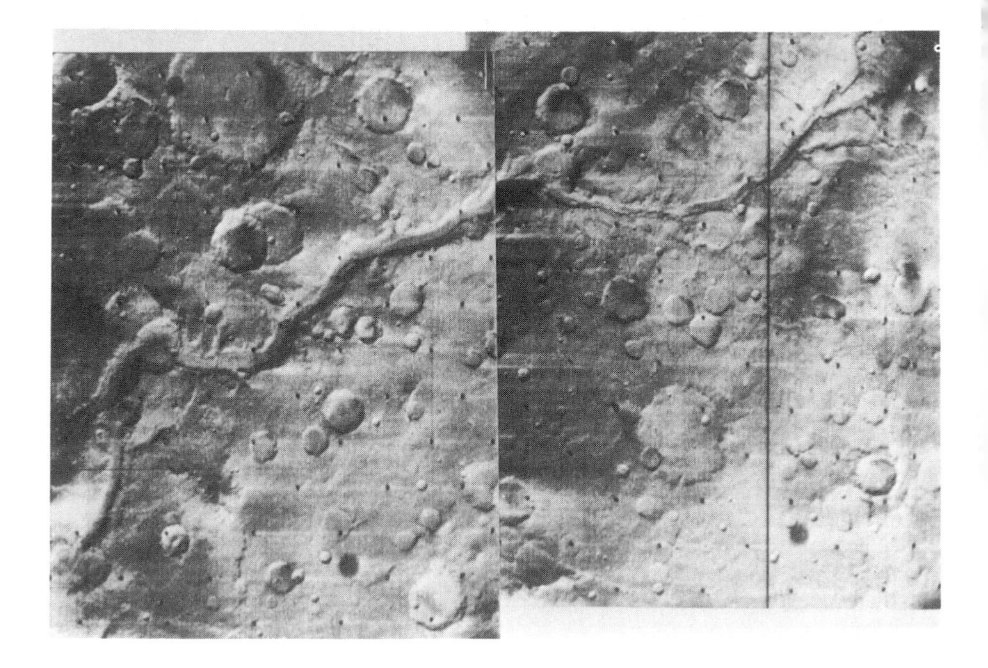

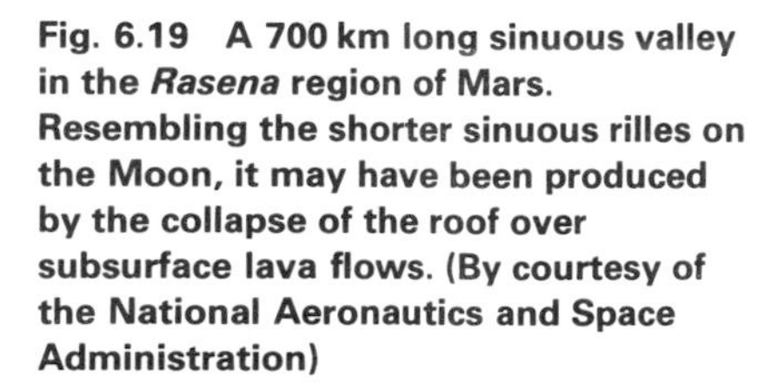

Fig. 6.19 A 700 km long sinuous valley in the *Rasena* region of Mars. Resembling the shorter sinuous rilles on the Moon, it may have been produced by the collapse of the roof over subsurface lava flows. (By courtesy of the National Aeronautics and Space Administration)

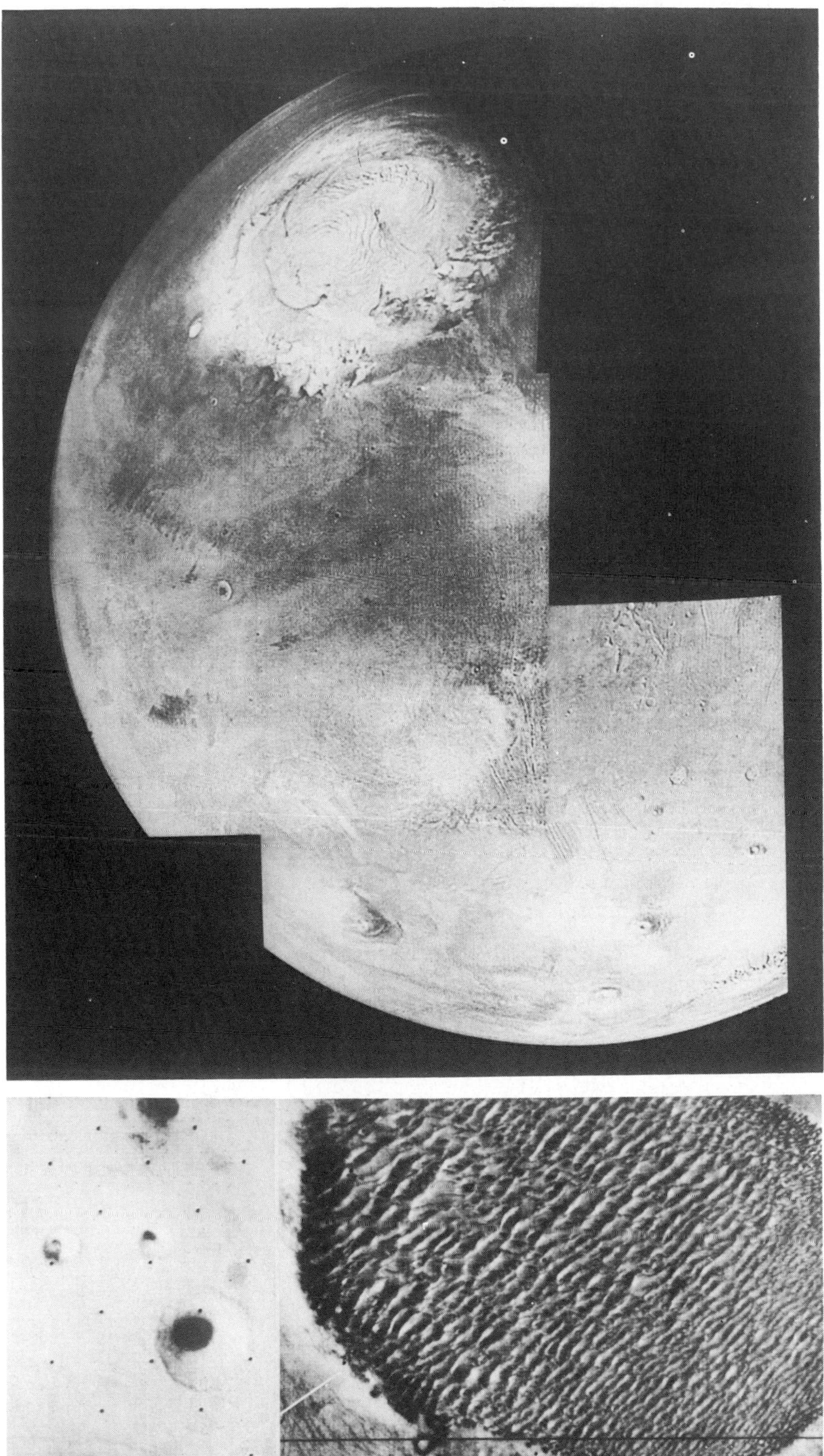

Fig. 6.20 The northern hemisphere of Mars from the polar cap to a few degrees south of the equator. The north polar ice cap is shrinking during the late Martian spring and the area shows complex sedimentary systems. Fractured terrains partially flooded by volcanic extrusions are visible in the centre of the disk. The huge Martian volcanoes are also visible. (By courtesy of the National Aeronautics and Space Administration)

Fig. 6.21 At right is a 128 × 64 km dune field of loose material in the floor of a 150 km wide crater in the *Hellespontus* region of Mars (Mariner 9 narrow-angle camera photograph). At left, the dune field appears as a black spot (wide-angle photograph) in the crater. The numerous long dunes are spaced about 1½ km apart. The similarity of size and direction of the individual dunes indicate that they are formed by strong winds blowing from a consistent direction, in this case from the south–west (By courtesy of the National Aeronautics and Space Administration)

Fig. 6.22 Clouds in the Martian atmosphere. A cyclonic storm created when the cold polar air flows under warmer air nearer the equator. (By courtesy of the National Aeronautics and Space Administration)

of the host of questions raised by the Mariner missions. The enigmatic results from the Viking I and II packages on the Martian surface, suggesting the presence of either exotic and highly unusual chemical reactions or life processes, show that the question "Is life present on the planet?" is certainly not to be answered with an outright negative. Martian surface conditions, though severe, are by no means completely inhospitable to life as we know it. In a later chapter, we will return to this matter.

Fig. 6.23 The surface of Mars near a Viking lander, a scene of rugged boulders, dust and desert. (By courtesy of the National Aeronautics and Space Administration)

6.4.2 The Martian Satellites The two satellites of Mars, Phobos and Deimos, are very small objects and revolve very close to Mars. They are seen more easily when Mars is close to opposition, but even then a large telescope is required. Their physical sizes were deduced by measuring their brightness and assuming a value for their albedo. These calculations give radii of the order of 8 km and 4 km for Phobos and Deimos respectively.

The inner satellite, Phobos, revolves around Mars in $7^h 39^m$, this being less than one-third of the rotation period of the parent planet. As a consequence, an observer on Mars would see Phobos in the same part of the sky on three occasions during a full Martian day. Its easterly orbital motion far outweighs the apparent westerly motion caused by the rotation of Mars. Phobos would uniquely be seen to rise in the west and set in the east. The more distant satellite, Deimos, has a revolution period of $30^h 18^m$.

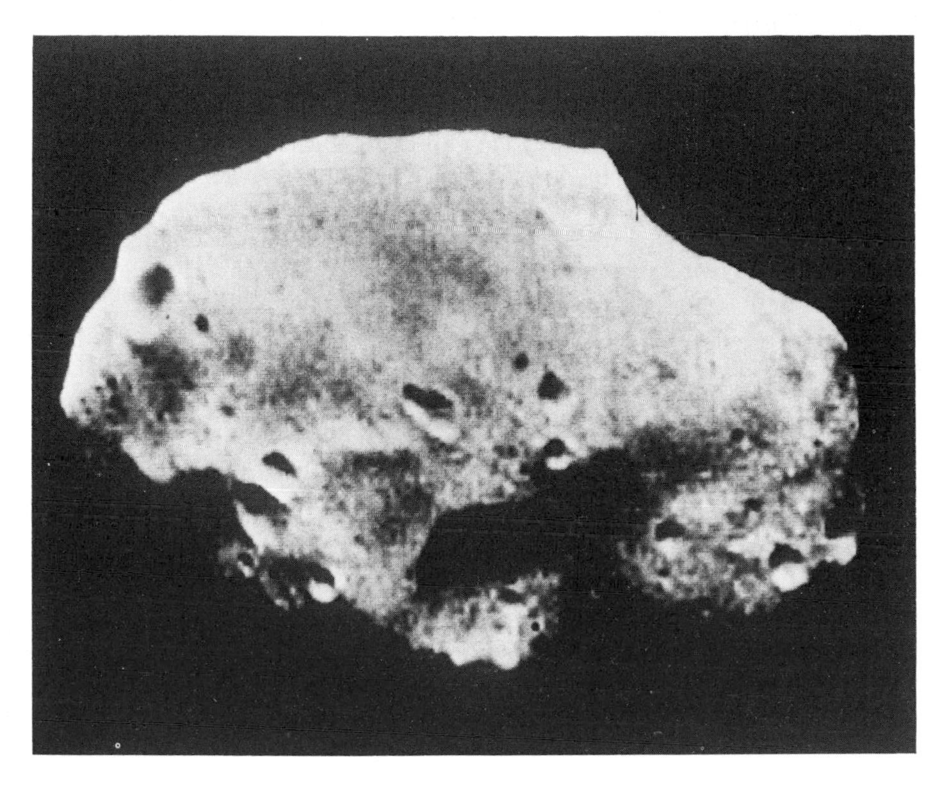

Fig. 6.24 Computer-enhanced photograph of Martian satellite Phobos, taken by Mariner 9. Many craters are visible, suggesting that Mars' inner moon is very old and possesses considerable structural strength. (By courtesy of the National Aeronautics and Space Administration)

Pictures of the two moons sent back to Earth by Mariner 9 (see Fig. 6.24) show the small satellites to be of irregular shape, their surfaces pitted with craters. These bodies are small enough for the structural strength of their materials to retain such irregular shapes against their own weak gravitational fields.

6.5 The Asteroids

Sometimes called the **minor planets**, these bodies form a numerous group of objects revolving about the Sun. Most of them have orbits lying between the orbits of Mars and Jupiter, the mean distance of all asteroid orbits being close to 2·8 AU. There are a few asteroids, usually of highly eccentric orbit, that can approach to within Mercury's orbit or recede as far as Saturn's. There are also two groups, the Trojans, that revolve in Jupiter's orbit.

The first asteroid to be discovered, Ceres, was found by Piazzi on January 1st, 1801. By 1891, when Wolf first used photography, 322 had been found by visual means, the stars in a given region of the sky being compared with a chart of the same region in the hope that a discrepancy would reveal one or more of the stars to be asteroids. The modern method is to take a time-exposure photograph of the region selected. If the telescope is guided on the stars, they will produce point-like images, but any asteroid image will appear as a streak (see Fig. 6.25); if the faintest asteroids are sought, however, the telescope is given a rate equal to the calculated angular drift of an asteroid so that its light will build up a point-image with the stellar images appearing as trails on the plate.

There are now about 1700 asteroids known in the sense that they have been discovered, numbered and named, and orbits have been calculated for them. Only one, Vesta, is visible at times to the unaided eye. The others are in general faint telescopic objects. In size these minor planets range downwards from about 770 km in diameter for Ceres to diameters for the most part less than 80 km. Many are less than 10 km in size. For the larger bodies, measurements of size have been made using the polarization properties of the light reflected from them. From a study of the polarization–phase curve, values of albedo may be predicted. Use of these values, together with knowledge of the bodies' apparent magnitudes and distances, give their sizes. Other estimates of size have been made by radiometric and speckle-interferometric techniques.

Asteroids are solid bodies, of densities probably not greatly different from that of the Earth's crust, and cannot possess any atmosphere. Careful photometric measurements show that many of them are highly irregular in shape and are rotating, usually in under 12 hours. Their combined mass would fall far short of the Moon's mass. It seems likely that there are hundreds of thousands of smaller ones, ranging in size down to rocks indistinguishable from meteors.

Study of the orbits of asteroids shows that most are almost circular and lie at small inclinations to the plane of the ecliptic. A minority have eccentricities much higher than those of the main planetary orbits and higher inclinations. Hidalgo, for example, has an orbital eccentricity of 0·66 and inclination of 43°. The distribution of the mean distances of the asteroid orbits shows a feature first discovered by Kirkwood—a distinct avoidance of certain regions within which the periods of revolution would be simple fractions of Jupiter's orbital period, for example, a half, a third, two-fifths, and so on. On the other hand, there is an accumulation of asteroid orbits at a distance where the period is approximately two-thirds that of Jupiter and, as mentioned, the Trojan asteroids move in Jupiter's orbit, so that their period of revolution is equal to that of Jupiter.

The avoidance by asteroids of the distances where their sidereal periods are commensurable with that of Jupiter is often explained as follows. If there is such a relationship in period, it means that the closest approaches of Jupiter and asteroid take place at very nearly the same heliocentric longitude over successive *synodic* periods of the two bodies. The net disturbance of the asteroid by Jupiter in a synodic period is largely cumulative from period to period and so shifts the asteroid into a new orbit. Jupiter thus has swept these regions clear of asteroids. Unfortunately, this "resonance" theory, although attractive, does not explain the accumulation of asteroids at distances where the sidereal periods are almost two-thirds that of Jupiter. R. B. Hunter's work has in fact indicated that any asteroids that previously existed in orbits for a range on either side of that distance will have been removed, but that the two-thirds period orbit may be a stable one. The Trojans themselves are illustrations of the stable solution of the problem of three bodies, first

Fig. 6.25 Three minor planet trails amongst the point images of stars. (By courtesy of the Royal Astronomical Society)

discovered by Lagrange (see Roy & Clarke, *op. cit.*, Section 13.3), which states that a small body can remain at a corner of an equilateral triangle, the other two corners being occupied by two massive bodies in orbit about each other. There are obviously two cases where the Sun and Jupiter are the two massive bodies and the Trojans are divided between them, one group oscillating about the equilateral triangular point with heliocentric longitude 60° less than Jupiter's, the other group behaving likewise about the point with heliocentric longitude 60° more than Jupiter's. From analysis of the various orbital parameters, additional associations or families of asteroids have been found. Important groupings include the Themis, Koronis and Eos families.

Asteroids on occasion pass near the Earth and some must have collided with our planet in the past. Weathering, except in a few cases, will have removed all visible traces of such events. Among those that have drawn near to the Earth recently are Eros (22 000 000 km), Apollo (4 500 000 km), Adonis (1 500 000 km), Hermes (1 000 000 km). The close approach of Icarus (5 900 000 km) in 1968 allowed radar contact to be made with an asteroid for the first time.

Two theories are put forward to account for the asteroid belt. One is that one or more larger bodies were disrupted in successive collisions. The other is that the proximity of Jupiter during the process of planetary formation robbed the region of most of its material and also successfully prevented the remaining material from forming one fair-sized body.

6.6 Jupiter

6.6.1 The Planet The planet Jupiter is the largest in the Solar System by mass and diameter, being 318 times the mass of the Earth; its equatorial radius is 71 370 km. The disk of the planet seen in a small telescope shows a high degree of flattening. Jupiter revolves about the Sun at a mean distance of 5·2 AU and in a period of 11·86 years. The planet is covered with a dense atmosphere arranged in belts of different shades and colours parallel to its equator. Like the Sun's atmosphere, the Jovian atmosphere has different periods of rotation depending upon the latitude, the periods being longer the higher the latitude. Values range from $9^h\ 50^m$ near the equator to $9^h\ 56^m$ in high latitudes. The transition in velocity between neighbouring belts is to all intents and purposes discontinuous. Dark and bright spots, often well-defined, are seen in these belts. Some spots persist for months. One such marking, the Great Red Spot (see Fig. 6.26), has certainly been visible since 1831 and may have been observed by Hooke in 1664. Dull red in colour, oval in shape, with dimensions 40 000 km long by 13 000 km broad, it has a varying period of revolution about the planet, somewhat longer than that of the belt in which it is immersed. The view that it and the other spots are essentially atmospheric disturbances of a semi-stable cyclonic nature is reasonable.

Fig. 6.26 Jupiter's Great Red Spot and large white oval. Both are cyclonic structures in Jupiter's atmosphere. (By courtesy of the National Aeronautics and Space Administration)

The spectrum of Jupiter's atmosphere reveals the presence of methane and ammonia in large amounts. Hydrogen is also present. The methane is in a gaseous state, but much of the ammonia must be in the form of crystals since the temperature at the level of the obscuring clouds is about −130 °C and ammonia freezes at −78 °C.

Jupiter emits radio waves. Some of its radio output is thermal in origin. Short non-thermal outbursts come from the planet's ionosphere. Other radio waves possibly come from radiation belts, similar to the Earth's van Allen belts, surrounding the planet. Jupiter possesses a strong magnetic field, the angle between the planet's magnetic axis and axis of rotation being 15°. The polarity is opposite to that of the Earth. The field and trapped particles in the radiation belts share Jupiter's rotation rate out to a distance of 35 planetary radii.

Various models have been constructed of the interior of Jupiter. Such models must fit the observable data such as mass, radius, Jupiter's marked oblateness, and the changes in its satellites' orbits caused by its internal constitution. Early models (around 1934 or before) proposed a heavy core (density $= 5{\cdot}5 \times 10^3$ kg m^{-3}) surrounded by a high-pressure modification of ice (density $= 10^3$ kg m^{-3}), which in turn was covered by a shell of solidified hydrogen and helium (density $= 3{\cdot}5 \times 10^2$ kg m^{-3}). Above that lay the atmosphere. Later views have developed from models of planets consisting of pure hydrogen, or hydrogen with a proportion of helium and just 1 or 2% of the other elements. The chemical composition of such planets closely resembles that of the Sun, though the interior conditions are vastly different. The physical theory of solid hydrogen shows that it assumes a metallic form under very high pressure and, in the case of a hydrogen planet, a metal-type core would exist if its mass exceeded about 80 times the mass of the Earth. The mass of Jupiter is 318 times that of the Earth. Above the core would be a region in a different state, while above this would lie a layer behaving more like an ocean than an atmosphere. It is calculated that the top layer of atmosphere in a mass the size of Jupiter would have a mass less than one-tenth that of the Earth.

The total thermal output from the planet is about $2\frac{1}{2}$ times the quantity received from the Sun. A possible explanation for this source of heat is that Jupiter is still contracting gravitationally, changing potential energy into heat energy, a process in fact believed by astronomers to be an early phase in the lives of stars. In this respect it is interesting to note that Jupiter is about as massive as a planet can be while still remaining a planet.

6.6.2 Jupiter's Satellite System Jupiter has sixteen known satellites. Their mean distances from the planet's centre range from 128 000 km to 24 000 000 km. The four large moons, Io, Europa, Ganymede and Callisto, were discovered by Galileo and are often referred to as the Galilean satellites. They move in almost circular orbits very close to the plane of Jupiter's equator. The other twelve are much smaller and though they have names are often numbered in order of discovery. The fifth, Amalthea, revolves nearest to the planet and may be only 160 km in diameter. Jupiter VI, VII, X and XIII form a separate group, all having orbits about 11 500 000 km from the centre of Jupiter. Their orbits are markedly elliptical and are highly inclined to the planet's equatorial plane. Although the mean distances of these satellites are almost equal, the orbits are so orientated that the chance of a collision is slight. Because they are so far from Jupiter, the Sun's gravitational pull strongly perturbs their orbits.

Of the remaining seven, Jupiter VIII, IX, XI and XII, form a third group in much larger retrograde orbits, even more strongly perturbed by the Sun.

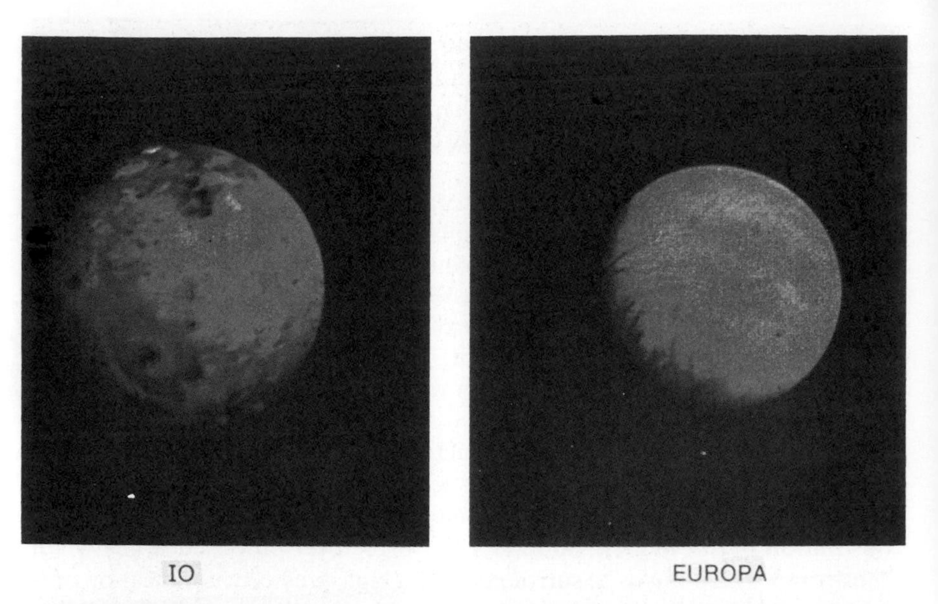

Fig. 6.27 Pictures of the four large Galilean satellites of Jupiter, showing their correct relative sizes. While Ganymede and Callisto are about the size of the planet Mercury, Io and Europa are about the size of Earth's moon. (By courtesy of the National Aeronautics and Space Administration)

Indeed, calculation shows that if the orbits were direct at such distances, Jupiter could not retain these objects as satellites for more than a short time. They would be pulled away by the Sun to become asteroids pursuing independent orbits about the Sun. The reverse course of events can also take place, with Jupiter capturing asteroids and holding them as satellites for an indefinite time interval. The outer four Jovian satellites may be captured asteroids that in future, under the right conditions, could escape from the Jovian system. In such an event they need not return to the region of the asteroid belt. Theoretical studies have shown that they could be sent into quasi-stable orbits midway between the orbits of Jupiter and Saturn.

Two of the remaining satellites, Jupiter XV and XVI, orbit much closer to Jupiter than Jupiter V (Amalthea) while Jupiter XIV follows an almost circular orbit between those of Amalthea and Io.

All the small satellites are probably under 80 km in diameter. At their distances from Sun and Earth, and at their close angular proximity to Jupiter, they are difficult objects to photograph, so that it is quite possible that Jupiter

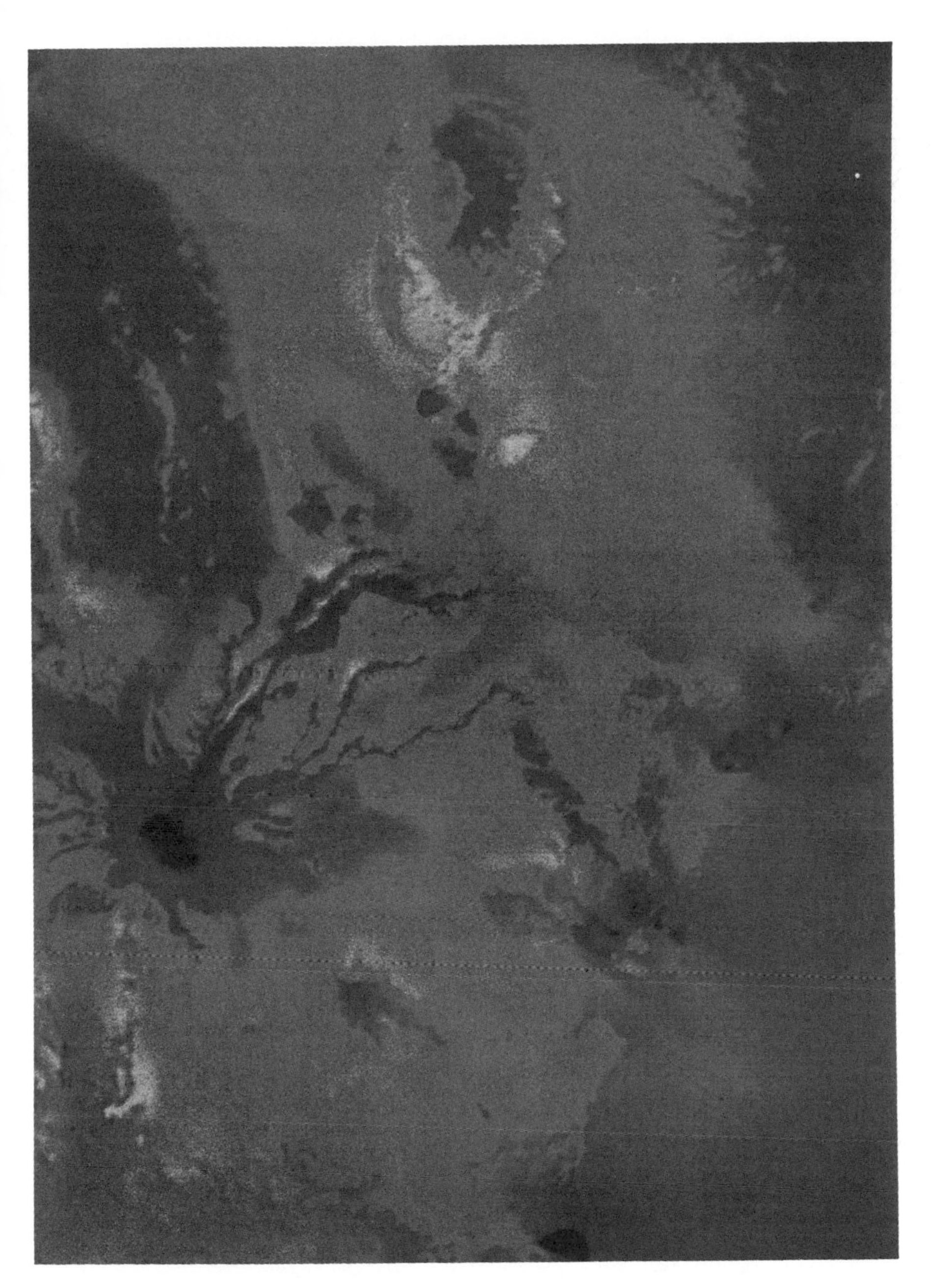

Fig. 6.28 The surface of Io showing surface deposits of sulphur compounds and salts. Possible lava flows are also visible radiating from a volcanic crater. (By courtesy of the National Aeronautics and Space Administration)

may yet possess undetected moons. In contrast to these small objects, the four Galilean satellites are between the Moon and Mercury in size and are visible with binoculars.

With a small telescope it is possible to watch the eclipses, occultations and shadow transits of the Galilean satellites. Their periods of revolution about the planet are so small, and their orbital planes are so close to the plane in which Jupiter journeys round the Sun that such phenomena are frequent. *The Astronomical Almanac* carries predictions of these happenings. Thus in Fig. 6.29, the satellites as seen from Earth exhibit the following phenomena:

Io (I) precedes its shadow across Jupiter's disk.
Europa (II) is emerging from occultation.
Ganymede (III) is eclipsed.
Callisto (IV) is occulted.

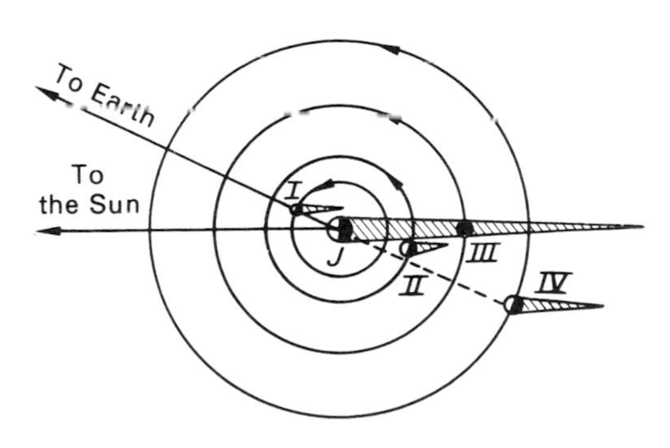

Fig. 6.29 A configuration of Jupiter's Galilean satellites showing I in transit, II emerging from occultation, III eclipsed and IV occulted

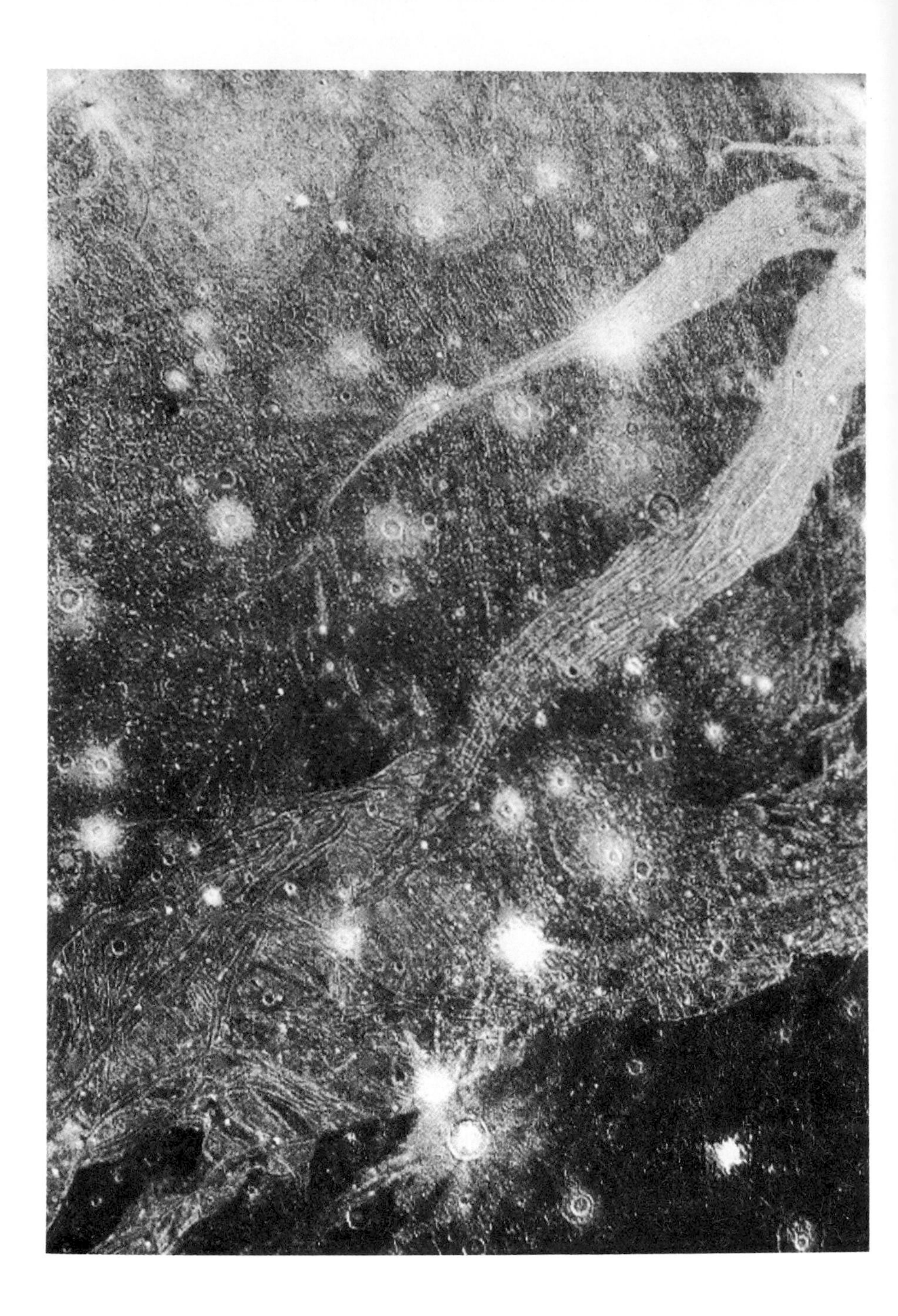

Fig. 6.30 Surface features of Ganymede, showing many impact craters, some old, some comparatively new. Also shown are features revealing large-scale faulting and slippage of terrain. (By courtesy of the National Aeronautics and Space Administration)

It will be remembered that it was by a comparison of the observed times of these events with predicted times that Roemer was able to deduce that the velocity of light is finite.

Although accurate orbital data concerning the Galilean satellites had been acquired prior to the Voyager flypasts, data concerning the physical make-up of these moons were sketchy in the extreme. The situation has now changed remarkably and the four Galilean moons have become Solar System worlds in their own right, providing a wide variety of geophysical, geological and cosmogonic problems that will continue to exercise the wits of researchers long after Voyager 2 has left the Solar System.

Io, the innermost Galilean moon, has a sidereal period of revolution of 1·76914 days and an orbital radius of 422 000 km. Its diameter is 3530 km, and its mean density 3500 kg m^{-3}, figures not much different from the corresponding ones of 3476 km and 3330 kg m^{-3} for the Earth's Moon. But there the resemblance ceases. Whereas the Moon is a dead, inactive world, Io is a world of continuously active volcanoes, its changing surface covered by multi-hued deposits of sulphur and sulphur dioxide. Its interior is probably molten silicate. Some of the products of vulcanism escape into space forming a doughnut-shaped ring or torus surrounding Jupiter and coinciding roughly with Io's orbit. The heat energy creating this activity comes from tidal imbalances within Io produced by Io's orbital relationship with Jupiter.

Europa is smaller than Io, having a diameter of about 3120 km and a mean density of 3000 kg m^{-3}. Its distance from Jupiter's centre is 671 000 km giving it an orbital period of 3·55118 days, just over twice that of Io. Its surface carries a vast network of light and dark markings, possibly faults or fractures, many of them more than 1000 km long. It is probable that the satellite has a thick icy crust about 100 km thick overlaying a rocky interior.

Ganymede is the largest of the Galilean satellites with a diameter of 5220 km and a mean density of only 1900 kg m^{-3}. Its orbital radius and period are respectively 1071 000 km and 7·15455 days. A model of the satellite fitting the observed data suggests an ice crust some 75 km thick over a water or ice mantle below which lies a rocky core. The surface features include grooved and cratered terrains that intermingle in places indicating a complicated geological history.

The outermost Galilean satellite Callisto resembles Ganymede in make-up rather than the inner two, Io and Europa. Its mean density is 1800 kg m^{-1} while its diameter is 4900 km. Like Ganymede, it may be composed of a rocky core carrying on top of it a water or ice mantle with an outer layer or crust made up of ice and/or rock. Its surface is the darkest of the four satellites and is also the most densely cratered. It has been described as resembling a dirty, pock-marked frozen slush.

Callisto orbits at a distance from Jupiter's centre of 1884 000 km with an orbital period of 16·68902 days.

All four satellites have rotational periods equal to their orbital periods so that, like our Moon, they always keep the same hemisphere turned towards their planet.

In addition to the satellite system Jupiter possesses a ring in orbit about it, about 6500 km wide, its outer edge only 57 000 km above the outermost limits of Jupiter's cloud tops. It is probably only a few kilometres thick and is composed for the most part of particles no larger than a few microns in diameter.

6.7 Saturn

6.7.1 The Planet The second largest planet of the Solar System is Saturn, appearing as a bright yellowish object to the unaided eye. Its equatorial radius is 60 400 km compared with a polar radius of 54 600 km. Saturn's mean distance from the Sun is 9·54 AU and it has an orbital period of 29·46 years.

From observations of Saturn's inner satellites the mass has been determined to be 95·2 times that of the Earth. Calculation of the mean density gives a value of 7×10^2 kg m^{-3}, i.e. less than water. Saturn has the lowest density of all the major planets except, possibly, Pluto.

Telescope views of Saturn reveal markings similar to Jupiter's, but less distinct. The markings do not allow an accurate rotational period to be measured. However, this has been done by measuring Doppler shifts in the spectral lines in the light from the planet's limb. The measurements show that the period of rotation at the equator is $10^h 14^m$ and that this increases progressively towards the poles where the value is $10^h 38^m$.

Measurements of thermal radio radiation are in good agreement with the temperature predicted by considering the heat received by Saturn from the Sun. The temperature of Saturn is of the order of −150 °C.

The presence of ammonia in the spectrum is not as strong as in the case of Jupiter. This may be a reflection of Saturn's lower temperature—at −150 °C practically all the ammonia present would be frozen out and the bands corresponding to its gaseous state would be very weak. Methane bands are stronger than for Jupiter.

The internal structure of Saturn is probably similar to that of Jupiter. The Voyager spacecraft infrared scans measured a ratio of 9 : 1 for the hydrogen-to-helium abundance, exactly the same as the corresponding value for Jupiter. In addition Saturn, like Jupiter, radiates more energy into space than it receives from the Sun.

6.7.2 Saturn's Ring System Saturn was the first member of the Solar System observed to have a ring system in the plane of its equator. The phenomenon was first observed by Galileo but it was not until 1655 that the true nature of the system was explained by Huygens. Shortly afterwards in 1675, Cassini recorded that the ring had a division, the outer ring now being known as the A ring and the inner, the B ring. Further structure was found by Bond in 1850 when a third, hazy or crepe ring, now known as the C ring, was discovered inside the perimeter of the main rings. In 1969 Guerin discovered that the crepe ring also has structure; the inner part of it has now been labelled D.

Since then the Voyager missions have revealed other rings together with a wealth of structural detail in the classical rings.

Saturn's equator is inclined to the plane of its orbit by 26° 44′ and, as the rings are exactly in the equatorial plane of the planet, the aspect they present to us depends greatly on the position of Saturn in its orbit, this being demonstrated in

Fig. 6.31 The best available photographs of (*a*) Mars, (*b*) Jupiter, (*c*) Saturn, (*d*) Pluto taken with the Mt. Wilson telescope show the limitations of earth-based photography. (By courtesy of the Mt. Wilson and Palomar Observatories)

Table 6.1 The radii of Saturn's rings†

Ring	km	Ring	km
A Outer	136 600	D	Uncertain
A Inner	119 800	E Outer	294 600
B Outer	117 100	E Inner	210 000
B Inner	90 500	F	139 200
C Inner	74 600	G	168 000

† Each of the rings A, B and C consists of hundreds of ringlets, some elliptical.

Fig. 6.32. As the plane of Saturn's orbit is not quite in the plane of the ecliptic, the aspect of the rings also varies a little, according to the Earth's position within its orbit.

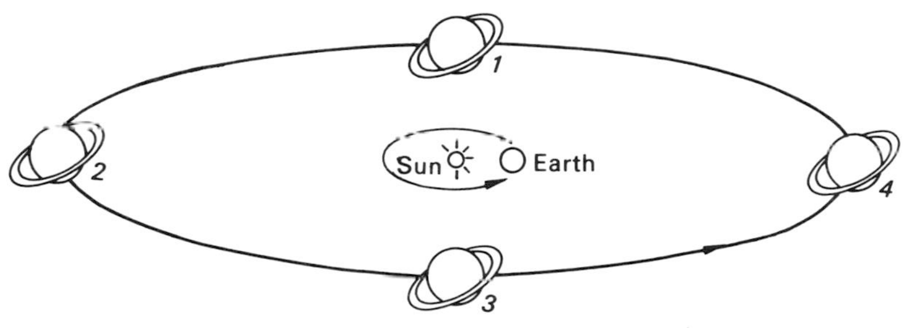

Fig. 6.32 The aspects of Saturn's rings

It will be seen in Fig. 6.32 that there are two positions (marked 1 and 3) some fifteen years apart, when, depending on the exact position of the Earth in its orbit, the rings are seen exactly edgeways. At other times (marked 2 and 4) the full aspect of the ring system is seen. The change in aspect over the years causes Saturn's apparent brightness to vary by some 70%.

The virtual disappearance of the rings when in the edgeways configuration shows that the rings are thin, probably no thicker than 20 km. On some occasions it is possible to see bright stars through the rings, thus demonstrating their small optical thickness.

In the 19th century, Maxwell proved that the rings could not be solid or liquid if they were to remain in stable orbits around Saturn. Gravitational forces across them would cause the rings to fragment. Maxwell also proved, however, that the rings would have stability if they were composed of many small particles.

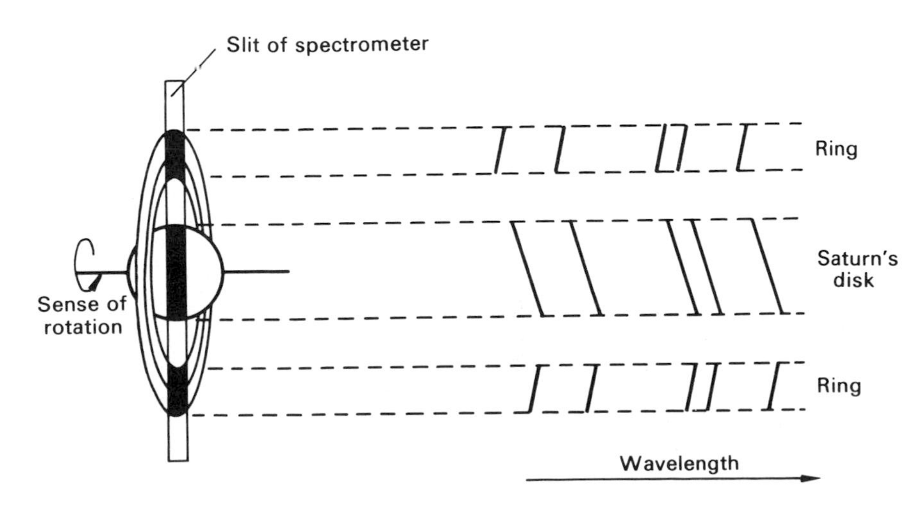

Fig. 6.33 The Doppler shifts in Saturn's spectrum

In 1895 Keeler used spectroscopy to demonstrate that the rings are not solid or liquid. By placing the entrance slit of a spectrometer at right angles to a line through the poles of the planet, a spectrum is produced which has a central part from the planetary disk and smaller parts on either side corresponding to sections of the ring system (see Fig. 6.33). The Doppler displacement is zero for light from the centre of the disk. According to Fig. 6.33 the lower limb is receding from the observer (red-shifted lines) while the upper limb is approaching the observer (blue-shifted lines). The continuous range in Doppler shift from the centre of the disk to the limb shows that the body is solid. The upper spectrum from the rings indicates that the parts of the rings have different relative velocities, but with the inner part being more blue-shifted than the outer and consequently having a higher velocity of approach. In the lower spectrum, the inner part of the ring has the greater velocity of recession. Thus the inner part of the ring has a higher velocity of rotation about Saturn than the outer part and the ring system therefore cannot be solid.

In 1850 Roche studied the effect of a planet's tide-raising forces on a finite-sized liquid satellite. He showed that if the satellite remained beyond a certain distance from the planet's centre, the satellite would merely be distorted; if, however, it orbited the planet within this limiting distance, now known as *Roche's limit*, the tide-raising forces would overcome the mutual gravitational forces of the satellite and tear it asunder. This limit r is given by the expression

$$r = 2{\cdot}44(\rho_S/\rho_P)^{1/3}R_P,$$

where ρ_S and ρ_P are the average densities of satellite and planet respectively and R_P is the primary's radius.

Forgetting for the moment that the law was derived for a liquid satellite we find that Janus, Saturn's innermost satellite at a distance 2·80 times the planet's radius, is just outside Roche's limit for that planet; on the other hand, the outer radius of the ring system of Saturn is 2·30 times that of the planet so that the whole ring system lies within Roche's limit. If its material was originally liquid it would have been disrupted; even if it were originally composed of solid particles, Saturn's tide-raising forces would have prevented them from joining into larger objects.

It is a straightforward matter to derive an approximate expression for Roche's limit for a solid satellite.

Suppose the planet P, of radius R_P, has a satellite S, of radius R_S moving in a circular orbit of radius r about it. Let the masses of planet and moon be M and m respectively. Suppose also that $M \gg m$ so that we can assume the centre of mass of the system to be at the centre of the planet.

The gravitational acceleration of planet on satellite is GM/r^2 while the centrifugal acceleration of the satellite in its orbit is $\omega^2 r$. They must balance each other so that

$$\omega^2 = GM/r^3. \tag{6.1}$$

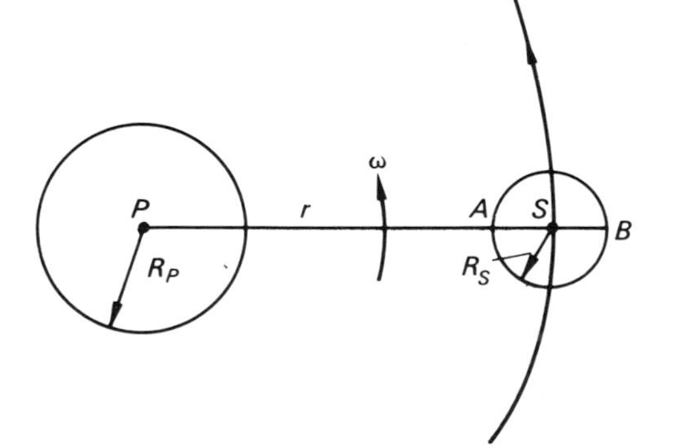

Fig. 6.34 Approximate derivation of Roche's limit

The difference between the gravitational accelerations of the planet on points S and B of the satellite (Fig. 6.34) is Δg, given by

$$\Delta g = GM/r^2 - GM/(r+R_S)^2,$$

or, approximately, since $R_S \ll r$,

$$\Delta g = 2GMR_S/r^3.$$

Similarly the differential centrifugal acceleration between S and B, $\Delta\alpha$, is given by

$$\Delta\alpha = \omega^2(r+R_S) - \omega^2 r = \omega^2 R_S.$$

Fig. 6.35 Voyager 2 picture of Saturn from a distance of 21 million kilometres. Three of Saturn's moons are visible: in order of increasing distance from the planet they are Tethys, Dione and Rhea. Tethys' shadow is seen on Saturn's southern hemisphere. (By courtesy of the National Aeronautics and Space Administration)

But by Equation (6.1), $\omega^2 = GM/r^3$, hence

$$\Delta\alpha = GMR_S/r^3 = \Delta g/2.$$

If the satellite is not to be torn apart, therefore, the sum of these accelerations, namely $3GMR_S/r^3$, must be less than the satellite's own gravitational acceleration on a particle at its surface, namely Gm/R_S^2.

Hence in the limit we can write

$$Gm/R_S^2 = 3GMR_S/r^3. \tag{6.2}$$

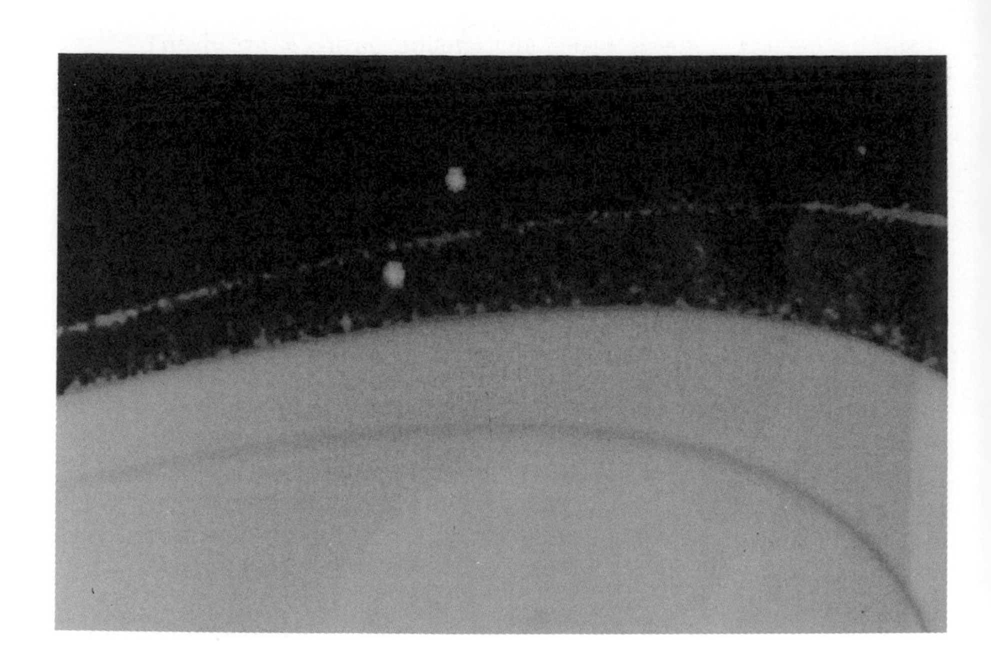

Fig. 6.36 Saturn's F ring with its two "shepherd" satellites. The F ring lies outside the A ring. The satellites' gravitational forces keep the F ring in position. (By courtesy of the National Aeronautics and Space Administration)

Fig. 6.37 Saturn's rings seen from Voyager 2. Note the many spoke-like features in the B ring. It is not certain that Newtonian gravitation can account for all those features. (By courtesy of the National Aeronautics and Space Administration)

But the mean densities of planet and satellite, ρ_P and ρ_S are given by

$$\rho_P = \frac{M}{\frac{4}{3}\pi R_P^3}; \qquad \rho_S = \frac{m}{\frac{4}{3}\pi R_S^3} \tag{6.3}$$

Combining (6.2) and (6.3) we therefore obtain

$$r = 3^{1/3}(\rho_P/\rho_S)^{1/3} R_P \sim 1{\cdot}44(\rho_P/\rho_S)^{1/3} R_P.$$

The Encke and Cassini divisions in the rings were explained by Kirkwood as caused by the perturbations of Saturn's innermost satellites, in particular Mimas, on particles originally within the gaps. Cassini's division contains distances where the period of revolution of a hypothetical particle would be 1/2 that of Mimas and 1/3 and 1/4 those of the next two satellites, while the boundary between rings B and C lies at a distance where the period of revolution would be 1/3 that of Mimas. Encke's division contains an orbital period 3/5 that of Mimas. It can be shown that perturbations on particles with such commensurable periods would lead to a dearth of small bodies at these distances.

It used to be thought that the situation was therefore analogous to that of the Kirkwood gaps in the asteroid region and to a first approximation the analogy probably is valid. The Voyager–Saturn missions, however, have shown that the dynamical problems presented by the fine structure of Saturn's rings are far more complicated.

Each of the major rings A, B and C consists of hundreds of individual ringlets, some elliptical, with even the classical gaps containing ringlets; the Cassini division contains at least five. Transient, spoke-like features exist, radiating outwards across the B-ring system.

Additional rings have been discovered, for example the D, E, F and G rings. The F ring, discovered by the spacecraft Pioneer 11 in 1979, is formed by a number of ringlets. The two "shepherd" satellites, Pandora and Prometheus, discovered in 1980, orbit respectively on the outside and inside of the F ring. They control the outer and inner limits of the F-ring system. It seems probable that the F ring and its shepherding satellites exist at distances such that the systems Saturn–ring particle–Pandora, Saturn–Prometheus–ring particle, Saturn–Pandora–Prometheus form three-body systems that are stable dynamically.

Indeed four possible explanations have been put forward for the rings' extraordinary multiplicity: resonances with external satellites, gravitational instabilities within the rings, density waves moving through the rings and small additional satellites sweeping up particles along their orbits. A satisfactory tying-in of theory to data has yet to be achieved.

6.7.3 Saturn's Satellite System Saturn is now known to have at least 17 satellites, the largest of which, Titan, was discovered by Huygens in 1655. Four more, Tethys, Dione, Rhea and Iapetus, were found by J. D. Cassini between 1671 and 1684 while Sir William Herschel, the discoverer of Uranus, found two, Mimas and Enceladus, in 1789. W. C. Bond found the eighth, Hyperion, in 1848 and W. H. Pickering discovered the ninth, Phoebe, in 1898 by photography.

These pre-Voyager satellites have a wide range in size of orbit, mass and body. The inner five move in nearly circular coplanar orbits almost in the plane of the rings. Titan and Hyperion, while also in orbits near to the plane of the rings, have more elliptical orbits. Iapetus has an orbital plane lying between the planet's equatorial and orbital planes. Phoebe has a retrograde

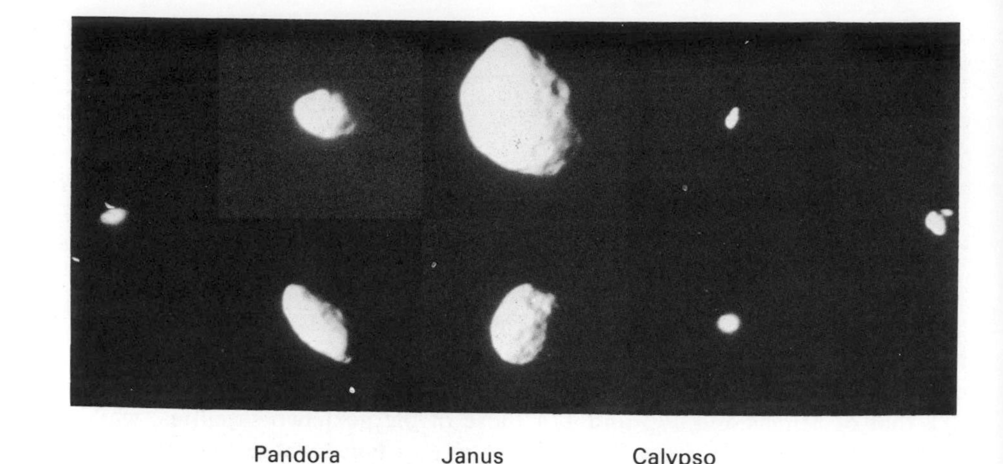

Fig. 6.38 Eight of the smaller moons of Saturn as seen from Voyager 2. The images are scaled to display the true relative sizes. The bodies may be fragments of larger satellites broken up by collision. Their dimensions (in km) and positions are given below the figure. (By courtesy of the National Aeronautics and Space Administration)

motion. The orbital radii range from 186 000 km for Mimas to 12 960 000 km for Phoebe with corresponding periods of revolution ranging from 0·94242 days to 550·45 days. Titan is the biggest and most massive with a radius of 2400 km and a satellite-to-planet mass ratio of 1/4150; the smallest, Mimas, has a radius of 240 km and a satellite-to-planet mass ratio of about 1/15 000 000.

The Voyager missions have added eight satellites to the list of Saturn's family of moons and also shown us what those discovered pre-Voyager look like.

Saturn's satellite system is listed in Table 6.6.

Of the pre-Voyager satellites, all, with the exception of Titan, are without atmospheres and show the cratered features characteristic of the chaotic, violent early era in the life of the Solar System when collisions with massive pieces of debris or other smaller bodies were frequent. A number of the moons also show sinuous valleys or crevasses fracturing the surfaces. Enceladus seems to have had a complicated history with traces of at least five distinct evolutionary eras visible on its surface. Some of its features resemble the grooved terrain on Jupiter's satellite Ganymede. It is possible that Enceladus is still geologically active, probably through tidal heating induced by Saturn and Dione.

Iapetus presents a different problem. It had been known observationally as far back as J. D. Cassini's time that part of Iapetus' surface is much darker than the rest, in fact six times as dark. The Voyager pictures confirm this and show that the dark side is the hemisphere that always faces the direction in which Iapetus orbits Saturn. It is not known at present if the dark material has erupted onto the predominantly icy surface, whether it has been exposed as the original ice layer over it has been eroded or whether it has been laid down from space, possibly from Phoebe.

The biggest satellite, Titan, is surrounded by a dense orange-coloured atmosphere with a surface pressure estimated to be more than $1\frac{1}{2}$ times that

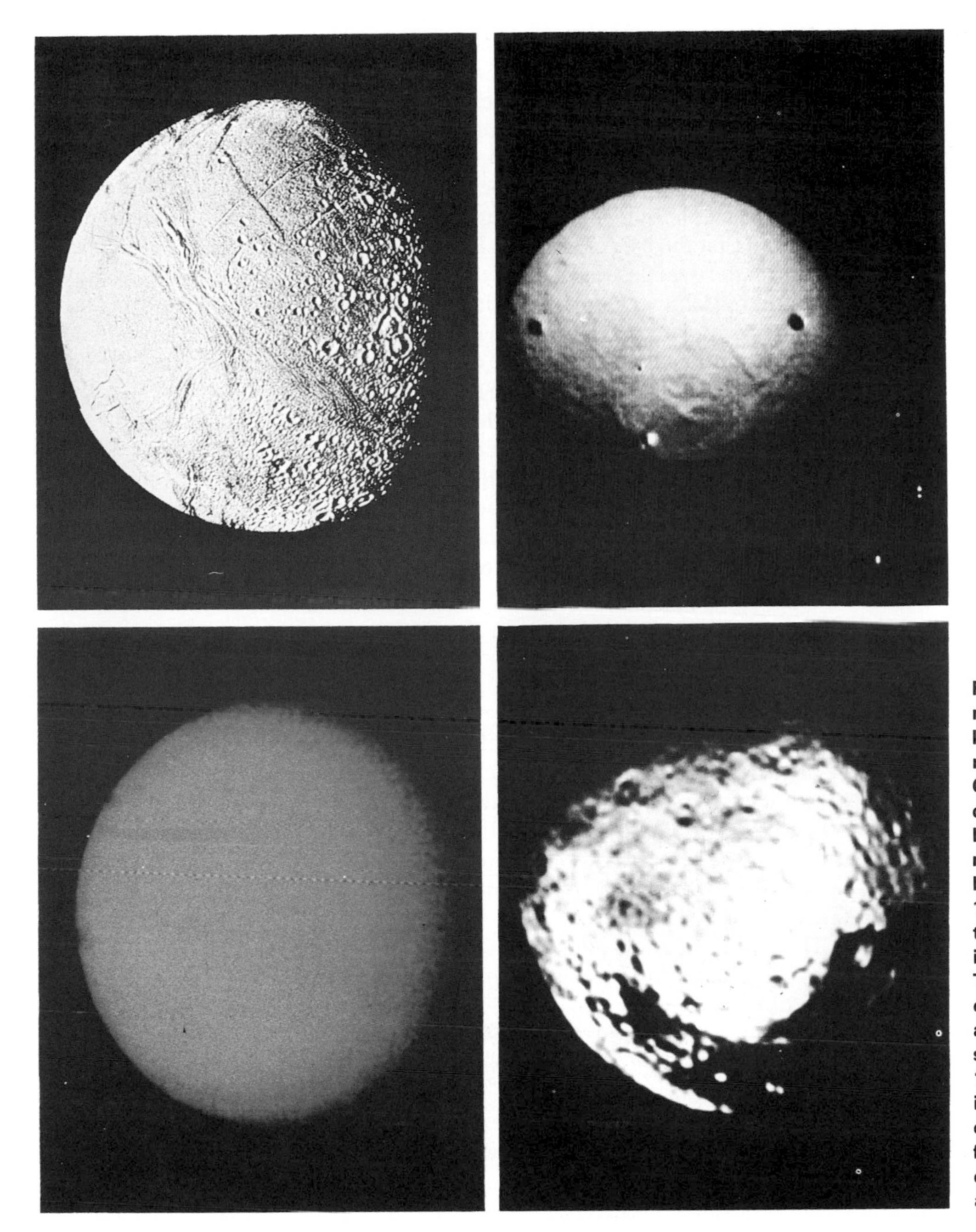

Fig. 6.39 Four of Saturn's larger moons, Enceladus, Tethys, Titan and Iapetus. Enceladus bears a marked resemblance to Jupiter's moon Ganymede. Enceladus (about 500 km in diameter) shows older terrain crossed by strips of younger terrain. Some regions are smooth; others are covered by sets of grooves. Tethys, about 1050 km in diameter, is remarkable for the presence on it of an enormous impact crater some 400 km in diameter. Titan, Saturn's largest moon, with a diameter of 4800 km, has a dense atmosphere completely obscuring its solid surface. Iapetus, of diameter 1450 km, is covered by impact craters; it is also noted for the extraordinarily dark material that covers a considerable fraction of the satellite's ice crust. (By courtesy of the National Aeronautics and Space Administration)

at the Earth's surface. The atmospheric constituents are predominantly nitrogen and methane but there may be as much as 12% of argon and a tiny amount of molecular hydrogen. Organic compounds such as propane, acetylene and hydrogen cyanide are also present.

Of the nine "classical" satellites, Hyperion possesses the most irregular shape, resembling a battered hamburger and suggesting, in its cratered, almost shattered condition, innumerable damaging collisions with large bodies in the past. Its present orbit is locked in a most remarkable way to that of Titan, its chief gravitational disturber. Hyperion's conjunction line with Titan swings at a rate just sufficient to keep conjunctions of Saturn, Titan and Hyperion occurring near Hyperion's aposaturn. In this way the close approaches of

Titan and Hyperion are as large as possible thus minimising the former's disturbing effect.

The other newly-discovered satellites of Saturn are small and in many cases even more irregular in shape than Hyperion. They also exhibit remarkable orbital properties.

Atlas coasts just 800 km outside the outer edge of the A ring and possibly acts in a "shepherding" capacity to the A-ring particles. It has already been remarked that the pair of satellites Prometheus and Pandora act as "shepherd" moons on the outside and inside of the F ring, controlling its width and position. Two other satellites, Janus and Epimetheus, share a common orbit. In Saturn's satellite system are also found two more examples of the Lagrange equilateral triangle solution to the problem of three gravitating bodies (see Roy and Clarke, *op. cit.*, Section 13.3 and Section 6.5 of the present chapter). Calypso and Telesto occupy respectively the leading and trailing corners of the equilateral triangles whose other two corners are occupied by Saturn and its satellite Tethys. Thus Calypso precedes Tethys in its orbit about Saturn, while Telesto follows it, both of these small bodies' radius vectors keeping 60° ahead or behind Tethys' radius vector. Similarly the satellite Helene (Dione B) precedes Saturn's satellite Dione, keeping 60° ahead of Dione and so forming with it and Saturn an equilateral triangle.

6.8 Uranus

As a result of systematic searches of the sky, a seventh planet, originally named *Georgium Sidus* after George III but now known as Uranus, was discovered by Sir William Herschel in 1781. In searching through records made by previous observers it was later found that Uranus had in fact been seen but unidentified on some twenty occasions. These observations stretched over a time greater than one sidereal period of this planet and so allowed its orbit to be determined fairly well.

The disk presented by Uranus is small (3·75 arc sec) but, converted to physical size, the planet has an equatorial radius of 25 600 km, being nearly four times the size of the Earth. It appears greenish in colour and there are no distinct markings that would allow its rotation to be measured; however, it shows a relatively high degree of flattening, indicating a large speed of rotation.

The mean solar distance of Uranus is 19·18 AU and at this distance it is interesting to note that the Earth would never appear to be more than 3° away from the Sun.

Much of our knowledge about the planet's physical properties comes from the fly-past of the planet by Voyager 2 in January 1986. It was known that the rotational period (deduced from spectrometric measurement) was approximately $10^h\ 49^m$. This figure has now been revised to almost 17 hours. The axis of rotation is at an angle of 98° with respect to the plane of its orbit and, to be strictly correct, we must say that the sense of rotation is opposite to all the other planets except Venus. During its 84 year orbit, one pole, then the equator and then the other pole have long periods when they are turned towards the Sun.

The planet's mean density is $1200\ kg\ m^{-3}$, similar to that of Jupiter and Saturn. Its atmosphere is rich in hydrogen and also contains methane, identified from its infrared spectrum. Clouds were detected deep in the atmosphere, which has a temperature as low as −221° C. The planet possesses a magnetic field stronger than Earth's, the magnetic axis lying about 60° from the axis of rotation. It seems likely that Uranus consists of a superdense atmosphere

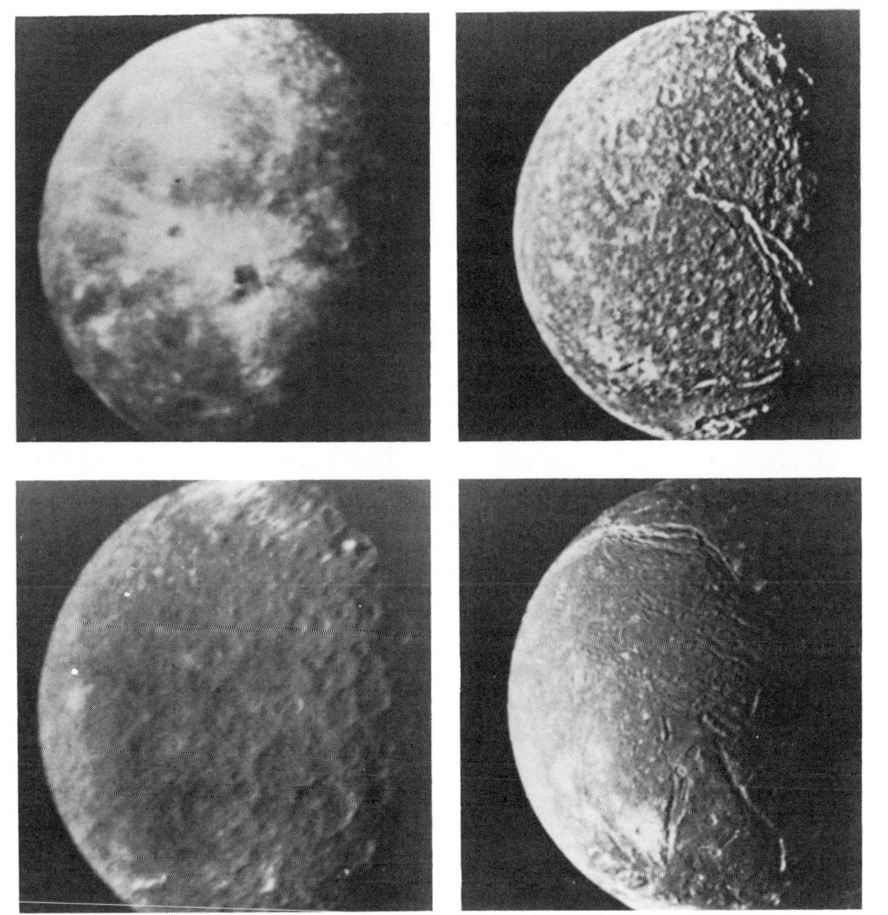

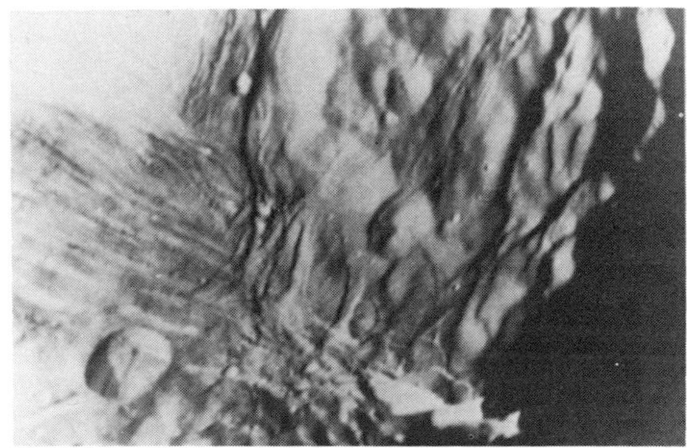

Fig. 6.40 Satellites of Uranus recorded by Voyager 2. (Top left) Oberon, the outermost moon of Uranus. (Top centre) Titania, the largest moon of Uranus. The giant cracks, as much as 5 km deep and 1600 km long, expose light-coloured material. (Top right) Valley walls on Miranda catch the sunlight while night blankets the surrounding area, revealing the steepness of the cliffs. Miranda is only of the order of 480 km in diameter but has canyons more than ten times deeper than the Grand Canyon on Earth. (Bottom left) Umbriel, darkest of the Uranian moons, shows craters but little geologic activity—except for two bright features near the top. Light-coloured material is visible at the bottom of a crater 80 km in diameter and on the slope of the central peak in a nearby crater. (Bottom right) Ariel has fault valleys up to 30 km deep that extend most of the way around the moon. (By courtesy of NASA/Jet Propulsion Laboratory)

forming a mixture of hydrogen, helium, methane and other constituents overlying a rocky core.

Soon after the planet's discovery, Herschel thought that he had also discovered six satellites. Only two, Titania and Oberon, proved to be real. However, two more were discovered by Lassell in 1851 and the fifth by Kuiper in 1948. All five satellites move in the equatorial plane of the planet and revolve in the same direction as the planet rotates. Measurement of their orbits has allowed an accurate mass to be be assigned to Uranus.

These five satellites have radii ranging from 242 km (Miranda) to 805 km (Titania). They have densities about 1·5 that of water. Voyager 2 gave the first-ever detailed views of these five moons. All have their individual characteristics, in some cases astonishing and perplexing differences. Oberon displays many large craters scattered everywhere whereas Titania, though it shows signs of cratering, lacks the large craters so evident on Oberon. In addition Titania possesses a web of giant crevasses exposing lighter-coloured material. Umbriel, the darkest of the five, also shows craters but seems somehow less battered than the others. Ariel reveals fault valleys extending most of the way around its orb but is possibly the least cratered. It is Miranda, the smallest of the five satellites, that displays the most surprising structure. It is almost as if chunks of moons had been selected from all over the Solar System to display different geological processes and had then been squeezed together to form a conglomerate body.

In addition to these five moderately sized satellites, Uranus is now known to possess ten small moons. Two, about 40 to 50 km in diameter, lie one on each side of the so-called Epsilon ring and probably act as shepherd moons on that ring in the same way that Saturn's two moons Prometheus and Pandora confine Saturn's F ring in width and position.

On 10 March 1977 the occultation of a star by Uranus led to the discovery of a ring system about the planet. The regular diminution of the star's brightness a number of times both before and after it was occulted by the planet revealed the presence of a number of rings. Direct observation of the ring system from sunlight reflected by them has been achieved in infrared wavelengths.

There are nine narrow rings that lie between 42 000 km and 51 000 km from the planet's centre. From inner to outer boundary they are designated 6, 5, 4, α, β, η, γ, δ and ε, lying at distances 41 980, 42 360, 42 663, 44 844, 45 799, 47 323, 47 746, 48 423 and 51 000 km respectively. Each ring is probably less than 10 km wide. The outermost ring, Epsilon, is distinctly elliptical, precessing about the planet at a rate of $1^\circ\!.4$ per day.

6.9 Neptune

After the orbit of Uranus had been calculated, it soon became evident from regular observations that its motion deviated from its predicted path. The cause of the anomalous motion of Uranus was a puzzle for quite a few years. One suggestion was that the inverse square law of gravitation was not strictly exact. The eventual conclusion of the study suggested that the orbit of Uranus was being perturbed by another planet at a greater distance from the Sun. The analysis allowing prediction to be made of the apparent position of the disturbing planet was commenced in 1841 by Adams in England, but his work was neglected by Airy who was then the Astronomer Royal. At that time Airy belonged to the school of thought which suspected the exactness of the inverse square law of gravitation. A belated attempt, however, was finally made at his suggestion to find the planet suspected by the work of Adams.

Unaware of Adams' work, the problem was also tackled by the French astronomer Le Verrier in 1845. After allowing for perturbations caused by Jupiter and Saturn, he also concluded that there must be another planet beyond the orbit of Uranus. He predicted its position in the sky and sent the information to Galle in Berlin and the new planet, to be called Neptune, was found immediately. The discovery of Neptune illustrates very well the power of the scientific method.

Neptune appears telescopically as a small greenish disk with an angular size of just over 2 arc sec, giving it an equatorial radius of 22 300 km, close to that of Uranus. No markings on the disk are visible. The size of the disk does not allow measurements of flattening to be made with any accuracy. However, with a rotation period of $15^h\,40^m$, obtained by spectrometric measurements, a relatively large degree of flattening is predicted.

Two satellites are known to be orbiting Neptune. Triton was discovered by Lassell in 1846 and Nereid by Kuiper in 1949. Triton's orbital plane is inclined at 160° to the planet's equator. (The value 160° rather than 20° is used to indicate that the satellite's motion is retrograde.) The orbit precesses with a period of 580 years. This motion is similar to the regression of the Moon's nodes caused by the Earth's equatorial bulge and is supporting evidence that Neptune is oblate.

Nereid is remarkable for the high value of its orbital eccentricity (0·76), being the highest of all known natural satellites in the Solar System. Its distance from Neptune varies by nearly a factor of 8, between $1{\cdot}3\times10^6$ km and $9{\cdot}8\times10^6$ km.

6.10 Pluto

Some time after the discovery of Neptune, it was found that its presence could not explain fully the perturbations in the orbit of Uranus and once again the possibility of there being a more distant undiscovered planet was entertained. The theoretical analysis was principally performed by Lowell and it was he who started the search. The problem of detecting what would inevitably be a faint, slowly moving object was very difficult but, after many patient years of research, the discovery of an object subsequently named Pluto, was made by Tombaugh in 1930. Again, it is ironical that once the discovery had been

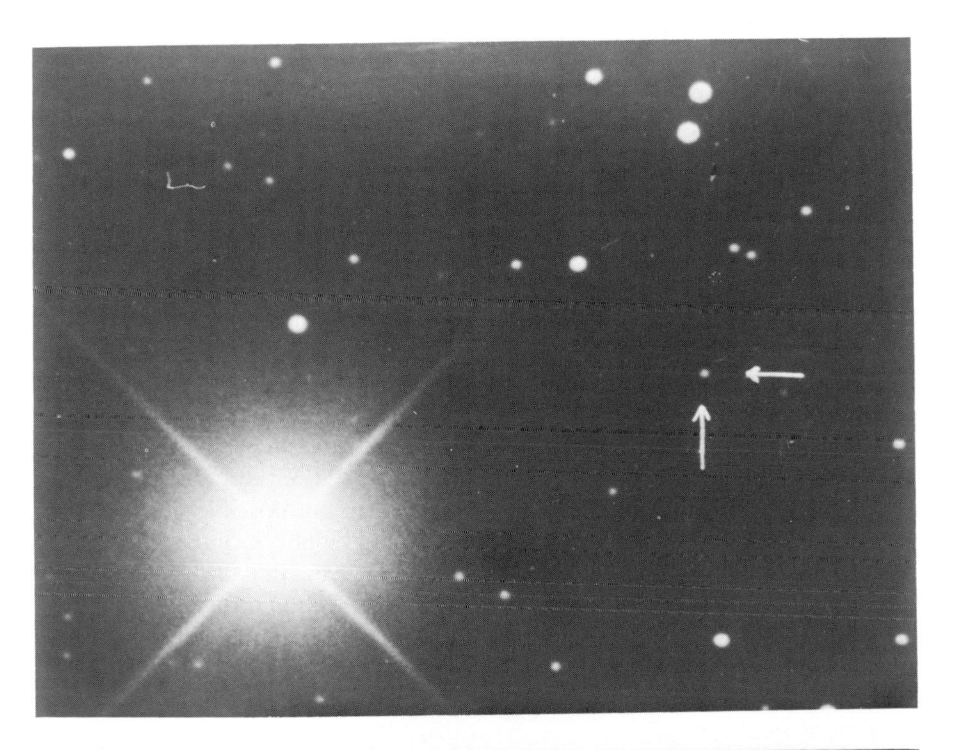

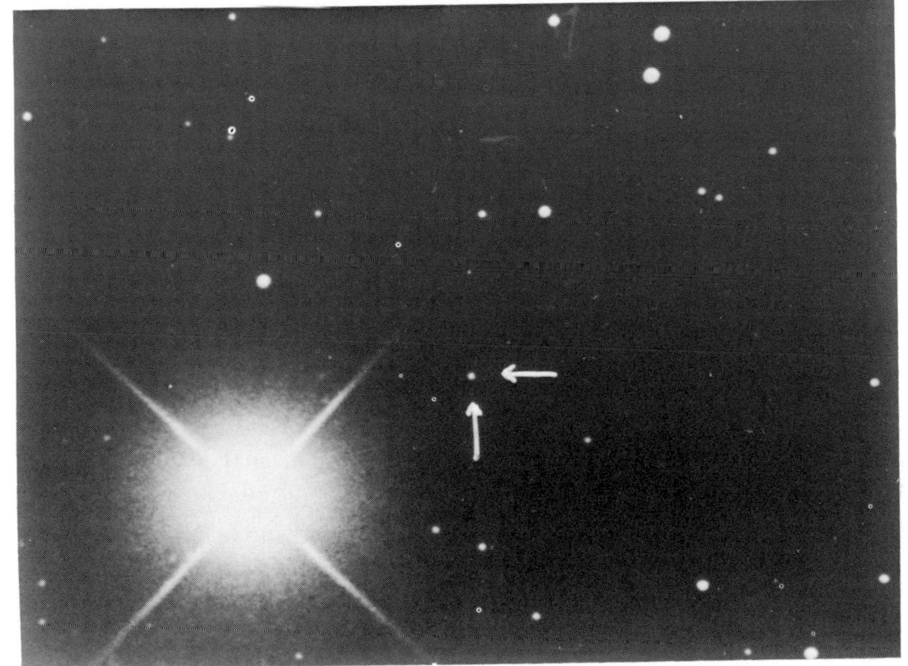

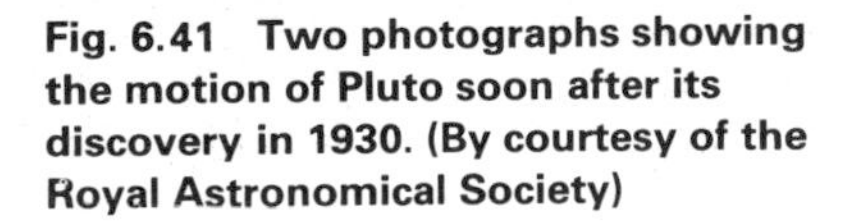

Fig. 6.41 Two photographs showing the motion of Pluto soon after its discovery in 1930. (By courtesy of the Royal Astronomical Society)

made, reference to photographs taken some fifteen years earlier as part of the search showed that Pluto had already been recorded but unnoticed (see Fig. 6.41).

Pluto's mean distance from the Sun at 39·4 AU makes it the most distant member of the Solar System so far known. However, its orbit has a large eccentricity and at perihelion it is some 50 000 000 km inside Neptune's orbit. At the time of writing, it is nearer to the Sun than Neptune and this situation will persist until the close of the 20th century. Its orbital period is 247·7 years.

Pluto is so far distant that it hardly presents a disk which is resolvable by telescopes. In 1950, however, Kuiper managed to see a disk using the 200-inch (5·08 m) telescope, and measured it to be 0·23 arc sec, giving it a diameter just less than half that of the Earth.

Obtaining a reliable value of Pluto's mass had to wait until the discovery in 1978 that Pluto possessed a satellite, now named Charon. Charon, about 1300 km in diameter, moves in an orbit about Pluto of radius 20 000 km with a period of 6.39 days. A fluctuation in Pluto's brightness of exactly the same period reveals that both planet and moon have rotational periods equal to Charon's orbital period.

The diameter and mass of Pluto are now deduced to be 3000 km and 0.0023 Earth masses respectively. Charon's mass is estimated to be 0.0002 Earth masses so that in terms of sizes and mass, Charon, compared to its planet, is the largest satellite in the Solar System. It might indeed be more correct to talk about the double planet system Pluto–Charon.

However we describe it, Pluto is not the planet predicted by Lowell's calculations as the disturber of Uranus' orbit. Nor is it likely, as has been suggested, that it is a "lost satellite" of Neptune, the remarkable features of the orbits of Neptune's satellites (Section 6.9) being a possible consequence of the dynamical changes in Neptune's system that could have occurred when Pluto was released. Against this idea is the work on the orbits of the five outer planets carried out by Cohen, Hubbard and Oesterwinter. On numerically integrating these orbits over a period of one million years, they found no evidence that Pluto made close approaches to Neptune; if anything, their work suggests that Pluto and Neptune's orbits evolve together to avoid close encounters.

In addition, it seems unlikely that the Pluto–Charon system could have escaped as a system from Neptune or formed after such an escape.

Very little is known of the planet's surface or structure. Spectra show indications of a methane atmosphere though at Pluto's distance from the Sun, its surface temperature of 50 K is below the freezing point of methane. It may be that at this period of its existence when it is at its nearest to the Sun, the surface ices and frosts have a slight tendency to sublimate. The planet's mean density of about 1000 kg m^{-3} suggests a resemblance to the gas giant planets rather than to the inner terrestrial planets.

6.11 Comets

At any one time, there are likely to be one or two comets in the sky and in order to see them a long-exposure photograph would be required. However, on some occasions a comet may be sufficiently bright to be seen with the naked eye and on others its appearance may be so spectacular as to attract the attention of the non-astronomer.

Comets vary appreciably in appearance but generally they are seen to have a bright nucleus or head, containing most of the mass, and a translucent tail. It is now thought that most of the comets have elliptical orbits about the Sun. When any comet is close to perihelion its appearance changes from night to night, even from minute to minute. The growth and decay of the tail as the comet first approaches and then recedes from the Sun is very noticeable (see Fig. 6.42).

Cometary tails may take one of two forms and for many comets both types of tail may be present simultaneously. Both in general point away from the Sun. The *Type I* tail provides a spectrum revealing ionized molecules and radicals, e.g. molecular carbon (C_2) hydroxyl radical (OH) and cyanogen (CN). It is usually fairly straight but may be photographed with "knots" in it which, over a period of hours or days, seem to work their way along the tail. For this type of gaseous tail, the direction lies within a few degrees of the radius vector from the Sun.

The *Type II* tail is generally broader and more curved than the gaseous tail. It provides a spectrum similar to that of the Sun, revealing that its light is essentially solar radiation scattered by dust particles.

From a series of photographs it is sometimes possible to determine the accelerations on various portions of the tail. Typical movements within Type I tails indicate that there are repulsive forces acting on the gas which are of the order of one hundred times the attractive force of solar gravitation. The mechanism of these forces is not well understood, but perhaps involves a magnetic coupling of the solar wind and the cometary magnetic field.

The curvature of the Type II tail shows that the repulsive force pushing the tail outwards is only just larger than solar gravitation. It appears as though the force is mainly that of radiation pressure and that, in this case, the dust particles must be tiny, being of the order of one micron in diameter.

The nuclei of comets are not solid but are collections of particles and dust, cemented together by ice or frozen gases. Occasionally, when passing close to the Sun, the solar gravity causes the head to split into components. As an example, Biela's comet, with a period of nearly 7 years, was seen in 1772, 1806, 1826 and 1832 without any major changes in its structure. However, in 1846 the head became more bulbous and then separated into two parts. The two components were seen again in 1872 to be well separated, but neither of them has been seen since.

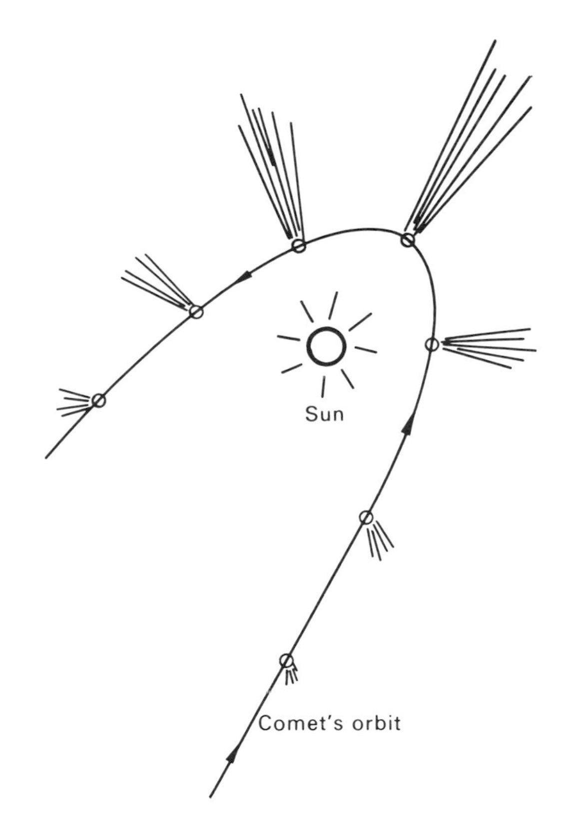

Fig. 6.42 The growth and decay of a comet's tail

(*a*)

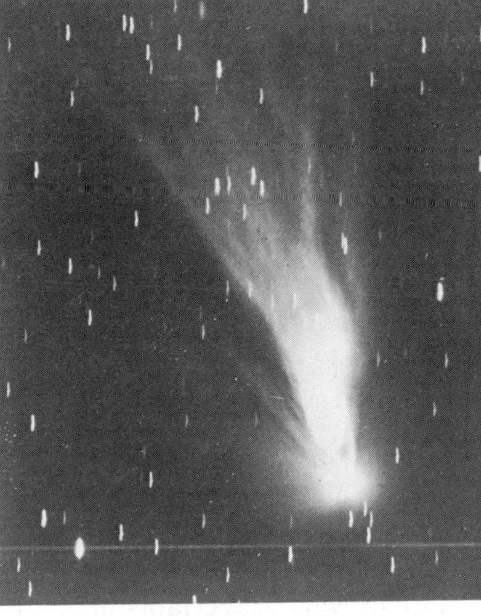

(*b*)

Fig. 6.43 Comet Humason in 1962 showing distinct changes over 24 hours. (*a*) July 9th, 1962; (*b*) July 10th, 1962. (By courtesy of the Royal Astronomical Society)

That some of the comets had orbits round the Sun and were periodic was first demonstrated by Halley. He showed that the comets which appeared in 1531, 1607 and 1682 were the same one and he predicted its return in 1758. This comet, perhaps the most famous, bears his name. It has, in fact, been traced back over 2000 years. Data of some of the more important periodic comets are listed in Table 6.8 at the end of this chapter.

Altogether there are some 40 comets with periods less than 100 years, which are likely to become sufficiently bright to be detectable close to perihelion. Nearly all of these revolve round the Sun in the same direction as the planets, a notable exception being Halley's comet which is retrograde.

All the other periodic comets have names in honour of their discoverer or of the person who first calculated the orbit. Two or three names may be used if the discovery is made independently by different observers over a short period of time. As comets are discovered, they are designated by the year in which they are found and a letter. Thus, 1953*a* was the first comet to have been discovered in 1953; 1953*b* was the second, etc. At a later date, the comets are re-catalogued and they are rearranged in a new order according to their dates of perihelion passage, using Roman numerals. The comets 1953*a* and 1953*b* are now known as 1953 III and 1953 V.

A few decades ago, it was thought that not all the comets were periodic. It was then believed that some of them had hyperbolic orbits. In other words, some of them must have originated in interstellar space, been swept up by the Sun and then returned to continue travelling outside the limits of the Solar System. The problem of deciding whether an orbit is hyperbolic, parabolic or elliptical with a high eccentricity is very difficult. When a comet is newly discovered, the orbit that is calculated is based on the gravitational action of the Sun only. As the comet has small mass, the orbit may, just prior to the comet's discovery, have been disturbed significantly by the action of either or both of the major planets, Jupiter and Saturn. Unless such planetary perturbations are calculated and allowed for, its exact original orbit of approach will remain uncertain. Indeed, observed perturbations of many comets suggest that none of these objects had a mass greater than 10^{-6} that of the Earth.

There is also positive evidence that comets belong to the Solar System, the most important being that there is no preferential direction of arrival. If comets were swept up from interstellar space, a preferential direction would be expected, corresponding to the Sun's motion. The current theory suggests that a body of comets surrounds the Solar System in a shell extending some 150 000 AU from the Sun, most of the comets spending much of their time well away from the planetary system.

The return of comet Halley in 1985/1986 provided the stimulus for scientific international cooperation on a grand scale, the most important achievements being the experiments carried by spacecraft for *in situ* measurements. The scene was set for the Halley excursions by a preliminary mission—International Cometary Explorer—to comet Giacobini-Zinner in August 1985 using a vehicle from an earlier, non-related mission—International Sun–Earth Explorer-3. Five vehicles were launched specially to intercept comet Halley, these being two Vega spacecraft from the USSR, two Japanese vehicles (MS-T5 and Planet A) and Giotto under the European Space Agency. This latter spacecraft left Earth on 2 July 1985 and travelled 7×10^9 km to comet Halley with the encounter being achieved on 13 March 1986. Experiments on board included a dust impact detection system, apparatus for measuring solar wind interactions and a camera. Several thousand dust impacts were recorded and large grains were found to be more abundant than expected, one of them having sufficient momentum to upset the orientation of the module and prevent transmitted signals being received for a time. Pictures of the comet's nucleus (see Fig. 6.44) verified the general concept of the icy conglomerate model and revealed a surface area of the dirty snowball of about 80 km^2.

Fig. 6.44 This image of the nucleus of comet Halley is composed of seven pictures taken by the Halley Multicolour Camera during the encounter of the ESA Giotto spacecraft on 13 March 1986 over a range of distances of 25 630 to 2730 km from the nucleus. (© 1986 Max-Planck-Institut für Aeronomie, courtesy of H. U. Keller)

6.12 Meteors

Meteors make themselves apparent to us when they happen to penetrate the Earth's gravitational field and enter the atmosphere at great speed. Because of the resistance of the atmosphere, the frictional forces cause the meteoritic particle to heat up and become incandescent. We see the event as a flash of light in the form of a trail in the night sky. They are popularly known as shooting stars. The events can be recorded photographically and, because of the ionization they produce, they can also be detected by radar, this being particularly convenient for recording daytime meteors.

The numbers of meteors which are detected can vary greatly, depending on whether the Earth's orbit happens to be passing through a meteor stream. Generally the meteors arrive at random and the best time to observe them is after the hours of midnight, when the blackest part of the sky is in the direction of the Earth's orbital motion.

By estimating the amount of energy which is released during the burn-up, this can be related to the kinetic energy of the particle. From measurements of the meteor velocity, the mass of the particle is then calculated. A typical meteor is likely to have a mass of only a few milligrams.

For bright meteor trails it is sometimes possible to record the spectrum and perform a chemical analysis. For some larger particles which are not completely burned up and which arrive at the Earth's surface, direct chemical analysis may be performed. These bodies are known as **meteorites**. From such studies, there appear to be two kinds of meteor. One kind, the stony meteor, has a predominance of calcium and silicon while the second variety, the metallic meteor, contains larger percentages of iron and nickel.

There have been several investigations of the velocities of random meteors. At one time it appeared that a sizeable percentage had velocities which were greater than the velocity of escape of the Solar System, i.e. some meteors were being swept from interstellar space by the Sun's motion. The most recent results now show that this is not so and it appears that all the meteors recorded belong to the Solar System.

On some occasions, a large number of meteors may be recorded, all apparently arriving from a particular point or **radiant** in the sky. Such meteor showers, lasting perhaps over a few days, are often annual events. For example, one particular shower occurs every November; its radiant is in the constellation of Leo and the shower is known as the Leonids (see Fig. 6.45).

Meteor showers result from the Earth passing through a swarm of particles which are also in motion round the Sun, their orbit cutting through the plane of the ecliptic close to the Earth's orbit. The direction of the radiant corresponds to the relative velocity vector of the swarm with respect to the Earth. The particles are spread round the orbit and the numbers of meteors recorded in

Fig. 6.45 Members of the Leonid meteor shower, 1966 Nov 17th. Photograph by D. Mclean, Kitt Peak. (By courtesy of the Royal Astronomical Society)

any one year depends on the distribution of the particles. In some years, a particular shower may be very spectacular with counts of a few thousand meteors per hour. Data concerning the important meteor showers are listed in Table 6.9 at the end of this chapter.

It appears likely that meteor swarms originate from some previous comet having the same orbit and may indeed be the debris from a disintegrated comet. For example, the orbit of the Perseid shower is very nearly the same as that of the comet 1861 I.

6.13 Zodiacal Light

Soon after the fading of twilight, it is sometimes possible to see a faint glow of light in the western sky in the shape of a tongue extending along the ecliptic. It is known as the **Zodiacal Light**. Its brightness decreases with elongation from the Sun but the general impression is that it is about as bright as the Milky Way. The same phenomenon may also be seen in the eastern sky some two hours before sunrise, fading away as dawn breaks. With sensitive photometers, the Zodiacal Light can be traced along the complete path of the ecliptic. It is best seen at high altitude stations well above the Earth's dust layer and at sites close to the equator where the ecliptic makes an angle of almost 90° with the horizon.

The exact nature of the Zodiacal Light is not known with certainty. There is good reason for believing that it is due to the scattering of sunlight by interplanetary dust particles between the Earth's orbit and the Sun. Photometric studies indicate a smooth transition between the solar corona and the interplanetary dust and the suggestion is that the interplanetary dust is just the continuation of the extended solar atmosphere.

Another theory, particularly held by Russian astronomers, considers some of the light to be caused by scattering from a tail of dust attached to the Earth. Just as a comet's tail always points away from the Sun, so also does the Earth's dust tail. This theory explains the slight brightening of the Zodiacal Light at a position 180° away from the Sun, the patch of light directly opposite the Sun being known as the **gegenschein** or **counterglow**.

However, recent data provided by Pioneer 10 *en route* to Jupiter suggest that the gegenschein is not attached to the Earth, but is the effect of a strong back-scatter from the interplanetary dust cloud out as far as the asteroid belt.

Analysis of data from IRAS (Infrared Astronomical Satellite) has revealed that the dust is also found in bands above and below the ecliptic. Some of the particles are in inclined orbits with nodes which are distributed in ecliptic longitude. As more time is spent by the particles at perihelion, their distribution is biased to these positions, so providing the apparent banding. Finer analysis reveals that the bands have structures which are related to the asteroid families Themis, Koronis and Eos.

6.14 The Age and Origin of the Solar System

This is a problem that has attracted the attention of scientists and philosophers for 300 years. Before considering their suggested solutions, we will summarize the nature of the problem. It is a situation reminiscent of the classical detective story. The detective arrives, views the body, hunts for significant clues and weaves a hypothesis to fit the facts and solve the crime. Some clues are more important than others; some may turn out to be red herrings, but whatever conclusion the investigator comes to, it must be one that leaves no important fact unexplained.

Some of the facts have already been mentioned, namely the make-up of the Solar System and its dynamical properties. Others reside in the distribution and abundance of the elements within the System. Additional pieces of evidence are the ages found for some of the bodies in the Solar System—the Earth, the Moon, the Sun, meteorites.

We recall that the Solar System consists of the Sun, which is a hot, gaseous body, the nine major planets, fifty-four known satellites, thousands of asteroids or minor planets, innumerable comets and meteors and the tenuous interplanetary medium.

The major planets move in almost circular coplanar orbits about the Sun, all orbits being direct. With the exceptions of Venus and Uranus, the rotations of these objects, as well as the Sun, are also direct, their equatorial planes being not too far removed from the plane of the Solar System.

The larger satellites of the Solar System follow this scheme in rotation about their axes and revolution about their primaries. Most of the smaller satellites, most of the asteroids, the comets and the meteors also have direct orbits, but there are exceptions.

The distribution of planetary distances from the Sun and satellite distances from their planets shows another feature. This is referred to as **Bode's law**. The planetary form of Bode's law is often written as

$$r_n = 0{\cdot}4 + 0{\cdot}3 \times 2^n,$$

where r_n is the distance of a planet from the Sun, n being $-\infty$, 0, 1, 2, 3, etc., in turn. If we give n these values, we obtain the second column in Table 6.2 below. In the third column are given the actual distances of the planets from the Sun, where the distance of Earth is taken to be 1 AU. It is seen that the agreement is very good except where Neptune and Pluto are concerned, though it is interesting to note that Pluto is near the distance of 38·8 predicted for Neptune by Bode's law.

Table 6.2

	Distance from Sun (AU)	
Name	From Bode's law	Actual
Mercury	0·4	0·39
Venus	0·7	0·72
Earth	1·0	1·0
Mars	1·6	1·52
Asteroids	2·8	2·8
Jupiter	5·2	5·2
Saturn	10·0	9·54
Uranus	19·6	19·2
Neptune	38·8	30·06
Pluto	77·2	39·4

It is also interesting to note that when the law was first published, the asteroids, Uranus, Neptune and Pluto were undiscovered. When Uranus was discovered in 1781, it fitted Bode's law. The gap in the Solar System, according to Bode's law, between the orbits of Mars and Jupiter, drew the attention of a number of astronomers so that a search was made for the missing planet. The mean distance of the asteroids was found to be almost precisely that predicted by Bode's law.

Similar laws can be found for the major satellite systems. In addition there exists in the Solar System a remarkable number of near-commensurabilities in mean motion between pairs of bodies in the planetary and satellite systems.

Table 6.3 The Sun–Earth–Moon system*

1. THE SUN
Radius = 696 000 km = 432 000 miles
Mass = $1{\cdot}99 \times 10^{30}$ kg
Mean density = $1{\cdot}41 \times 10^{3}$ kg m^{-3}
Surface gravity = $2{\cdot}74 \times 10^{2}$ m s^{-2} = $27{\cdot}9\ g_{\oplus}$

2. THE EARTH
1 astronomical unit (AU) = 149 600 000 km = 92 960 000 miles
Equatorial radius = 6378·16 km = 3963·31 miles
Polar radius = 6356·740 km = 3949·9 miles
Flattening (f) = 1/298·2 = 0·003352
Mass = $5{\cdot}98 \times 10^{24}$ kg
Mean density = $5{\cdot}52 \times 10^{3}$ kg m^{-3}
Normal gravity $g_{\oplus} = (9{\cdot}8064 - 0{\cdot}0259 \cos 2\phi)$ m s^{-2}, where ϕ is the geodetic latitude

3. THE MOON
Mean radius = 1738 km = 1080 miles
Mass = $7{\cdot}35 \times 10^{22}$ kg
Escape velocity from surface = 2·37 km s^{-1}
Surface gravity = 1·62 m s^{-2} = $0{\cdot}16\ g_{\oplus}$

* From *Explanatory Supplement to The Astronomical Ephemeris and The American Ephemeris and Nautical Almanac* and from the reference list on constants recommended by the International Astronomical Union.

For example, if n_J, n_S, n_N, n_P are the mean motions of Jupiter, Saturn, Neptune and Pluto respectively, then from Table 6.5,

$$n_J = 0{\cdot}083091,$$

$$n_S = 0{\cdot}033460,$$

$$n_N = 0{\cdot}005981,$$

$$n_P = 0{\cdot}003979,$$

in degrees per day.

It is then found that

$$2n_J - 5n_S = -0{\cdot}001118,$$

$$3n_P - 2n_N = -0{\cdot}000025,$$

showing how close the ratios of these pairs of mean motions are to simple fractions.

Perhaps the most remarkable case of this kind occurs among the Galilean satellites. If n_1, n_2, n_3 are the mean motions of Io, Europa and Ganymede, in degrees per day, then

$$n_1 = 203{\cdot}488\,992\,435,$$

$$n_2 = 101{\cdot}374\,761\,672,$$

and

$$n_3 = 50{\cdot}317\,646\,290.$$

We then have

$$n_1 - 2n_2 = 0{\cdot}739\,469\,091,$$

Table 6.4 Physical elements of the planets*

Planet	Equatorial radius (km)	Reciprocal of flattening	Mass (Earth = 1)	Mean density ($\times 10^3$ kg m^{-3})	Surface gravity (Earth = 1)	Velocity of escape (km s^{-1})	Rotation period	Inclination of equator to orbit
Mercury	2 420	∞	0·056	5·13	0·36	4·2	59^d	?
Venus	6 200	∞	0·817	4·97	0·87	10·3	$244^d.3$	174°
Earth	6 378	297	1·000	5·52	1·00	11·2	$23^h\ 56^m\ 04^s$	23° 27′
Mars	3 400	192	0·108	3·94	0·38	5·0	$24^h\ 37^m\ 23^s$	23° 59′
Jupiter	71 370	16·1	318·0	1·33	2·64	61	$9^h\ 50^m\ 30^s$†	3° 04′
Saturn	60 400	10·4	95·2	0·69	1·13	37	$10^h\ 14^m$†	26° 44′
Uranus	23 530	16	14·6	1·56	1·07	22	$10^h\ 49^m$	97° 53′
Neptune	22 300	50	17·3	2·27	1·41	25	14^h?	28° 48′
Pluto	1 500	?	0·002	1?	0·04	1·0	$6{\cdot}39^d$	?

* From *Explanatory Supplement to the Astronomical Ephemeris and the American Ephemeris and Nautical Almanac.* Recent data are given for Pluto.

† In these planets' cases, the period of rotation depends upon the latitude, increasing towards the poles. In the case of Pluto, a question mark is added to indicate that the values given are very uncertain.

and

$$n_2 - 2n_3 = 0{\cdot}739\ 469\ 092,$$

giving

$$n_1 - 3n_2 + 2n_3 = 0,$$

this being correct to one part in 7×10^8, the limit of observational accuracy.

Corresponding to this remarkable commensurability in the mean motions of the satellites, there is an equally exact one in their mean longitudes. Thus

$$l_1 - 3l_2 + 2l_3 = 180°.$$

It may be noted that Fig. 6.24 represents Jupiter's Galilean satellites at a real configuration and that their positions agree with this commensurability.

Let us now consider the chemical elements that form the bodies in the Solar System. The elements most common in the Sun are hydrogen and helium. They probably form about 98% of the solar mass, hydrogen alone accounting for about 80%. Jupiter, Saturn, and to a lesser degree Uranus and Neptune, are similar in chemical composition. In the Earth's case, however, the percentage of heavier elements is much greater. Silicon, oxygen, iron, nickel, calcium, sodium are abundant and it is believed that the composition of the Earth as a whole resembles the chemical composition of meteorites and the Moon. The planets Mars, Venus and Mercury in all likelihood have compositions that are similar to Earth's.

With respect to the ages found for the Earth, the Moon, the Sun and meteorites, there is substantial agreement among them. Radioactive studies of the rocks in the Earth's crust suggest that the oldest rocks are at least $3{\cdot}3 \times 10^9$ years in age, though recently a figure of $4{\cdot}5 \times 10^9$ years was obtained for the age of the Earth's mantle. Similar techniques put the age of the lunar samples at much the same figure; meteorites are found to be just as old. Measurement of the solar chemical composition and the solar energy output give, with the use of theories of stellar structure and evolution, an age for the Sun of around 5×10^9 years. We may therefore with some confidence put the formation of the Solar System at a period prior to 4·5 to $5{\cdot}0 \times 10^9$ years ago.

The distribution of angular momentum in the System must also be considered. Let a particle in the Sun at distance r from its centre have velocity v and

Table 6.5 Mean elements of the planetary orbits*
[Epoch 1960, January 1.5, Ephemeris Time (ET)]

Planet	Mean distance *a* (AU)	Mean distance *a* (10^6 km)	Sidereal period (tropical years)	Synodic period (days)	Mean daily motion *n* (deg/day)	Orbital velocity (km s^{-1})
Mercury	0·387099	57·9	0·24085	115·88	4·092339	47·8
Venus	0·723332	108·1	0·61521	583·92	1·602131	35·0
Earth	1·000000	149·5	1·00004	—	0·985609	29·8
Mars	1·523691	227·8	1·88089	779·94	0·524033	24·2
Jupiter	5·202803	778	11·86223	398·88	0·083091	13·1
Saturn	9·538843	1426	29·45772	378·09	0·033460	9·7
Uranus	19·181951	2868	84·01331	369·66	0·011732	6·8
Neptune	30·057779	4494	164·79345	367·48	0·005981	5·4
Pluto†	39·43871	5896	247·686	366·72	0·003979	4·7

Planet	Inclination *i* (deg)	Eccentricity *e*	Mean longitude (deg) Of node	Of perihelion	At epoch
Mercury	7·00399	0·205627	47·85714	76·83309	222·62165
Venus	3·39423	0·006793	76·31972	131·00831	174·29431
Earth	0·0	0·016726	0·0	102·25253	100·15815
Mars	1·84991	0·093368	49·24903	335·32269	258·76729
Jupiter	1·30536	0·048435	100·04444	13·67823	259·83112
Saturn	2·48991	0·055682	113·30747	92·26447	280·67135
Uranus	0·77306	0·047209	73·79630	170·01083	141·30496
Neptune	1·77375	0·008575	131·33980	44·27395	216·94090
Pluto†	17·1699	0·250236	109·88562	224·16024	181·64632

* From *Explanatory Supplement to the Astronomical Ephemeris and the American Ephemeris and Nautical Almanac.*
† The elements for Pluto are osculating values for epoch 1960 September 23.0 ET.

mass m. Then its angular momentum is l, where

$$l = mvr.$$

The Sun's total angular momentum is obtained by summing l over every particle in the Sun.

Similarly the planets and other bodies in the Solar System will possess angular momentum because of their rotation but, in addition, they possess a great deal of angular momentum by virtue of their revolution about their primaries—the Sun in the case of the planets, the planets in the case of their satellites. When this sum is carried out, it is found that the rotation of the Sun accounts for only 2% of the total angular momentum of the Solar System. Jupiter's share alone is 60%, most of the remaining angular momentum being distributed among Saturn, Uranus and Neptune.

Any theory of the origin of the Solar System must account for all of these properties. Not all of them, of course, were known to the early investigators of the problem.

Descartes suggested in 1644 a vortex theory of the origin of the Solar System. Buffon in 1745 thought that a comet striking the Sun obliquely might have removed sufficient material to create the planets, the slanting collision not only setting the planets in orbit round the Sun but also producing their rotation and the Sun's. He considered that the planets may have rotated so fast that in their turn they threw off material to form the satellite systems. In Buffon's day comets were supposed by many to be almost as massive as the Sun. His

Table 6.6 Satellite elements and dimensions

Planet and satellite		Mean distance (10^3 km)	Sidereal period (days)	Inclination of orbit to planet's equator or planet's orbit†	Eccentricity of mean orbit	Radius of satellite (km)	Reciprocal mass (planet = 1)
Earth							
	Moon	384·4	27·32166	5° 08′†	0·05490	1738	81.3
Mars							
I	Phobos	9·4	0·31891	0° 57′	0·0210	11	?
II	Deimos	23·5	1·26244	1° 18′	0·0028	6	?
Jupiter							
XVI	Metis	128	0·295	~0°	<0·004	20	?
XV	Adrastea	128	0·297	~0°	~0	12	?
V	Amalthea	181	0·49818	0° 24′	0·003	138 × 78	?
XIV	Thebe	222	0·678	0° 48′	0·015	40	?
I	Io	422	1·76914	0°	0	1826	21 300
II	Europa	671	3·55118	0°	0	1560	39 000
III	Ganymede	1 071	7·15455	0°	0	2500	12 700
IV	Callisto	1 884	16·68902	0°	0	2450	17 800
VI	Himalia	11 480	250·57	27° 38′†	0·15798	85	?
VII	Elara	11 740	259·65	24° 46′†	0·20719	40	?
X	Lisithea	11 860	263·55	29° 01′†	0·13029	10	?
XIII	Leda	12 300	292	?	?	4	?
XII	Ananke	21 200	631·1	147°†	0·16870	10	?
XI	Carme	22 600	692·5	164°†	0·20678	10	?
VIII	Pasiphae	23 500	738·9	145°†	0·378	10	?
IX	Sinope	23 700	758	153°†	0·275	10	?
Saturn							
XV	Atlas	137·7	0·60192	0° ?	~0	20 × 10 × ?	$2{\cdot}2\times10^{-11}$*
XVI	Prometheus	139·4	0·61300	0° ?	~0·0024	70 × 50 × 40	$1{\cdot}0\times10^{-9}$*
XVII	Pandora	141·7	0·62854	0° ?	~0·0042	55 × 45 × 35	$6{\cdot}4\times10^{-10}$*
X	Janus	151·4	0·69433	0° ?	~0·007	110 × 90 × 80	$6{\cdot}5\times10^{-9}$*
XI	Epimetheus	151·5	0·69467	0° ?	~0·009	70 × 60 × 50	$1{\cdot}5\times10^{-9}$*
I	Mimas	185·6	0·94221	1° 31′	0·0201	195	15 000 000
II	Enceladus	238·1	1·36908	0° 01′	0·00444	255	7 000 000
III	Tethys	294·7	1·88521	1° 06′	0	525	910 000
XIV	Calypso	294·7	1·88521	0° ?	~0	17 × 11 × 11	1.5×10^{-11}*
XIII	Telesto	294·7	1·88521	0° ?	~0	17 × 14 × 13	$2{\cdot}2\times10^{-11}$*
IV	Dione	377·5	2·73313	0° 01′	0·00221	560	490 000
XII	Helene	377·4	2·73921	0°	0·005	18 × 16 × 15	$3{\cdot}1\times01^{-11}$*
V	Rhea	527·2	4·51075	0° 21′	0·00098	765	250 000
VI	Titan	1 221·6	15·91029	0° 21′	0·0289	2575	4 150
VII	Hyperion	1 483·0	21·28121	0° 26′	0·104	205 × 130 × 110	5 000 000
VIII	Iapetus	3 560·1	79·15492	14° 43′	0·02828	720	300 000
IX	Phoebe	12 950·0	549·14775	150°†	0·16326	100	?
Uranus							
	1986 U7	49	0·33	~0°	~0	20?	?
	1986 U8	53	0·37	~0°	~0	25?	?
	1986 U9	59	0·44	~0°	~0	25?	?
	1986 U3	62	0·47	~0°	~0	30?	?
	1986 U6	63	0·48	~0°	~0	30?	?
	1986 U2	64	0·50	~0°	~0	40?	?
	1986 U1	66	0·52	~0°	~0	40?	?
	1986 U4	70	0·56	~0·	~0	30?	?
	1986 U5	75	0·62	~0°	~0	30?	?
	1985 U1	86	0·77	~0°	~0	80?	?
V	Miranda	124	1·414	0°	<0·01	150?	1 000 000
I	Ariel	192	2·52038	0°	0·0028	400	67 000
II	Umbriel	267	4·14418	0°	0·0035	300	170 000
III	Titania	438	8·70588	0°	0·0024	600	20 000
IV	Oberon	587	13·46326	0°	0·0007	500	34 000
Neptune							
I	Triton	354	5·87683	159° 57′	0	2000	750
II	Nereid	5 570	359·4	27° 27′†	0·76	150	3 000 000
Pluto							
I	Charon	20	6·39	0° ?	?	650	11·5

* Assuming mean densities of 10^3 kg m^{-3}.
? Figure unknown or uncertain.

Table 6.7 Approximate elements of the four major asteroids*
(Epoch 1957 June 11, 0^h ET referred to mean equinox and ecliptic of 1950·0)

Asteroid	a (AU)	n (deg)	i (deg)	e	Ω (deg)	$\varpi(=\Omega+\omega)$ (deg)	Mean anomaly at epoch M(deg)	Diameter of asteroid (km)
Ceres	2·7675	0·21408	10·607	0·07590	80·514	152·367	279·880	770
Pallas	2·7718	0·21358	34·798	0·23402	172·975	122·734	271·815	490
Juno	2·6683	0·22612	12·993	0·25848	170·438	56·571	329·336	190
Vesta	2·3617	0·27157	7·132	0·08888	104·102	253·236	79·667	390

* From *Explanatory Supplement to the Astronomical Ephemeris and the American Ephemeris and Nautical Almanac.*

promising ideas were developed by Chamberlin and Moulton, a star being substituted for the comet, a close encounter taking the place of the collision. Jeans in the first half of the present century further modified the theory, but today it has been largely discarded because of serious difficulties. For example, during the close encounter of the two stars, the tidal filaments drawn from them would be made up of gas at such a high temperature that it would spread out in space without giving rise to solid material. In addition, most of the gas would fall back into the stars from which it came.

The other major line of investigation has followed more directly from Descartes' original vortex theory. In 1755 Kant published his theory of the formation of the Solar System, postulating that in the beginning the Universe was filled with gas. The denser regions would attract more matter than the less dense and separation of these regions from each other would take place, especially as they would contract under their own gravitational forces. As each individual region, or nebula, contracted, it would rotate faster and faster to conserve angular momentum. It would therefore flatten into a disk at right angles to the axis of rotation and, according to Kant, would break up into a number of components. In the case of the nebula that developed into the Solar System, the central most massive component would become the Sun; the other smaller parts condensed into the planets.

Laplace, who seems to have been unaware of Kant's work, published a different form of the nebular hypothesis in 1796. He suggested that as the nebular disk cooled and contracted, the primitive Sun at the centre of the disk

Table 6.8 Periodic comets with 10 or more apparitions*

Designation	Name	No. of apparitions	Period (years)	Inclination i (deg)	Eccentricity e	Perihelion distance (AU)
1960 *i*	Encke	46	3·30	12·4	0·847	0·339
1961 *g*	Grigg–Skjellerup	10	4·91	17·6	0·703	0·858
1961 *b*	Temple 2	13	5·26	12·5	0·549	1·364
1951 VI	Pons–Winnecke	15	6·16	21·7	0·655	1·160
1950 II	d'Arrest	10	6·70	18·1	0·612	1·378
1960 *h*	Brooks 2	10	6·72	5·6	0·505	1·763
1961 *c*	Faye	15	7·38	9·1	0·576	1·608
1958 *c*	Wolf	10	8·43	27·3	0·395	2·507
1910 II	Halley	29	76·0	162·2	0·967	0·587

* From *The Nature of Comets* by N. B. Richter, Methuen & Co. Ltd. (1963).

Table 6.9 Night time meteor showers*

Shower name	Normal limits	Maximum	Radiant RA	Radiant Dec
Quadrantids	Jan 1–4	Jan 4	$15^h 28^m$	+50°
Corona Australids	Mar 14–18	Mar 16	16 20	−48
Lyrids	Apl 19–23	Apl 22	18 08	+32
η Aquarids	May 1–8	May 5	22 24	00
Ophiuchids	June 17–26	June 20	17 20	−20
Capricornids	July 10–Aug 5	July 25	21 00	−15
δ Aquarids	July 15–Aug 15	July 29	22 36	−17 to 00
Pisces Australids	July 15–Aug 20	July 30	22 40	−30
α Capricornids	July 15–Aug 25	Aug 1	20 36	−10
ι Aquarids	July 15–Aug 25	Aug 5	22 32 to 22 04	−15 to − 6
Perseids	July 25–Aug 17	Aug 12	03 04	+58
κ Cygnids	Aug 18–22	Aug 20	19 20	+55
Orionids	Oct 17–26	Oct 21	06 24	+15
Taurids	Oct 10–Dec 5	Nov 1	03 28 to 03 36	+14 to +21
Leonids	Nov 14–20	Nov 17	10 08	+22
Phoenicids	Dec 5 (one day)	Dec 5	01 00	−55
Geminids	Dec 7–15	Dec 13	07 28	+32
Ursids	Dec 17–24	Dec 22	14 28	+78

* Taken from *The Handbook of the British Astronomical Association for 1970.*

became unable to hold the outer parts of the disk, centrifugal force overcoming gravitational force. An equatorial ring of gas would therefore be left behind by the still-shrinking disk. The process would be repeated, creating a series of such rings, each successive ring of smaller radius than the previous one. Each ring would subsequently coalesce into a planet.

In fact the form of the nebular hypothesis advocated by Laplace contains a number of serious difficulties though it does explain many of the regularities found in the Solar System. Among the difficulties are the tendency of gas to disperse and the requirements of the theory that the Sun should possess most of the System's angular momentum and that the total angular momentum, moreover, should be about 200 times the amount found in the Solar System today.

Most recent developments of the nebular hypothesis begin with an acceptance that stars condense out of the interstellar dust and gas. They contract, their central temperatures rising until nuclear reactions begin to take place, resulting in a continual output of radiation. Surrounding the developing star is a primitive nebula containing dust and gas. Magnetic forces and radiation from the star, which we can now call the Sun, conspired to place most of the System's angular momentum in the planets developing out of the nebula. The radiation also removed most of the lighter elements from the inner planets, producing the small high-density terrestrial planets such as Mercury, Venus, Earth and Mars, and the low density outer planets. Once the main planetary bodies had reached their present sizes, very little change would take place in the planetary distances. It may be noted that in some modern versions of the formation of the Solar System the creation of the planets imposes a distance relationship rather like Bode's law.

It is not possible in this text to give details of any of these theories or to discuss their relative merits in explaining how the Solar System came into being. There are still many questions to be answered, but it is felt that the main picture has now been sketched out correctly. With the advent of space research and the possibility of investigating directly the surfaces of some of our nearest celestial neighbours, it is probable that long before the end of this century we will possess completely acceptable answers to the remaining problems that still defy solution in this field.

Possibly the most important consequence of these modern versions of the nebular hypothesis is that the formation of the Solar System is not a highly-improbable incident in the life of one star, the Sun. Planetary systems may well be common throughout the Universe; indeed several of the nearer stars, as we shall see, exhibit signs of possessing planetary companions.

Problems: Part 1

1. The orbits of asteroids are found mainly between those of
   (*a*) Saturn and Uranus,
   (*b*) Jupiter and Saturn,
   (*c*) Mars and Jupiter,
   (*d*) Earth and Mars.
   Which of these four statements, (*a*), (*b*), (*c*) or (*d*) is correct?

2. The following statements are true or false.
   (*a*) Of all the natural satellites in the Solar System only the Moon always turns the same face towards its primary.
   (*b*) The mass of a planet in the Solar System can be determined only if it possesses one or more satellites.
   (*c*) The planet with the largest apparent angular diameter when nearest the Earth is Venus.
   (*d*) Pluto is the planet farthest from the Sun.
   (*e*) A lunar eclipse may occur if the Moon is new.

3. Complete the following sentences.
   (*a*) The first man to land on the Moon was ....
   (*b*) Adams and Le Verrier, using celestial mechanics, predicted the existence of the planet now called ....
   (*c*) The .... tide occurs when the Moon is new or full.
   (*d*) The gaps in the asteroid belt are attributed to the gravitational disturbance of the planet ....

4. Identify the following planets:
   (*a*) Has two satellites one of which has an orbital eccentricity of about 0·7.
   (*b*) Has a mean density of around $0{\cdot}7 \times 10^3$ kg m^{-3}.
   (*c*) Exhibits phases like the Moon and has little or no atmosphere.
   (*d*) Is definitely known to have life on it.
   (*e*) Has a ring-system containing hundreds of rings.
   (*f*) Rotates in the retrograde direction with a period of rotation longer than its period of revolution.

5. Calculate the mean density of Jupiter from the following data, assuming the orbits of Earth and Jupiter to be circular and coplanar:

Angular semi-diameter of Jupiter at opposition	$= 21''\!.8$
Orbital radius of Jupiter	$= 5{\cdot}2$ AU
Mass of Jupiter/mass of Earth	$= 318$
Mean density of Earth	$= 5{\cdot}5$ kg m^{-3}
Sun's horizontal parallax	$= 8''\!.8$

6. The angular distance of the illuminated summit of a lunar mountain from the terminator is measured to be 27″; the Moon's elongation at the moment of measurement is 30° and its distance from the Earth's centre is 384 400 km. If the Moon's radius is 1738 km, calculate the approximate height of the mountain.

7. If the inclination of the Moon's orbit is $5° 09'$ and the obliquity of the ecliptic has the value $23° 27'$, calculate the north latitude of a place where the Moon passes through the zenith at its maximum possible declination north. If this occurs at new moon, *estimate* the meridian altitude of the next full moon.

8. The albedos of Saturn and Uranus are 0·50 and 0·65 respectively while the ratio of the planets' radii is 2·50, Saturn's being the larger. Their heliocentric distances (orbits being assumed circular and coplanar) are 9·5 and 19·2 AU respectively. Neglecting the effect of Saturn's rings, calculate the magnitude difference in the brightness of the two planets when both are observed at opposition.

9. Show that any station where the Moon is always above the horizon at upper culmination, the maximum and minimum possible altitudes at this instant differ by $2(\varepsilon + i)$, where ε is the obliquity of the ecliptic and i is the inclination of the lunar orbit.

10. Given that the Moon's synodic period is 29·53 days, find the phase of the Moon $4\frac{3}{4}$ days after new moon.

11. In a certain year the Moon was seen to be at its ascending node at 0^h October 15th and again at 0^h December 8th. Given that the Moon's longitudes at these times were respectively $239° 35'$ and $236° 45'$, calculate the synodic period of the node.

12. A sunspot on the centre of the Sun's disc subtends an angle of $5''$ at the Earth. Show that its linear diameter is approximately 3600 km. (Take the solar parallax to be $8''8$ and the Earth's radius to be 6372 km.)

13. The star *Alcyone* (η Tauri) has ecliptic longitude $59° 22'$ and ecliptic latitude $4° 02'$ N. Calculate the possible longitudes of the ascending lunar node when the lunar orbit passes through the star. (Inclination of lunar orbit to ecliptic = $5° 09'$.)

14. In a particular lunar eclipse the Moon's centre was displaced from the centre of the umbra

$$\text{in RA by } \Delta\alpha = 0 + 24'\, t,$$

$$\text{in Dec by } \Delta\delta = -20' + 13'\, t,$$

where t is the time in hours measured from the moment of opposition of Sun and Moon in RA. The Moon's declination was small during the eclipse so that

$$\Delta\alpha \cos\delta - \Delta\alpha$$

may be neglected. The radius of the umbra is $39'$ and the Moon's semi-diameter is $15'$. Show by a diagram, or otherwise, that totality lasts nearly 75 minutes.

Part 1: Further Study

If it is possible, arrange to make visual observations of the planets through a telescope. Check on the dates of forthcoming meteor showers (see Table 6.9) and make an effort to observe them.

Part 2
Astrophysics

Chapters 7–14

PROGRAMME: This part describes the condition of matter and how it behaves over a wide range of physical conditions by discussing various features and objects within the Galaxy.

Basic stellar measurements are first reviewed together with the facts and relationships which emerge from the observations. With this information some of the more simple theories of stellar atmospheres, stellar structures, stellar evolution and stellar relationships are explored.

All the other objects within the Galaxy are described. These include gaseous clouds, dust clouds, planetary nebulae, supernova remnants and pulsars.

7
The Stars: Observational Data

7.1 Introduction

A great deal of information concerning stars can be obtained by measuring their radiation. For example, from positional measurements the distances of some nearby stars can be calculated by parallax methods. In addition, the angular sizes of some of the stars can be measured and, with knowledge of their distances, their physical diameters can therefore be calculated. Also two- or multi-coloured photometry allows the effective temperatures of stars to be determined (see Roy & Clarke, *op. cit.*, Sections 9.7, 17.7, 21.5).

In this chapter, further means of obtaining knowledge of stars are presented. We shall also show the relationships between the parameters describing the stars. Some of these relationships are empirical or are deducible from theoretical standpoints. They allow us to make comparisons between different stars and give us insight as to why there is a range of stellar types and why the different types exhibit the properties they do. But firstly let us see how the observational information about the stars is catalogued.

In the early days, the stars were listed according to their places within a constellation and designated by a letter or number. The constellation zones are quite arbitrary as far as the stellar distribution is concerned, but their designations persist for the ease of identification of the zone of sky under observation.

To describe the philosophy behind the development in star catalogues we can do no better than to quote the relevant section of *Norton's Star Atlas*.

"The origin of most of the constellation names is lost in antiquity. Coma Berenices was added to the old list (though not definitely fixed till the time of Tycho Brahé), about 200 B.C.; but no further addition was made till the 17th century, when Bayer, Hevelius, and other astronomers, formed many constellations in the hitherto uncharted regions of the southern heavens, and marked off portions of some of the large or ill-defined ancient constellations into new constellations. Many of these latter, however, were never generally recognized, and are now either obsolete or have had their rather clumsy names abbreviated into more convenient forms. Since the middle of the 18th century, when La Caille added thirteen names in the southern hemisphere, and sub-divided the unwieldy Argo into Carina, Malus (now Pyxis), Puppis, and Vela, no new constellations have been recognized. Originally, constellations had no boundaries, the position of a star in the "head", "foot", etc., of the figure answering the needs of the time; the first boundaries were drawn by Bode in 1801.

The star names have, for the most part, been handed down from classical or early mediaeval times, but only a few of them are now in use, a system devised by Bayer in 1603 having been found more convenient, viz., the designation of the bright stars of each constellation by the small letters of the Greek alphabet, α, β, γ, etc., the brightest star being usually made α, the second brightest β—though sometimes, as in Ursa Major, sequence, or position in the constellation figure, was preferred. When the Greek letters were exhausted, the small Roman letters, a, b, c, etc., were employed, and after these the capitals, A, B, etc.,—mostly in the Southern constellations. The capitals after Q were not required, so Argelander utilized R, S, T, etc., to denote variable stars in each constellation, a convenient index to their peculiarity.

The fainter stars are most conveniently designated by their numbers in some star catalogue. By universal consent, the numbers of Flamsteed's *British Catalogue* (published 1725) are adopted for stars to which no Greek letter has been assigned, while for stars not appearing in that catalogue, the numbers of some other catalogue are utilized. The usual method of denoting any lettered or numbered star in a constellation is to give the letter, or Flamsteed number, followed by the genitive case of the Latin name of the constellation: thus α of Canes Venatici is described as α Canum Venaticorum.

Flamsteed catalogued his stars by constellations, numbering them in the order of their right ascension. Most modern catalogues are on this convenient basis (ignoring constellations), as the stars follow a regular sequence. But when right ascensions are nearly the same, especially if the declinations differ much, in time "precession" may change the order: Flamsteed's 20, 21, 22, 23 Herculis, numbered 200 years ago, now stand in the order 22, 20, 23, 21.

For convenience of reference, the more important star catalogues are designated by recognized contractions."

For example, HD 172167 is at once known by astronomers to denote the star numbered 172167 in the *Henry Draper Catalogue*. This particular star is bright and well-known, being *Vega* or α Lyrae. It also appears in all other catalogues and may, for example, be known as 3 Lyrae (Flamsteed's number), Groombridge 2616 and AGK_2+38 1711 (from *Zweiter Katalog der Astronomischen Gesellschaft für das Äquinoktum*, 1950).

"Bode's constellation boundaries were not treated as standard, and charts and catalogues issued before 1930 may differ as to which of two adjacent constellations a star belongs. Thus Flamsteed numbered in Camelopardus several stars now allocated to Auriga, and by error he sometimes numbered a star in two constellations. Bayer, also, sometimes assigned to the same star a Greek letter in two constellations, ancient astronomers having stated that it belonged to both constellation figures: thus β Tauri = γ Aurigae, and α Andromedae = δ Pegasi.

To remedy this inconvenience, in 1930 the International Astronomical Union standardized the boundaries along the Jan 1st, 1875, arcs of right ascension and declination, having regard, as far as possible, to the boundaries of the best star atlases. The work had already been done by Gould on that basis for most of the S. Hemisphere constellations.

The I.A.U. boundaries do not change their positions among the stars and so objects can always be correctly located, though, owing to precession, the arcs of right ascension and declination of today no longer follow the boundaries, and are steadily departing from them. After some 12 900 years, however, these arcs will begin to return towards the boundaries, and 12 900 years after this, on completing the 25 800-year precessional period will approximate to them, but not exactly coincide."

Nowadays, as well as recording the stars' positions for a particular epoch, a general catalogue will also list other observed parameters of each star. For example, the annual changes in right ascension and declination may be given. Other headings might include proper motion, annual parallax, radial velocity, apparent magnitude, colour index, and spectral type (see Section 7.7). Special peculiarities may be noted—for example, if the star is a binary system.

In some catalogues, quantities derived from the above data may be given. One example is that of absolute magnitude, a quantity describing the intrinsic brightness or luminosity of a star. This concept is discussed in the next section.

7.2 Absolute Magnitude

From measurements of stellar parallax, it is immediately obvious that the stars are scattered through space and may be found at distances within the range of a few parsecs to 100 pc. Beyond 100 pc the parallactic method is insensitive and is no longer applicable but other methods can be applied to measure the distances of the millions of stars away from the solar neighbourhood.

In order to be able to compare the total amounts of energy radiated by the stars, allowance must be made for their different distances. For stars whose distances are known, their magnitudes may be converted to values that would have been obtained if they had been placed at some standard distance. For the convenience of comparison, the chosen standard distance is 10 pc. The **absolute magnitude** of a star is defined as the magnitude that it would have if it happened to be placed at the standard distance of 10 pc. Thus, in order to compare the energy outputs of stars, their apparent magnitudes should be reduced to absolute magnitude values.

The apparent brightness of a star is defined by the amount of energy received per unit area per unit time. This quantity depends on the star's output but it is also inversely proportional to the square of its distance. Let us consider a star of apparent brightness B_r at a distance r pc and suppose that its brightness at a distance of 10 pc would be given by B_{10}. If none of the star's light is absorbed on its passage through interstellar space then, according to the inverse square law, we have

$$B_r/B_{10}=10^2/r^2.$$

By taking logarithms we have

$$\log_{10}(B_r/B_{10})=2-2\log_{10} r. \tag{7.1}$$

By letting the star's apparent magnitude be m and its absolute magnitude be M, the difference between these magnitude values can be obtained by using Pogson's equation

$$m=k-2{\cdot}5\log_{10} B.$$

Hence

$$m-M-5\log_{10} r-5. \tag{7.2}$$

The quantity $(m-M)$ is defined as the **distance modulus** since its value expresses the distance of the star (i.e. if m has been measured and M determined by other means, r could be evaluated from $(m-M)$). Rearrangement of Equation (7.2) allows the absolute magnitude to be expressed as

$$M=m+5-5\log_{10} r. \tag{7.3}$$

If P is the parallax of the star, measured in seconds of arc, we recall that by definition

$$P=1/r.$$

Hence an alternative form of Equation (7.3) is

$$M=m+5+5\log_{10} P. \tag{7.4}$$

If the apparent brightness of a star has been weakened by an absorbing interstellar cloud then both Equations (7.3) and (7.4) need modification to allow for this. By letting the absorption be equivalent to A magnitudes, these equations may be rewritten as:

$$M=m+5-5\log_{10} r-A \tag{7.5}$$

and

$$M=m+5+5\log_{10} P-A \tag{7.6}$$

Just as each star may be assigned an apparent magnitude according to the detection system used, so also values of absolute magnitudes and values of absorption may be expressed in different magnitude systems. For example, an absolute magnitude may be expressed in terms of the visual system as M_V or in the B band system as M_B. Perhaps the most useful absolute magnitude values are those expressed in the bolometric system, M_{bol}, representing the absolute magnitudes that would be obtained by a receiver which is equally sensitive to all wavelengths.

By measuring and comparing the amounts of energy received from the Sun and from the stars it has been calculated that the Sun's apparent visual magnitude, $m_{V\odot}$, is equal to $-26{\cdot}74$. Its mean distance from the Earth is approximately $1{\cdot}5\times10^8$ km, and since 1 pc is approximately equal to 3×10^{13} km, its mean distance from the Earth is given by

$$1{\cdot}5\times10^8/3\times10^{13}\ \text{pc}=5\times10^{-6}\ \text{pc}.$$

(Equally well, we could have remembered that 1 pc = 206265 AU or 1 AU ≈ 5×10^{-6} pc.)

Substituting $m_{V\odot}=-26{\cdot}74$ and $r=5\times10^{-6}$ in Equation (7.3), we have

$$\begin{aligned}M_{V\odot}&=-26{\cdot}74-5\log_{10}(5\times10^{-6})+5\\&=-21{\cdot}74-5(-6)-5(0{\cdot}699)\\&=+4{\cdot}76.\end{aligned}$$

Thus if we could place ourselves at a distance of 10 pc from the Solar System, the Sun would have an apparent magnitude not untypical of the stars which we see by the unaided eye.

Clearly, a value of absolute bolometric magnitude is directly related to the total energy radiated in all wavelengths by a star. Comparisons of absolute magnitudes are equivalent to comparisons of the energies radiated by stars.

7.3 Stellar Luminosity

Let us define the energy output of a star as being its intrinsic **luminosity**, L. This may be expressed in units of joule s^{-1} or more conveniently it may be expressed in terms of the Sun as unit ($L_\odot=3{\cdot}79\times10^{26}\ \text{J s}^{-1}$). When stars are considered to be at the standard distance of 10 pc, their brightnesses at this distance are related directly to their luminosities. If L and M_{bol} refer to a star and $L_\odot$ and $M_{bol\odot}(=+4{\cdot}75)$ refer to the Sun, then using Pogson's equation we have

$$M_{bol\odot}-M_{bol}=2{\cdot}5\log_{10}(L/L_\odot). \tag{7.7}$$

When absolute magnitudes are determined for different stars it is immediately apparent that they differ markedly and therefore so must their luminosities. Typical luminosities run from 10^2 times to 10^{-2} times that of the Sun. In fact the complete list of stars covers a much wider range; a magnitude range of more than 20 is required to include the extent of stellar absolute magnitudes.

Now according to Stefan's law, the total amount of energy radiated per unit area of a body is proportional to the fourth power of its absolute temperature. For a star of radius R, its surface area is equal to $4\pi R^2$. We may therefore express its luminosity as

$$L=4\pi R^2\sigma T_e^4 \tag{7.8}$$

where σ is Stefan's constant and T_e the effective temperature of the star.

It is apparent from Equation (7.8) that the observed range in stellar luminosities can be explained by there being stars of different temperatures and/or different radii.

Now, consider a star to be at a distance r from the observer. At this distance, the energy E_* received from the star per unit collecting aperture is given by the total output from the star divided by the surface area of a sphere of radius r, i.e.

$$E_*=L/4\pi r^2=4\pi R^2\sigma T_{e*}^4/4\pi r^2.$$

But the angular diameter, α_*, of a star is equal to $2R/r$ and therefore we may write the above expression for E_* as

$$E_*=\tfrac{1}{4}\alpha_*^2\sigma T_{e*}^4.$$

For the case of the Sun we have

$$E_{\odot} = \tfrac{1}{4}\alpha_{\odot}^{2}\sigma T_{e\odot}^{4}.$$

The quantity $E_{\odot}$ is known as the **solar constant** and from measurements made above the Earth's atmosphere it is found to have a value close to 1400 W m^{-2}.

By comparing the energy received from a star to that from the Sun we have

$$\frac{E_{*}}{E_{\odot}} = \frac{\alpha_{*}^{2}}{\alpha_{\odot}^{2}}\frac{T_{e*}^{4}}{T_{e\odot}^{4}}.$$

The energy received from a star is related to the star's bolometric magnitude by an expression similar to Pogson's equation. By using bolometric magnitudes, the comparison above may be written as

$$m_{bol\odot} - m_{bol*} = 5 \log_{10}(\alpha_{*}/\alpha_{\odot}) + 10 \log_{10}(T_{e*}/T_{e\odot}).$$

With knowledge of the values of $m_{bol\odot}$, $\alpha_{\odot}$ and $T_{e\odot}$, this equation may be used to investigate the relationship between m_{bol*}, α_{*} and T_{e*}.

As an example, let us consider what the angular size of the Sun would be if it were placed at a distance of 10 pc. At this distance its bolometric magnitude (i.e. its absolute bolometric magnitude) would be +4·75. For this example α_{*} corresponds to the angular size of the Sun at 10 pc, $m_{bol\odot} = -26{\cdot}82$, $m_{bol*} = +4{\cdot}75$, $\alpha_{\odot} = 1920$ arc sec $= 9{\cdot}3 \times 10^{-3}$ radians and $T_{e*}/T_{e\odot} = 1$. Hence

$$5 \log \alpha_{*} = m_{bol\odot} - m_{bol*} + 5 \log \alpha_{\odot}.$$

Therefore

$$5 \log \alpha_{*} = -31{\cdot}57 + 5(-2{\cdot}03) = -41{\cdot}72,$$

$$\log \alpha_{*} = -8{\cdot}34,$$

so that

$$\alpha_{*} = 4{\cdot}6 \times 10^{-9} \text{ radians},$$

which is approximately one hundred times smaller than a large telescope could hope to resolve. As we have already mentioned, only a few of the nearby bright giant stars have had their diameters measured and these by interferometric techniques.

7.4 The Hertzsprung–Russell (HR) Diagram

It is in fact well known that each star has a colour which is indicative of its temperature. The range of the observed colours therefore shows that the stars have a range of temperatures. By obtaining a colour index of a star its effective temperature can be calculated (see Roy & Clarke, *op. cit.*, Section 21.5).

In 1911 independent studies made by Hertzsprung and Russell showed that there is a relationship between the colour indices of stars and their absolute magnitudes. This relationship is schematically shown in Fig. 7.1 for the stars close to the solar neighbourhood. Such a plot is known as a Hertzsprung–Russell diagram or as an HR diagram, after its initiators. It appears that most stars are placed in a diagonal band, known as the **main sequence**, but that there are some red, cool stars which are more luminous than the main group and that there are some blue, hot stars which are less luminous than the main group.

The group of stars at the top right of Fig. 7.1 have a smaller absolute magnitude than those of the main group with the same colour index and hence with the same temperature. They are therefore more luminous. This, by Equation (7.8), must mean that their radii are larger. These stars are known as **red giants**.

By a similar argument, the stars in the group at the bottom left of Fig. 7.1 must be smaller than stars of the same colour index contained within the main sequence. These stars are known as **white dwarfs**.

Let us now consider quantitatively the radii of the range of stars depicted in Fig. 7.1.

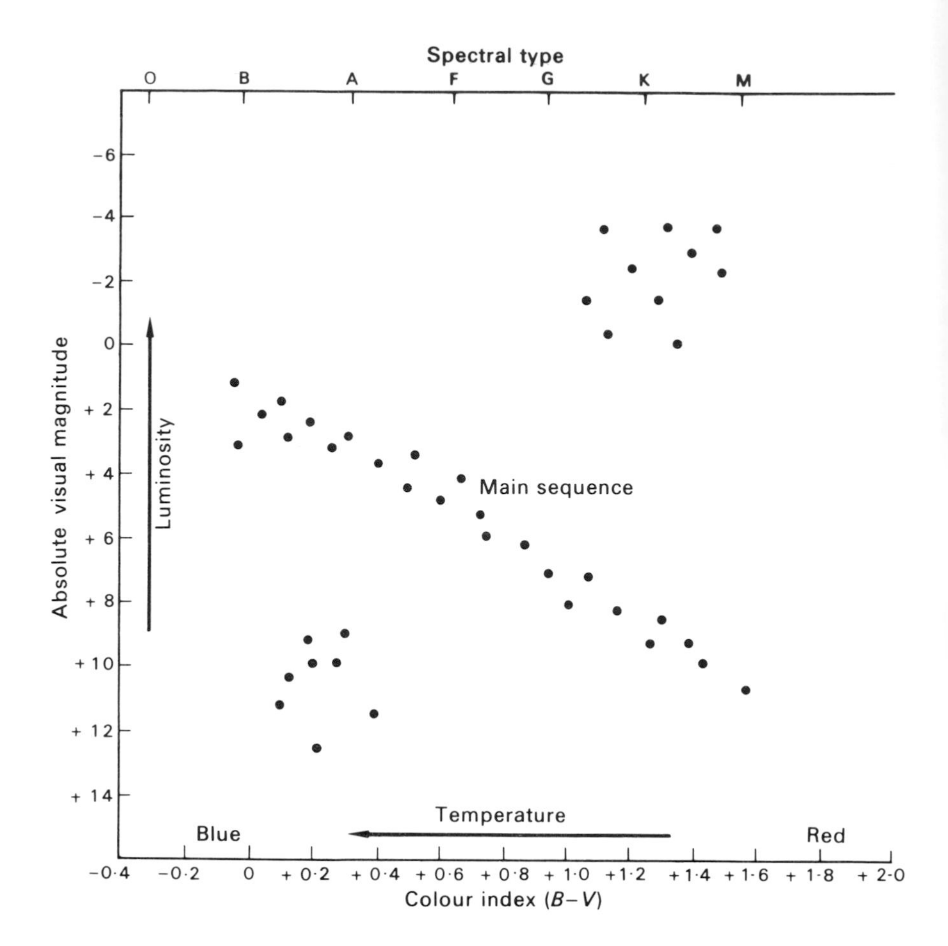

Fig. 7.1 The HR diagram for the local stars. (Spectral sequence is discussed in Section 7.7)

From Equation (7.7) it is obvious that the absolute magnitude scale of the HR diagram may be replaced with a scale given by $\log_{10}(L/L_\odot)$ and that, as the colour index is related to effective temperature, the colour index scale may be replaced by a scale of $\log_{10}(T_e/T_{e\odot})$. Now by taking logarithms of Equation (7.6) we have

$$\log_{10} L = 4 \log_{10} T_e + 2 \log_{10} R + \log_{10}(4\pi\sigma). \tag{7.9}$$

By inserting values of L, R and T_e for the Sun, Equation (7.9) gives

$$\log_{10} L_\odot = 4 \log_{10} T_{e\odot} + 2 \log_{10} R_\odot + \log_{10}(4\pi\sigma). \tag{7.10}$$

Subtracting Equation (7.10) from (7.9), we have

$$\log_{10}(L/L_\odot) = 4 \log_{10}(T_e/T_{e\odot}) + 2 \log_{10}(R/R_\odot). \tag{7.11}$$

Now the equation for a straight line in the x–y plane is

$$y = mx + c,$$

where m is the gradient and c is the intercept made by the line on the y-axis.

Thus if we take a particular chosen value of $(R/R_{\odot})$ in Equation (7.11), we can put

$$y = \log_{10}(L/L_{\odot}),$$
$$m = 4,$$
$$x = \log_{10}(T_e/T_{e\odot}),$$

and

$$c = 2\log_{10}(R/R_{\odot}).$$

We see then that Equation (7.11) represents a straight line, where $\log_{10}(L/L_{\odot})$ is plotted against $\log_{10}(T_e/T_{e\odot})$. Indeed Equation (7.11) represents a family of straight lines characterized by the parameter $\log_{10}(R/R_{\odot})$. Fig. 7.2 shows the HR diagram plotted with the new scales, and lines of constant radius are also indicated plotted with values of $\log_{10}(R/R_{\odot})$. When the solar neighbourhood stars are plotted on this diagram, it is immediately apparent that these stars cover an enormous range of sizes from about 10^2 times to about 10^{-2} times the size of the Sun. Thus we might say that the stellar sizes cover a range from the size of the Earth's orbit to the size of the Earth itself.

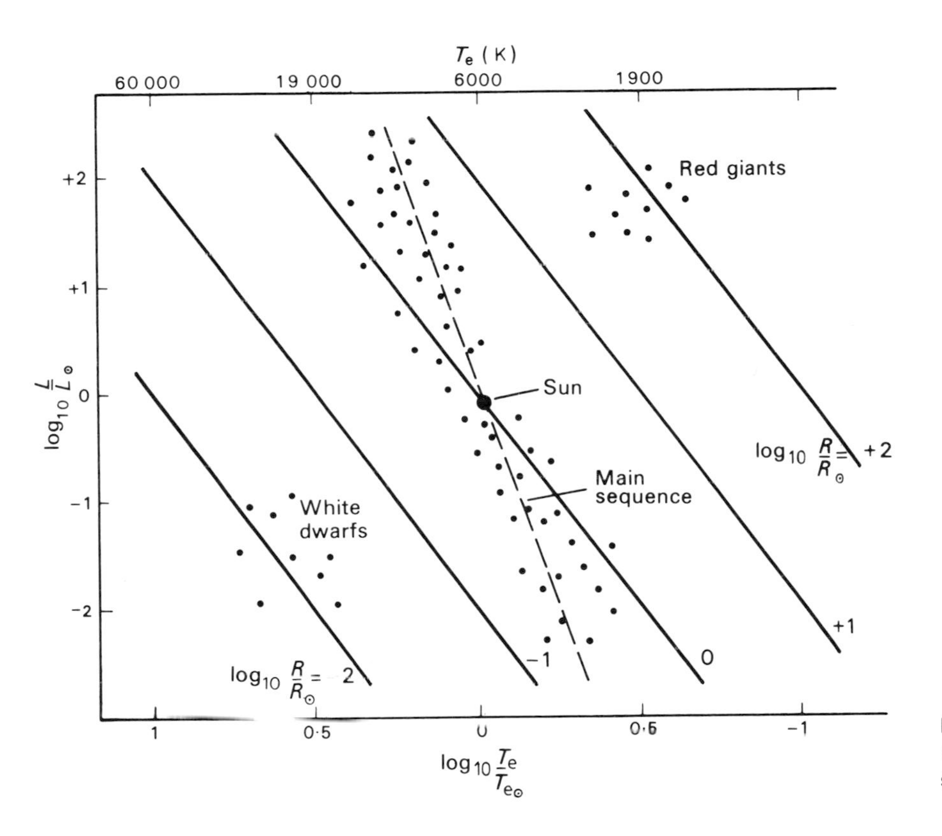

Fig. 7.2 HR diagram plotted with lines of constant $\log_{10}(R/R_{\odot})$ with main sequence indicated by the dashed line

Direct observation of stellar diameters by such means as the Michelson stellar interferometer has only been applied to stars whose sizes allow measurement, i.e. the giant stars. The large values obtained for such stars confirm the predictions of the simple theory.

When more stars are added to the HR diagram, the simple picture of a main sequence with giants and dwarfs on either side becomes more complicated by the presence of the other zones. In particular, the HR diagrams of stellar clusters vary significantly from one cluster to the next but the interpretation of these diagrams will be left until later (see Chapter 9).

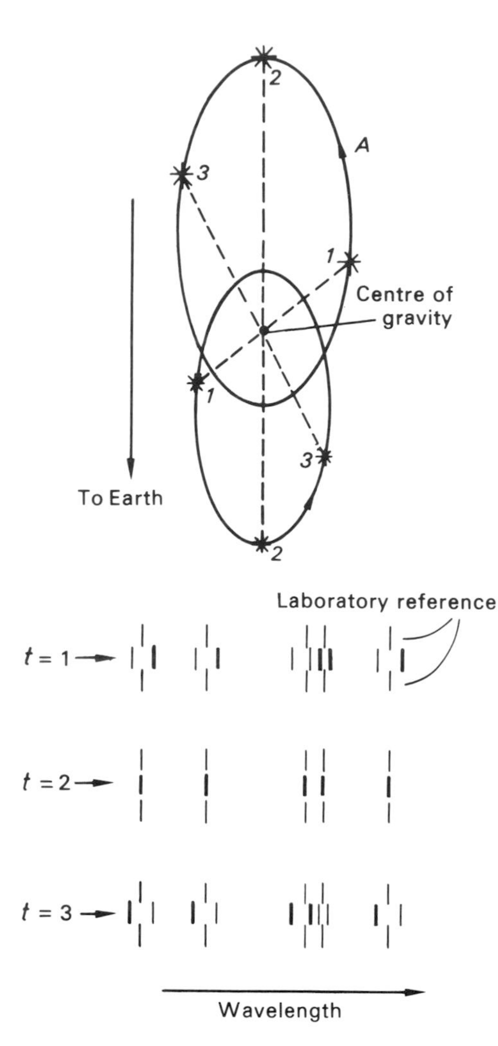

Fig. 7.3 The orbit and spectra of a spectroscopic binary, star *A* exhibiting the stronger spectrum

7.5 Binary Stars

7.5.1 General Description It has already been demonstrated that the masses of celestial bodies can be determined if they are observed to act gravitationally upon other bodies or possess satellites.

The fact that stars are frequently observed as physically-linked pairs has enabled us to determine many of their masses. It is only necessary to determine their orbits about each other for their masses to be deduced.

Stellar pairing exhibits itself in three different ways. In the first, the apparent closeness of some pairs of stars on the celestial sphere is statistically more frequent than might be expected from chance alignments of stars at different distances. In 1782, Sir William Herschel published a catalogue of the positions of many pairs which formed what are known as **optical doubles**. The aim of this work was to make regular observations of these stars and see if the brighter of the pair—and therefore supposed to be the nearer—suffered parallactic motion relative to the fainter. Re-observation of some of the pairs over a period of years revealed that the stars were in fact gravitationally connected and in orbit about each other. These pairs are therefore relatively close to each other in space, sufficiently close for the force of gravitation between them to be strong. They are known as **visual binaries**.

If it can be imagined that a pair of stars are brought progressively closer together (and at the same time speeding up their orbits) a situation will arise so that, at the distance of the observer, the two stars would be unresolvable. If, too, the stars are orbiting each other in a plane close to the one through which the line of sight runs, there will be times, according to the relative positions of the stars in their orbits, when one star will eclipse the other.

The eclipse would be registered by the observer as a diminution of the brightness of the apparent single star. Stars of variable brightness, with a pattern of variability which can be explained on the basis of eclipses, are not uncommon. An example is the star *Algol* which has a regular fluctuation with a period of $2^d\,20^h\,49^m$, this period being discovered by Goodricke in 1783.

Observations of the brightness changes allow a **light curve** to be obtained and from this curve orbital parameters and physical properties of the eclipsing pair may be deduced. The interpretation of the light curves of eclipsing binary systems therefore provides a second means of investigating the pairing of stars.

A third way is provided from the analysis of stellar spectra. Some stars, which otherwise might have been considered as being single, exhibit duplicity in their spectral lines. The simultaneous appearance of a particular spectral feature at two slightly different wavelength positions reveals that an apparent single star has two components and that the components are moving with different relative velocities with respect to the observer. Over a period of time, the relative positions of the lines are seen to change, showing that the velocities of the two stars change. This can only be interpreted by considering the two components to be revolving around each other. Fig. 7.3 illustrates the effect when the two stars are in orbit in a plane which contains the line of sight; typical spectra are presented for three epochs of the orbit.

At time $t = 1$, star A is receding from the Earth and star B approaching the Earth. Consequently the spectral lines of star A are red-shifted and those of star B, blue-shifted. At $t = 2$, both stars have no radial velocity with respect to the Earth and the spectral lines are superimposed. Later at $t = 3$, star A is approaching the Earth and exhibits a blue-shifted spectrum while star B recedes, exhibiting a red-shifted spectrum. From a collection of spectra it will be seen that the stars periodically reverse their sense of radial velocity and so a period can be ascribed to their orbits.

Such a system exhibiting periodic changes of the above nature is known as a **spectroscopic binary**. By plotting how the radial velocities of each component change with time, **velocity curves** are produced. Analysis of a velocity curve allows deduction of a star's orbit about the centre of mass of the system.

In some cases an apparent single star exhibits a single spectrum as expected, but it is found that the star has a radial velocity which exhibits periodic changes. This again is interpreted as the star being a component of a binary system but with the second star being too faint to contribute significantly to what would be the combined spectrum.

We have now seen that binary stars may be studied by three different means, the types of system being labelled visual binary, eclipsing binary and spectroscopic binary. Each class will now be discussed briefly in turn below, with the aim of showing the kind of information which can be gleaned from their study.

Fig. 7.4 Observations of a long period visual binary system with two possible orbits drawn

7.5.2 Visual Binaries: Stellar Masses Visual binaries are those binaries whose components have sufficient apparent angular separation to be resolvable by a telescope. Their separation may either be measured by eye with the aid of a rotatable micrometer eyepiece or their positions may be recorded photographically for subsequent measurement in the laboratory. By making regular observations, their apparent orbits may be determined. Typical orbital periods range from a few tens to hundreds of years. Some binaries have not yet been measured over a time sufficiently long for one complete orbit to have been observed and so considerable uncertainty arises about the orbital period. This is demonstrated in Fig. 7.4 where five observations are plotted for a visual binary. It will be seen that there is a large difference in the two possible orbits which have been drawn to fit the observed points. The real orbit will remain uncertain until time elapses and allows further points to be observed and plotted.

The most simple of measurements are made by choosing one star as reference. This is usually the brighter of the two and is known as the **primary star**; the other star is known as the **secondary star**. Observation is made at a chosen time, t, of the angular separation, ρ, of the stars and the position angle, θ, of the secondary star; the **position angle**, θ, is defined as the angle between the celestial north pole, the primary star and the secondary star, being measured positively in an anticlockwise direction (see Fig. 7.5).

The elliptical orbit which is obtained directly from observations by plotting them represents what is known as the **apparent orbit**. The **true orbit** is being performed in three-dimensional space and its plane may be tilted with respect to the observer. What the observer sees as the apparent orbit is the projection of the true orbit on a plane tangential to the celestial sphere. If the observer wishes to know all the parameters of the binary star orbit, he must allow for the tilt of the orbit with respect to himself. There are several standard mathematical procedures for doing this.

Any ellipse in a particular plane when projected on to another plane produces a figure which is again an ellipse, but with different characteristics. However, a focus in the first ellipse when projected does not appear at the position of the focus of the projected ellipse. Thus when the apparent orbit is examined it is generally found that the primary star does not sit at the position of the focus of the ellipse. The necessary change in perspective required to place the primary star at the focus can be determined by one of the standard methods, so giving the inclination of the true orbit with respect to the celestial sphere. After this has been determined, all the parameters describing the true orbit may be deduced. It must be pointed out, however, that the sign of the angle of inclination is indeterminate; a positive or negative tilt of the same amount produces the identical apparent orbit. If the radial velocity of the orbiting star can be measured, the sign ambiguity can be removed.

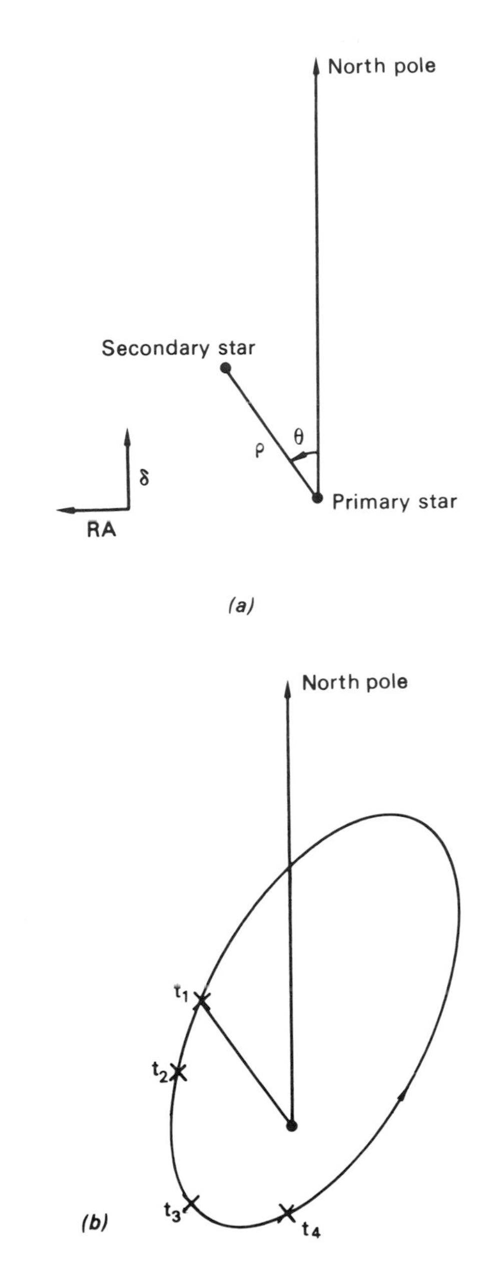

Fig. 7.5 A visual binary: (*a*) a single measurement at time t_1, (*b*) an apparent orbit plotted from observations over a number of years

Of immediate use are the orbital period, T in years, (which is available directly from the apparent orbit) and the size of the major axis, α, in arc sec. If the distance of the binary star is known, then we can determine the sum of the masses of the stars as follows.

If M_1 and M_2 are the masses of the primary and secondary stars, then the period of revolution, T, of the secondary about the primary is given by

$$T = 2\pi\left(\frac{a^3}{G(M_1+M_2)}\right)^{1/2} \tag{7.12}$$

where a is the semi-major axis of the orbital ellipse and G is the universal constant of gravitation. Now the corresponding formula for the Earth's orbit about the Sun is

$$T_{\oplus} = 2\pi\left(\frac{A^3}{G(M_{\odot}+M_{\oplus})}\right)^{1/2}.$$

If we express the periods of revolution in years and consider that $M_{\odot} \gg M_{\oplus}$, the last expression reduces to

$$1 = 2\pi(A^3/GM_{\odot})^{1/2}$$

therefore

$$(G)^{1/2} = 2\pi(A^3/M_{\odot})^{1/2}.$$

Substituting this into Equation (7.12) gives

$$T = \left(\frac{a^3}{M_1+M_2}\,\frac{M_{\odot}}{A^3}\right)^{1/2}$$

therefore

$$M_1+M_2 = \left(\frac{a}{A}\right)^3 \frac{M_{\odot}}{T^2}.$$

By letting the solar mass equal unity, this expression becomes,

$$M_1+M_2 = \left(\frac{a}{A}\right)^3 \frac{1}{T^2}. \tag{7.13}$$

Thus, if the period of revolution is determined and the size of the orbit is known, then the sum of the masses of the two stars may be deduced in terms of the solar mass.

If d is the distance of the binary star, the apparent angular size, α, of the semi-major axis is given by

$$\sin\alpha = a/d$$

and since α is a very small angle, this may be written as

$$\alpha = a/d. \tag{7.14}$$

Now the parallax, P, of the star is given by

$$\sin P = A/d$$

and since P is also a very small angle, this may be written as

$$P = A/d. \tag{7.15}$$

Solving for d by Equations (7.14) and (7.15) gives

$$\alpha/P = a/A.$$

By substituting this into Equation (7.13), we have that the sum of the masses of the two stars is given by

$$M_1+M_2=\left(\frac{\alpha}{P}\right)^3\frac{1}{T^2} \qquad (7.16)$$

(α and P are usually obtained in units of seconds of arc).

Example 7.1 *Sirius* is a visual binary with a time period of 50 years. Its parallax is 0·371 arc sec and the semi-major axis is 7·57 arc sec. Inserting these values into Equation (7.16) gives the sum of the masses of the stars as

$$M_1+M_2=\left(\frac{7{\cdot}57}{0{\cdot}371}\right)^3\left(\frac{1}{50}\right)^2=3{\cdot}40.$$

This binary system has a mass 3·4 times that of the Sun.

If it is possible to measure the stars' positions relative to the position of their centre of gravity then the ratio of the masses may be determined. This type of measurement requires very accurate positions of both stars to be observed against the distant star background over a long period of time. Such observations do in fact allow deduction of orbits for each star about their centre of gravity.

For a single star, prolonged observation over many years shows that it has a motion of its own with respect to the fainter background stars giving it a path which is part of a great circle on the celestial sphere. If it is a binary system, then it is the centre of gravity of the system which progresses along a great circle. The two stars forming the system follow curved paths with a slow oscillation about the centre of gravity (see Fig. 7.6). From the positional measurements of both stars, the path of the centre of gravity may be determined and then the separate orbits.

Suppose α_1 and α_2 are the angular distances of the primary and secondary stars from the apparent centre of gravity of the system. Then we have:

$$M_1\alpha_1=M_2\alpha_2$$

therefore

$$M_1/M_2=\alpha_2/\alpha_1. \qquad (7.17)$$

If observations allow parameters to be inserted into both Equations (7.16) and (7.17) then the masses of the individual stars may be evaluated.

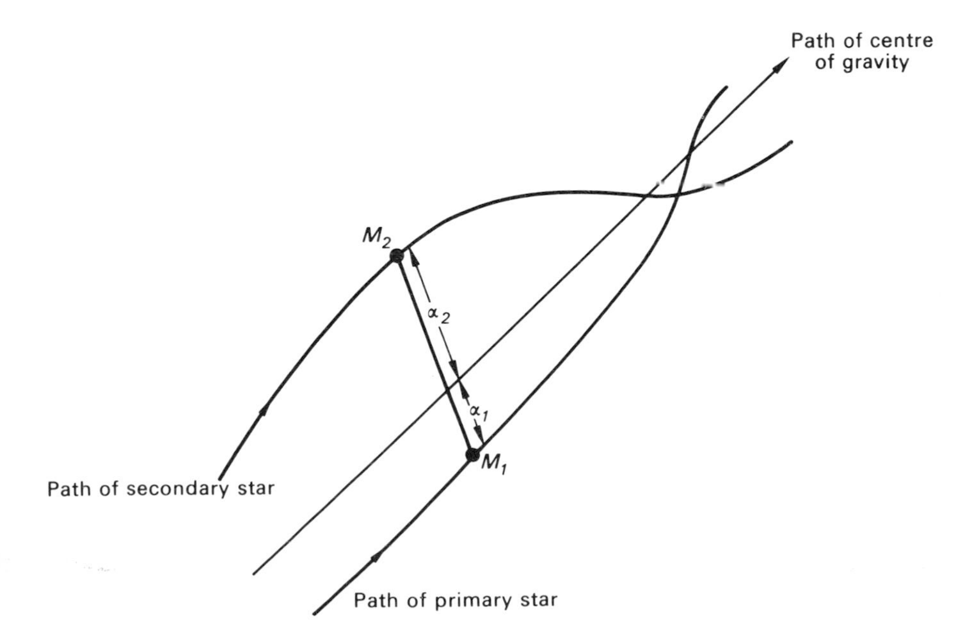

Fig. 7.6 Observed motion of a visual binary as a result of proper motion. (For simplicity, the effect of parallactic motion is not shown)

Typical masses obtained from the study of visual binary stars run from 0·1 to 20 times the mass of the Sun.

The fact that the masses of observed visual binaries do not cover a very wide range can be used to estimate the distances of those which cannot be measured by the usual parallax method. This method of distance determination is known as the method of **dynamical parallax**.

7.5.3 Dynamical Parallaxes The method involves a number of steps, repeated until a satisfactory answer is obtained.

Step 1. We assume as a first approximation that each star has solar mass. Then

$$M_1+M_2=2$$

and, by using Equation (7.16), in the form

$$P=\frac{\alpha}{T^{2/3}}\frac{1}{(M_1+M_2)^{1/3}}, \tag{7.18}$$

we can obtain a first approximation to the parallax by substituting observed values for α and T, and letting $M_1+M_2=2$.

Step 2. We now use the measured apparent magnitudes m_1 and m_2 of the binary components.

By Equation (7.4), we had

$$\mathcal{M}=m+5+5\log_{10}P.$$

If $\mathcal{M}_1$ and $\mathcal{M}_2$ are the absolute magnitudes of the components, then

$$\mathcal{M}_1=m_1+5+5\log_{10}P,$$

$$\mathcal{M}_2=m_2+5+5\log_{10}P.$$

Substituting the first approximation obtained in Step 1 for the parallax into these equations will give first approximations for the absolute magnitudes $\mathcal{M}_1$ and $\mathcal{M}_2$ of the components.

Step 3. Use is now made of the so-called *mass–luminosity relation* (see Section 7.6). This relation is given in Fig. 7.14. It shows that if the mass is known, the absolute bolometric magnitude can be determined and *vice versa*.

Using the first approximations found in Step 2 for the components' absolute magnitudes in this relation, we can read off improved (that is, more accurate) values of the masses M_1 and M_2 of the components.

Step 4. Use these values in the Equation (7.18) to derive an improved value P_2 of the parallax.

Step 5. Repeat Step 2.

Step 6. Repeat Step 3 and so on.

In practice it is found that the values of P settle down or converge very quickly. The reiterative process is halted when any difference between two successive approximations is less than one in the last significant figure to which the apparent magnitudes are known. For example, if the apparent magnitudes were 0·16 and 0·85 and it was found that $P_2=0{\cdot}15$ arc sec while $P_3=0{\cdot}14$ arc sec, it would be meaningless to carry the process any further.

It should also be noted that the quantity $\alpha T^{-2/3}$ in Equation (7.18) need only be calculated once.

Even if the first guess that $M_1+M_2=2$ is a poor one, the form of Equation (7.18) minimizes the error, since the quantity (M_1+M_2) is raised to the power one-third.

Thus, if in fact $M_1+M_2=20$ (an unusually large mass for a binary) and we put as a first approximation $M_1+M_2=2$, we see that $20^{1/3}=2{\cdot}714$, while $2^{1/3}=1{\cdot}260$. The factor of 10 in the sum of the masses is reduced immediately to a factor of about $2\frac{1}{2}$ in the term $(M_1+M_2)^{1/3}$.

Because of this, dynamical parallaxes are reliable, providing useful additions to our collection of stellar distances and masses.

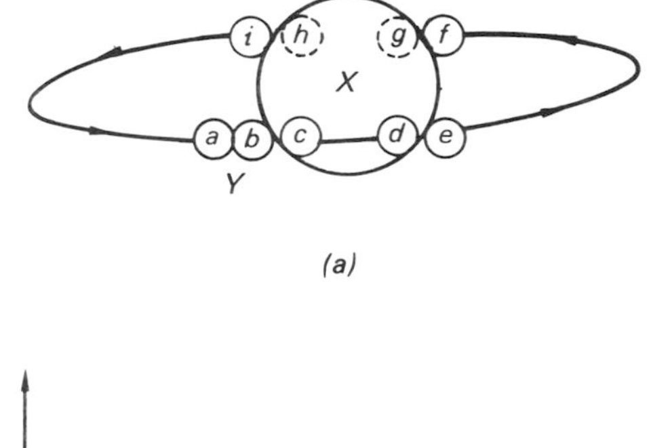

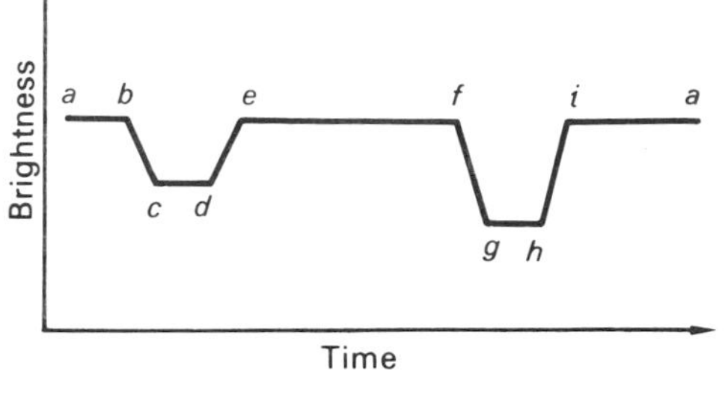

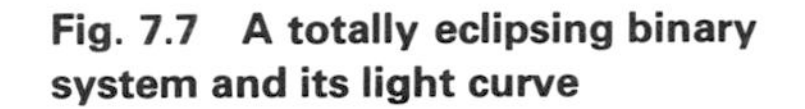
Fig. 7.7 A totally eclipsing binary system and its light curve

7.5.4 Eclipsing Binaries: Ratios of Luminosities and Radii The periodicity of light curves of eclipsing binaries is typically a few days, indicating that the components of this type of system are much closer together than in the cases of visual binaries. Actual shapes of light curves vary from one binary star to another, but the general characteristic of there being two falls in brightness within the period may only be interpreted by considering a system of two stars which are orbiting each other and presenting eclipses to the observer.

We shall consider the basic form of an eclipsing binary light curve as depicted in Fig. 7.7 where, during the periods of minimum brightness, the level remains constant to all intents and purposes. This particular form would indicate that the eclipses are total. Fig. 7.7(a) illustrates the configurations which produce the kind of light curve depicted in Fig. 7.7(b), representing the orbit that would be seen if it were possible to resolve the component stars.

By comparing Figs. 7.7(a) and (b), we see that when star Y is in position a, each component contributes fully to the total brightness. At position b this star is about to commence its passage across the disk of star X. In progressing from position b to c, light from X is blocked off and the total light level drops smoothly. It then levels off and remains at this brightness corresponding to an annular eclipse until star Y arrives at position d. In moving from d to e, more and more of the disk of star X is revealed until at position e the light level regains its full brightness.

Full brightness is then maintained until the motion brings star Y to position f. At this position it commences to be eclipsed by X and the light level falls. At position g star Y is fully eclipsed and remains so until it arrives at h. During the period from g to h the light level remains constant but, in general, not at the same level as the minimum produced between positions c and d, as the brightnesses of the component stars are usually different. On egress from the eclipse to position i, the light level rises until full brightness is recorded. This level is maintained until position b again when a new cycle of the light curve begins.

Let us now look at the light curve more quantitatively. Although the light curves may sometimes be expressed in terms of changes in stellar magnitude, it is more convenient here to consider them in terms of brightness changes. Suppose that star X has a luminosity L_X, and star Y a luminosity L_Y. Now the apparent brightness of the system is equal to the sum of the brightnesses of the two stars. They contribute to this total according to their luminosities. Out of eclipse we may write,

$$L_X+L_Y=kB, \tag{7.19}$$

where k is a constant related to the stars' distance from the observer and B is the full apparent brightness.

Suppose that at the primary minimum (usually defined as the deeper of the eclipses and hence corresponding to g, h in Fig. 7.7) the apparent brightness falls to B_1 and at secondary minimum the apparent brightness is B_2. It is easily seen that

$$L_X=kB_1 \tag{7.20}$$

and also that

$$L_X\left(\frac{A_X - A_Y}{A_X}\right) + L_Y = kB_2 \tag{7.21}$$

where A_X, A_Y are the projected areas of the two stars.

Let b_1, b_2 be the brightness losses at the two minima so that

$$b_1 = B - B_1$$

and

$$b_2 = B - B_2.$$

By dividing Equation (7.19) by L_X on the left-hand side and by kB_1 on the right-hand side (see Equation (7.20)), we have that

$$1 + L_Y/L_X = B/B_1$$

and hence

$$\frac{L_X}{L_Y} = \frac{B}{B - B_1} = \frac{B - b_1}{b_1}. \tag{7.22}$$

This simple analysis immediately shows that the ratio of the stars' luminosities may be obtained directly from the observed brightness out of eclipses and at the primary eclipse. Of course it has been assumed that the primary eclipse corresponds to the total rather than the annular one, implying that $L_X > L_Y$, i.e. the larger star is more luminous. In many systems this is not the case and for these, what, from observational data, has been designated as being the primary eclipse, refers to the annular eclipse. For such systems, the ratio of the luminosities would then be derived from

$$\frac{L_X}{L_Y} = \frac{B - b_2}{b_2}. \tag{7.23}$$

The question of which formula to apply is resolved by appealing to supplementary observations such as the behaviour of the spectrum during the eclipse phases.

By substituting Equations (7.19) and (7.20) into (7.21), it is easily shown that

$$\frac{A_X}{A_Y} = \frac{B_1}{B - B_2} = \frac{B - b_1}{b_2}. \tag{7.24}$$

Since the values of A_X and A_Y are proportional to the square of the stellar radii R_X and R_Y, Equation (7.24) can be rewritten as

$$\left(\frac{R_X}{R_Y}\right)^2 = \frac{B - b_1}{b_2}$$

and therefore

$$\frac{R_X}{R_Y} = \left(\frac{B - b_1}{b_2}\right)^{1/2}. \tag{7.25}$$

Thus by measuring the maximum brightness and the brightness loss at the minima, the ratio of the radii of the stars can be deduced. It may also be mentioned that the dimensions of the stars relative to the size of the orbit can be evaluated by measuring the lengths of time taken for the brightness to fall from full to the minimum and the length of the minimum in relation to the overall period.

Values of the ratios of luminosities and radii of the stars help us to compare the properties of stars which happen to be the components of an eclipsing binary system. The inclination of the orbit with respect to the observer may also be deduced. All this information is particularly useful if the eclipsing binary is also observed as a spectroscopic binary (see Section 7.5.5). However, the elegant methods that are applied to the light curve are beyond the scope of this text and will not be discussed here.

It may be noted though that the light curve described above represents a system which exhibits total eclipses. The fact that there are some systems which exhibit partial eclipses is clearly evident. For such systems, there is no extended period when the minima hold steady values; the light curve has two V-shaped minima, usually of different depths. Fig. 7.8(a) represents such a partially eclipsing system and Fig. 7.8(b) illustrates the light curve.

It will be seen in Fig. 7.8 that the maximum area of the larger star eclipsed by the smaller occurs at A. Because the eclipse is partial, the light curve immediately begins to rise again. It may easily be shown that the depths of the minima from such a light curve still allow the ratio of the luminosities to be determined. However, the ratio of the radii cannot be obtained by the simple expression (7.25). Other standard, but more complicated, ways are available for obtaining this information from the light curve.

The light curve can also provide knowledge of the eccentricity of the orbit of one star about another. As an example, an extreme case is illustrated in Fig. 7.9 where the major axis is at right angles to the line of sight. Now both stars are subject to the law of gravitation and therefore obey Kepler's three laws. The secondary star will therefore travel at its fastest when nearest the primary star, when it is said to be at **periastron**. Because of this, the secondary eclipse C occurs closer to the preceding primary eclipse A (Fig. 7.9(b)) than to the following primary eclipse, and the periods of maximum brightness (B and D) are not of equal length.

Besides providing orbital information, a detailed analysis of a light curve may provide knowledge about:

(i) departures from sphericity of the shapes of stars,
(ii) the uniformity of brightness across the stellar disks—i.e. limb darkening,
(iii) the effects of reflection—the light from one star being reflected by the other in the direction of the observer.

These are discussed briefly below:

(i) Some stars are so close together that they distort each other gravitationally, each star being elongated along the line joining their centres. Thus, as illustrated in Fig. 7.10, if two ellipsoidal stars revolve about each other in a plane such that eclipses occur, the light curve will contain no straight parts. It will change smoothly because the total area the stars present to the observer is never constant.

(ii) It is well known that the Sun does not have uniform brightness across its disk and that the brightness falls off towards the solar limb. This effect is known as **limb darkening**. From the light curves of eclipsing binaries, we know that some stars must exhibit the same effect. When the eclipse begins (see Fig. 7.11) the initial fall in brightness is slow as the less bright parts of the stellar disk at the limb are occulted first. The fall in brightness increases at a faster rate as the occulting star begins to cover the brighter parts of the eclipsed star. Thus the falls and rises in brightness are not linear when the stars exhibit limb darkening.

(iii) In this case the parts of the light curve between the minima are sloped and curved as shown in Fig. 7.12, so that even though neither star is entering eclipse nor emerging, the brightness of the system is altering. What is happening is that the smaller star is showing phases analogous to those exhibited by

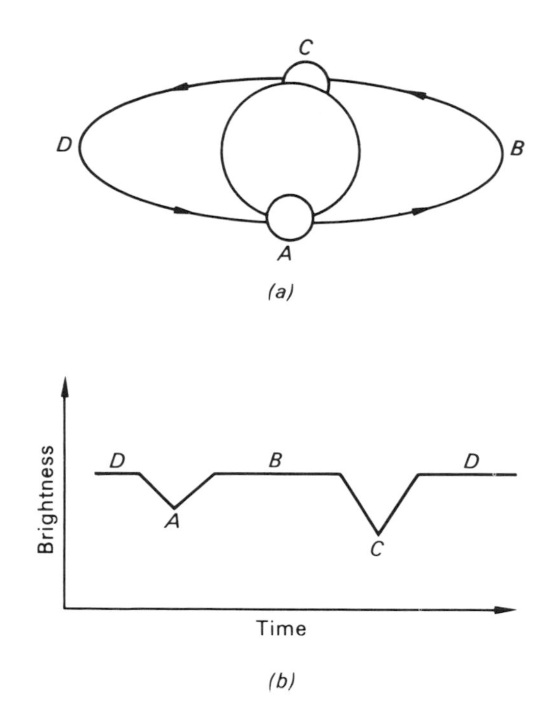

Fig. 7.8 A partially eclipsing binary star and its light curve

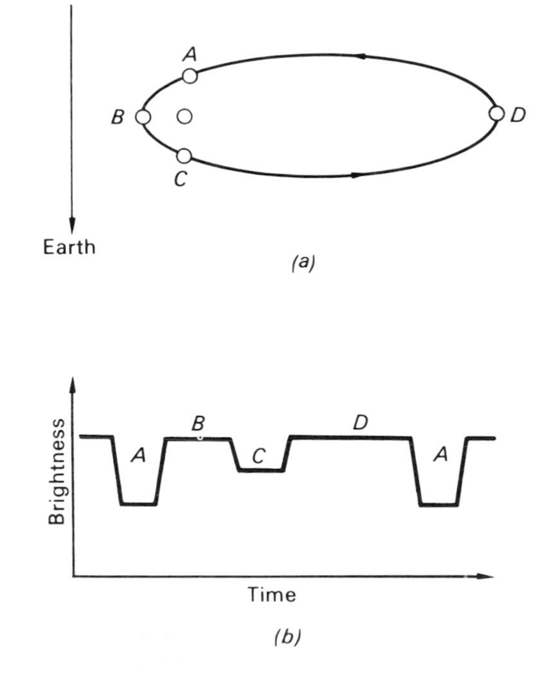

Fig. 7.9 The orbit and light curve of an eccentric binary star

Venus or the Moon. The side presented to the larger star appears brighter than the side turned away from it. It must be remembered however that, unlike Venus and the Moon, the smaller star is self-luminous as well.

7.5.5 Spectroscopic Binaries: Masses and Mass Functions An idea of the shapes that can be expected for a radial velocity curve can be obtained by considering three different types of orbit. For simplicity let us consider the orbit of one star about the centre of gravity and suppose the orbit to be in a plane which contains the line of sight. We shall consider the orbit as being (*a*) a circle, (*b*) an ellipse with its major axis at right angles to the line of sight, and (*c*) an ellipse with its major axis along the line of sight. The orbits are illustrated in order in Fig. 7.13(*a*), (*b*) and (*c*), together with their associated radial velocity curves.

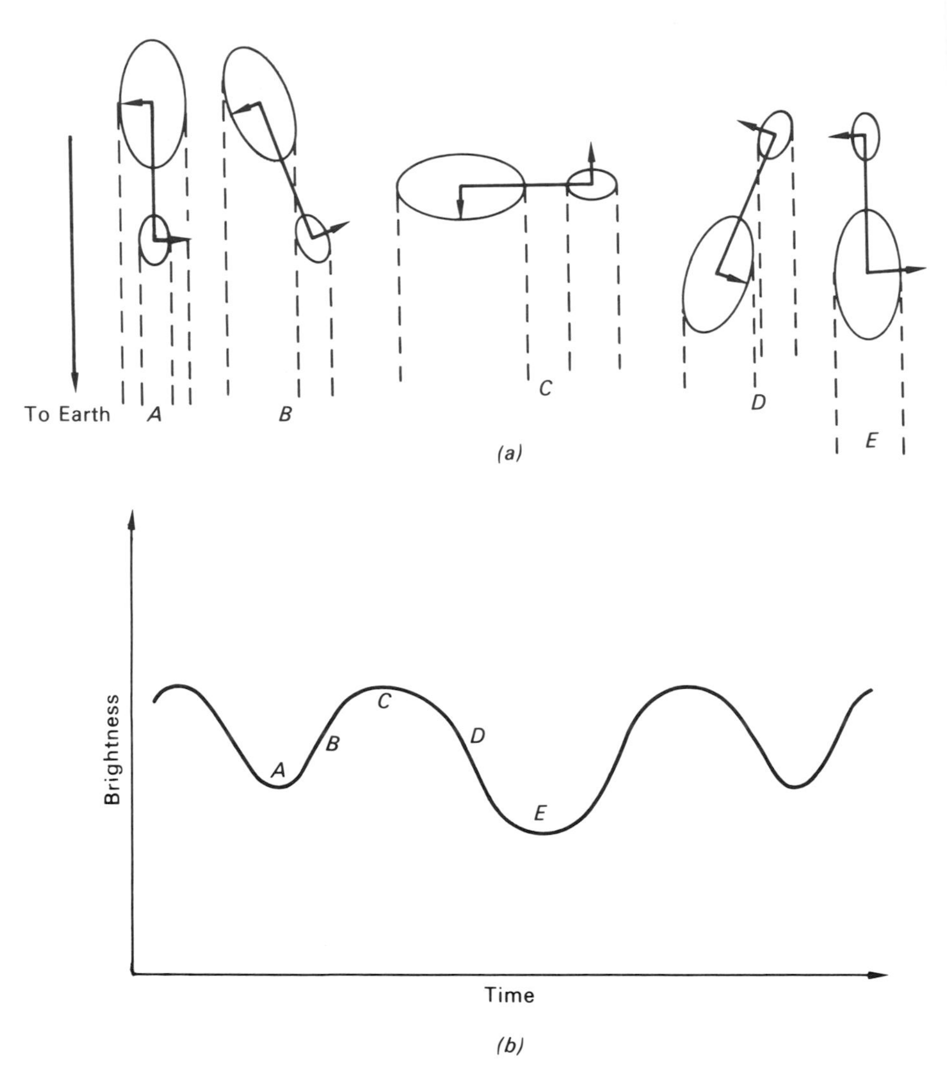

Fig. 7.10 The orbit and light curve of an eclipsing binary whose components show tidal distortion

It will be noted in all cases that for positions 1 and 3 the motion is transverse and the radial velocity is zero. Any measured radial velocity at these points represents the motion of the whole system with respect to the Earth.

For the circular orbit the radial velocity curve is symmetrical. The motion of the star towards and away from the observer is similar to that of simple harmonic motion and hence the velocity curve is in the form of a sine wave.

For the elliptical orbit with its major axis at right angles to the observer, Kepler's law predicts that the velocity of the star is greatest at periastron and it consequently spends a relatively short time in this part of its orbit. The velocity curve shows a sharp peak for the period through the points 1, 2 and 3. It spends a longer time with a motion which is nearly transverse. This corresponds to the orbit from point 3, through 4 and on to 1.

For the elliptical orbit with its major axis along the line of sight, the velocity changes its direction from negative to positive very quickly at point 1 near periastron. At point 3, the orbital speed is much slower than at 1. The crossover from a positive to a negative radial velocity is consequently much slower than the opposite crossover at point 1.

These three examples are all special cases. When it is considered that the orbit may be set with its major axis at a different angle and the plane inclined to the observer, then the shape of the curve must reflect these facts.

Since the net orbital velocity over one period is zero and since the velocity curve is a plot of velocity against time, a line of constant velocity can be drawn on the curve so that the area above the line is equal to the area below. The velocity indicated by this line represents the constant radial velocity of the binary system as a whole with respect to the Sun.

When both components contribute to the spectrum, two velocity curves may be obtained, corresponding to the orbits of each star about the centre of gravity of the system.

It goes without saying that any determined radial velocity must be corrected for the Earth's orbital motion about the Sun before the value can be plotted on the radial velocity curve.

If any binary star orbit is considered, it is possible to derive the expressions for the value of the radial velocity of each component at any particular time. Appearing in the radial velocity expression for the primary star is the product

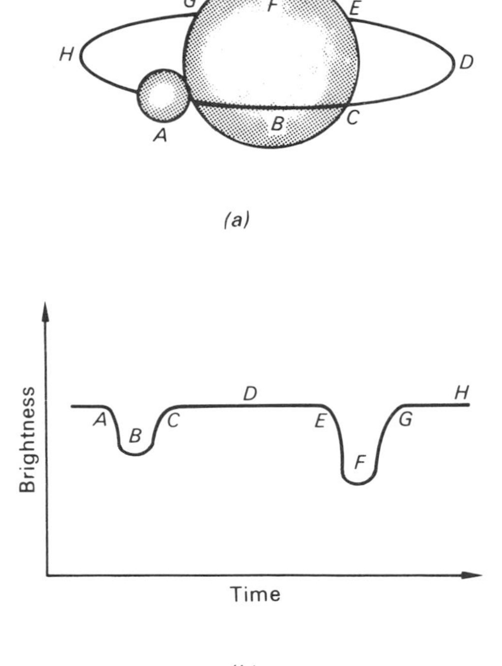

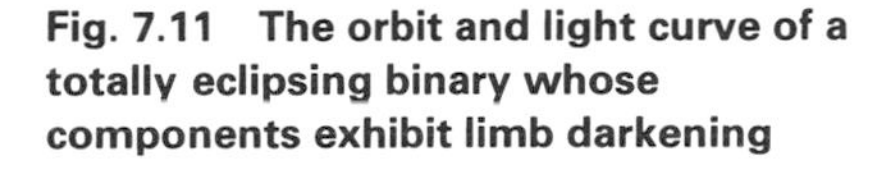
Fig. 7.11 The orbit and light curve of a totally eclipsing binary whose components exhibit limb darkening

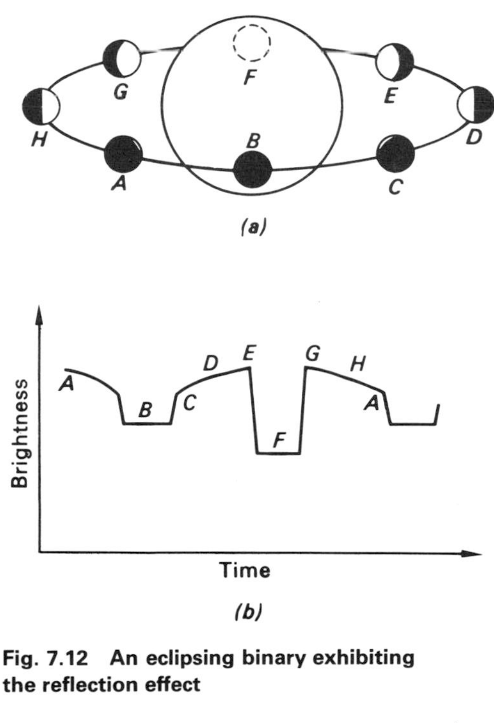

Fig. 7.12 An eclipsing binary exhibiting the reflection effect

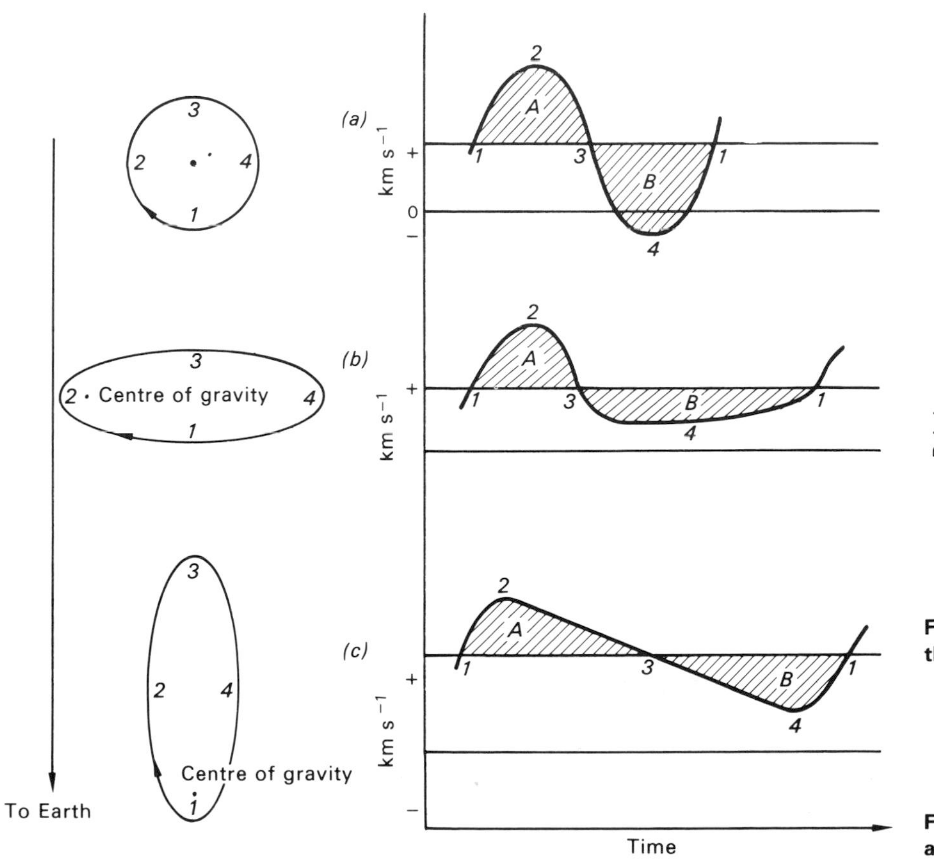

Fig. 7.13 Three binary stars and their associated velocity curves

$a_1 \sin i$ and in the expression for the secondary star is the product $a_2 \sin i$, where a_1 and a_2 are the semi-major axes of the orbits about the centre of gravity; the two products are the projections of these axes on the plane at right angles to the line of sight, i.e. i is the inclination of the plane of the orbit relative to the tangent plane on the celestial sphere. From the analysis of the two radial velocity curves, these products may be determined. The parameters a_1, a_2, however, cannot be separated from $\sin i$ using the radial velocity data alone.

From the definition of the centre of gravity we have that

$$M_1 a_1 = M_2 a_2. \tag{7.26}$$

We may multiply both sides of this equation by $\sin i$ giving,

$$M_1 a_1 \sin i = M_2 a_2 \sin i$$

and therefore

$$M_1/M_2 = a_2 \sin i / a_1 \sin i. \tag{7.27}$$

The numerator and denominator of the right-hand side of Equation (7.27) are the very quantities which may be determined from analysis of radial velocity curves.

When both curves are obtained, it is seen that one curve is a reflection of the other about the zero velocity line, though perhaps with a different amplitude. The ratio of the amplitudes of the two velocity curves is inversely proportional to the ratio of the masses of the stars. Thus if both curves are available, the ratio of the component masses can in fact be determined directly from the curves.

In Equation (7.13) we have already shown the relationship between the sum of the masses of two stars, the size of the major axis of the orbit of one star about the other and the period of revolution. By expressing distances in terms of the astronomical unit, this equation reduces to

$$M_1 + M_2 = \frac{a^3}{T^2}. \tag{7.28}$$

By substituting the value of M_2 obtained from Equation (7.26), the above equation may be written as

$$M_1 = \frac{a^3}{T^2[1+(a_1/a_2)]}. \tag{7.29}$$

In relating the two orbits about the centre of gravity to the one referred to the primary star, we have that

$$a = a_1 + a_2$$

and we may multiply each of these terms by $\sin i$ to give

$$a \sin i = a_1 \sin i + a_2 \sin i.$$

Now, as we have seen, the analysis of the radial velocity curves allows $a_1 \sin i$ and $a_2 \sin i$ to be deduced, so we are also able to calculate a value for $a \sin i$. By expressing the right-hand side of Equation (7.29) in terms of quantities which can be deduced we have

$$M_1 \sin^3 i = \frac{(a \sin i)^3}{T^2[1+(a_1 \sin i / a_2 \sin i)]} \tag{7.30}$$

thus showing that a value for $M_1 \sin^3 i$ may be determined. In a similar manner a value for $M_2 \sin^3 i$ may also be determined.

If only one curve is available then a quantity known as the **mass function** can be obtained.

Suppose then it is the primary star which provides the spectrum for measurement. We are therefore able to determine $a_1 \sin i$ but not $a_2 \sin i$. By adding $M_2 a_1$ to both sides of Equation (7.26) we have

$$M_1 a_1 + M_2 a_1 = M_2 a_1 + M_2 a_2,$$

$$\frac{a_1}{M_2} = \frac{a_1 + a_2}{M_1 + M_2}.$$

Since $a_1 + a_2 = a$, this may be rewritten as

$$\frac{a_1}{M_2} = \frac{a}{M_1 + M_2}.$$

Eliminating a from this equation by means of Equation (7.28) gives

$$a_1^3/T^2 = M_2^3/(M_1 + M_2)^2.$$

Multiplication of both sides of this equation by $\sin^3 i$ allows the left-hand side to be expressed in terms of measured and deduced quantities. Thus,

$$(a_1 \sin i)^3/T^2 = (M_2 \sin i)^3/(M_1 + M_2)^2. \qquad (7.31)$$

The right-hand term of Equation (7.29) is known as the mass function of the spectroscopic binary.

7.5.6 Combination of Deduced Data A summary of the information about the physical nature of binary stars which can be deduced from observations is given in Table 7.1.

Table 7.1 Deduced information from the measurement of binary stars

Visual binary	Eclipsing binary	Spectroscopic binary
The angular size of the major axis. The eccentricity of the ellipse. The modulus of the inclination of the orbital plane. If the parallax is known: The linear size of the major axis. The sum of the masses. If the centre of gravity is known: The ratio of the masses and hence the mass of each star	Ratio of luminosities. Ratio of radii to the radius of the relative orbit, considered as being a circle. The inclination of the orbital plane. The shape of the stars. The eccentricity of the orbit. Limb darkening.	The product of the size of the major axis and the sine of the inclination of the orbital plane. The eccentricity of the ellipse. From knowledge of both velocity curves the product of each mass and the cube of the sine of the inclination of the orbital plane.
	If the binary is both eclipsing and spectroscopic, the masses of the components and absolute values for the radii can be deduced.	

It can be seen that very complete information about the components of a binary system can be obtained if the system is an eclipsing binary and both the light curve and velocity curve are available. (It will be appreciated that for the binary system to exhibit eclipses, i must be close to 90°.)

Analysis of the light curve allows the ratio of the star's radius to the size of its orbit to be determined and also provides a value of i. If this latter value is combined with data from study of the radial velocity curves, e.g. $M_1 \sin^3 i$ (see Equation (7.30)), then the masses of the stars may be determined. A value for the size of the orbit (a) can also be evaluated and when this is applied to the data provided by the light curve, values of the individual radii can be deduced.

Once a star's mass and radius have been deduced, it is then a simple matter to calculate a value of the mean density of the stellar material.

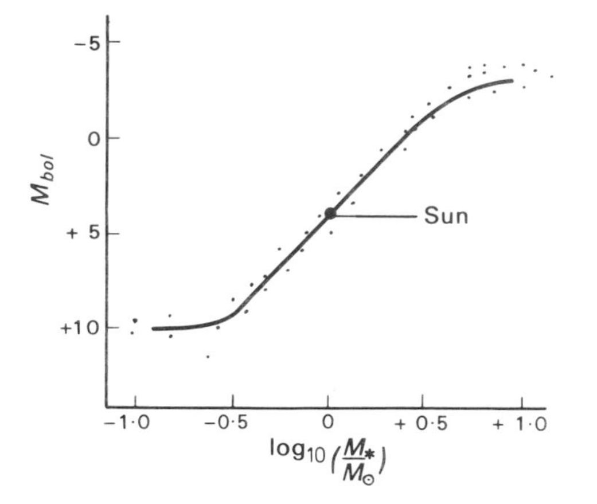

Fig. 7.14 The mass–luminosity relation

7.6 The Mass–Luminosity Relation

From the data concerning the masses and luminosities of stars, it is found that there is a fairly well-defined relation. This empirical law is known as the **mass–luminosity relation**. The law, as given directly from observational material, is depicted in Fig. 7.14. For convenience the relation is presented by plotting the absolute bolometric magnitudes (directly related to the luminosities) against the logarithms of the masses of the stars, the solar mass being taken as unity. It will be seen that the Sun with an absolute bolometric magnitude of 4·75 and a value of $\log_{10}(M_*/M_\odot)$ equal to zero lies on the curve.

Over the central part of Fig. 7.14, we see that there is a linear relation, these data corresponding to stars which lie on or close to the main sequence of the HR diagram. It will also be seen that the greater the stellar mass, the greater is the star's luminosity. At the extremes of the relation, the linearity breaks down. Based on the linear relation, red giants tend to be over-luminous for their masses while white dwarfs are decidedly under-luminous for their masses.

Now the data suggest that for the main sequence stars, the mass–luminosity relation can be expressed in the form

$$M_{bol} = -a \log_{10} M + b, \tag{7.32}$$

where a and b are constants.

From Equation (7.7) we have

$$M_{bol} = k - 2{\cdot}5 \log_{10} L,$$

where k is a constant.

Insertion of this last equation into Equation (7.32) allows the relation between the luminosity and the mass to be written as

$$L = AM^K,$$

where A and K are constants.

In other words, we may say that the luminosity of a star is proportional to the Kth power of its mass. From the most reliable data, the index K is a little greater than 3 and, to a good approximation, we may write

$$L \propto M^{3{\cdot}1}. \tag{7.33}$$

The mass–luminosity relation can obviously be used to assign a mass to a star if its luminosity is known. Of more importance, however, is that it is one of the relationships to be satisfied by theoretical models which describe the structures of stars and the relationships between the range of stars.

7.7 The Spectral Sequence

7.7.1 Secchi's Classification When the spectral features of stars were first investigated, it was immediately obvious that stars could be categorized into different types. It was found that stellar spectra were continuous, much like those of black bodies, with absorption and/or emission features superimposed.

After an analysis of some 4000 stars, Secchi in 1864 presented the stars as being in one of four broad groups as follows:

I – White	– Spectra with only hydrogen lines.
II – Yellow	– Spectra resembling that of the Sun.
III – Red	– Spectra bearing some resemblance to that of the Sun, but with bands shaded towards the red.

IV – Very Red – A small group, having strong bands quite different from those of group III, shaded to the violet.

7.7.2 The Harvard Classification At the beginning of the century, under the direction of Pickering, the Harvard classification system was set up. Initially the spectra of 225 000 stars were analysed from objective prism photographs. This classification was published as the *Henry Draper Catalogue*. Any star from this catalogue is described by the letters HD, followed by the number of the star in the catalogue. The work originally presented as the HD catalogue has now been extended and close to 400 000 stars have been classified at Harvard.

Letters are used to designate the types of spectra. Some distinctions are found to be very much smaller than others and do not merit the separation of a whole class. Any one class is therefore subdivided into 10; e.g. class B is divided into B0, B1, B2, . . . , B9.

At the time of the original cataloguing, it was soon recognized that the variations in spectra seemed to be a naturally changing sequence with temperature of the star according to its colour index. Bearing this in mind, the spectral sequence is now presented in an order dictated by temperature and the letters designating the spectral types are no longer in alphabetical order. The positions of the spectral types on the HR diagram are illustrated in Fig. 7.1.

Prior to the classification of Wolf–Rayet stars, designated by the letter W, the spectral sequence could be remembered by the mnemonic:

O Be **A** **F**ine **G**irl, **K**iss **M**e **R**ight **N**ow, **S**mack.

This may still be used, remembering to put the W classification at the beginning.

The spectral sequence is not, in fact, continuous, i.e. at the cooler end of the sequence, a particular temperature may allow more than one spectral type.

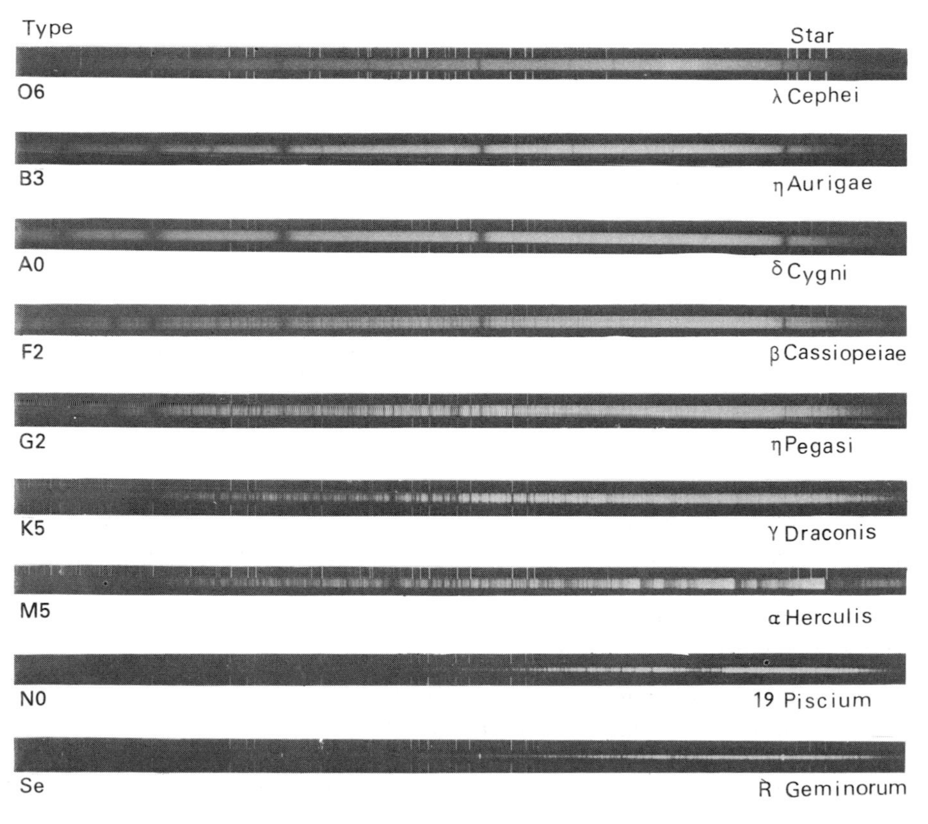

Fig. 7.15 Spectrograms of the major principal stellar spectra. (By courtesy of the Mt. Wilson and Palomar Observatories)

Diagrammatically, the sequence should be represented by:

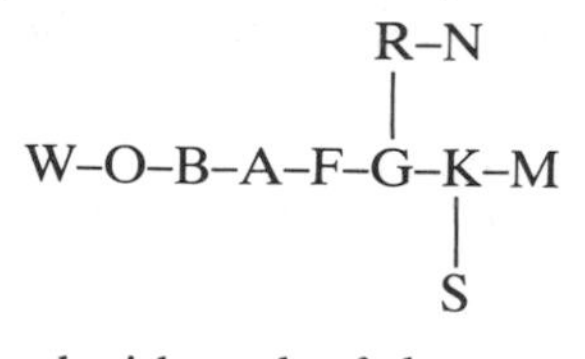

The main points associated with each of the spectral types are presented in Table 7.2. Examples of the major stellar spectral types are presented in Fig. 7.15.

Table 7.2 The spectral sequence

Type	Comments
W	Added since the original HD catalogue, Wolf–Rayet stars are characterized by broad bright bands of hydrogen and helium in emission. There is evidence of carbon and nitrogen being mutually exclusive—hence two classes WC and WN.
O	Average temperatures around 30 000 K. The absorption lines are characterized by elements which are highly ionized. These include helium, nitrogen and oxygen. Neutral hydrogen and helium are also present.
B	Less hot than O stars with a temperature range from about 13 000 K to 20 000 K. They are bluish-white in colour—a notable example being *Rigel* in the constellation of Orion. Hydrogen lines are stronger than in the O type and the strength continues to increase from B_0 to B_9. Neutral helium is at its strongest at B_2 but decays when B_9 is reached. Lines of ionized silicon, oxygen and magnesium are often visible.
A	The temperatures of these stars are close to 10 000 K. *Sirius*, *Vega* and *Altair* are examples of A-type stars. Helium lines are no longer present and those due to ionized elements are weaker. At A_0, the Balmer series of hydrogen is at its strongest, dominating the whole spectrum. The K line of calcium increases throughout the class.
F	These are yellowish stars with temperatures in the range 7000 K to 9000 K—an example being *Procyon*. The lines due to hydrogen decline in strength through the decimal sub-classification; the K line increases. Metals such as iron, sodium, manganese reveal their presence either as neutral or ionized atoms.
G	Stars in this class are yellow having average temperatures in the range 5000 K to 6000 K. The best known star of this class is the Sun, being type G_2. Hydrogen lines continue to weaken across the type whereas metallic lines increase. The K line is very strong.
K	These stars are distinctly reddish with temperatures of the order of 4000 K—*Arcturus* and *Aldebaran* belong to this class. Hydrogen lines fade into insignificance and the spectrum is dominated by the H and K lines of calcium. Lines due to molecules appear in this classification.
M	Such stars have temperatures of around 3000 K, well known ones being *Antares* and *Betelgeuse*. Molecular bands are stronger, especially those due to titanium oxide—metallic lines are still present. (The M classification is equivalent to Secchi's group III.)
R† N† S	Only a small percentage of stars belong to these classes. Those in classes R and N are orange-red with temperatures about 2300 K to 2600 K. Type N stars (Secchi's group IV) exhibit bands due to carbon compounds. S type stars have complex spectra containing bright lines due to hydrogen; there are also bands due to zirconium oxide.

† R and N are sometimes united as a single class designated by C.

There are also two spectral types which do not form part of the continuous sequence. These are designated by the letters P and Q and they do not have decimal subdivision. Class P is reserved for stars whose spectra suggest that the nuclei are contained within expanding shells and class Q refers to nova type stars.

At one time it was thought that the spectral classification revealed a certain stage in a star's life. According to this outdated theory stars commenced their life at the blue end of the sequence and gradually burned their way down the main sequence of the HR diagram to become red stars. That this theory no longer holds is demonstrated in Chapter 9, but the terms **early type** and **late type** are still applied respectively to stars at the left-hand and right-hand sides of the spectral sequence.

Even when stars of the same spectral class and decimal sub-classification are compared, it is found that the detailed shapes or profiles of the spectral lines vary from one star to another. In order to describe these details, a further prefix or suffix is sometimes used. These designations are discussed briefly below.

Prefixes:

c – The spectral lines of these stars are very sharp with a narrow profile. We shall see later that these stars are larger than the giants which have already been mentioned. They are known as supergiants (e.g. *Deneb* is a cA_2 type star).

g – These stars have spectral features peculiar to giant stars.

d – These stars have spectral features peculiar to dwarf stars.

Suffixes:

n – These stars have broad shallow lines, indicating that the star is rotating with a high speed.

s – Denotes sharp lines without exhibiting the fineness denoted by the prefix c.

e – These stars display emission lines in their spectra.

v – These spectra are known to be variable.

ev – The emission lines exhibit variability.

k – The spectrum shows the presence of interstellar absorption features.

p – A peculiar spectrum—usually with strong metallic lines.

pq – The spectrum shows a peculiarity similar to spectra of novae.

7.8 Spectroscopic Parallax

The researches leading to the production of the first HR diagram were based on absolute magnitude determinations which were obtained for stars whose trigonometric or dynamic parallaxes had been measured. The results, for what must be nearby stars, showed that there were stars in distinct groups on the luminosity/spectral type diagram (i.e. the HR diagram). It might be expected that stars of the same spectral type but with different luminosities would show some differences in their spectra, even if only very minor.

This problem has been tackled by considering stars of the same spectral type and with similar apparent brightness. If a pair of stars is chosen in this way but with proper motions which are very different, then there is a high probability that the star with the smaller proper motion is more distant. As the apparent brightnesses are approximately the same, the star with the smaller proper motion must be more luminous.

From close examination of spectra of such pairs of stars many distinct differences have been found. For example, in stars of spectral classes F to M some particular metal lines are strong in stars of high luminosity, but weak in stars of the same spectral type but with low luminosity. Other metal lines show the opposite effect.

This type of technique has been extended to include A and B spectral-type stars whose spectra do not exhibit metal lines. It has been shown that the detailed profiles of the hydrogen lines depend on a star's luminosity. These lines are relatively narrow for the more luminous stars. (See note concerning the use of prefix c in stellar classification.)

Thus, if a detailed study is made of a stellar spectrum, the star can be assigned a luminosity or absolute magnitude. If then its apparent brightness is measured, its distance modulus can be evaluated (see Equation (7.2)). This means of evaluating stellar distances is called the **spectroscopic parallax** method. Relying on the interpretation of features in a stellar spectrum, this method provides a sensitive way of determining stellar distances. It is one of the major methods employed.

7.9 The Complete HR Diagram and Yerkes Classification

The early HR diagrams were obtained by evaluating the absolute magnitudes of stars from knowledge of either their trigonometrical or dynamical parallaxes. The numbers of stars that could be placed on the diagram was consequently small.

After the development of the spectroscopic method of determining absolute stellar magnitudes, many more stars could be placed on the HR diagram. It soon became evident that the diagram did not just represent the three categories of main sequence, giants and dwarfs, but that a larger range in stellar sizes is represented (see Fig. 7.16).

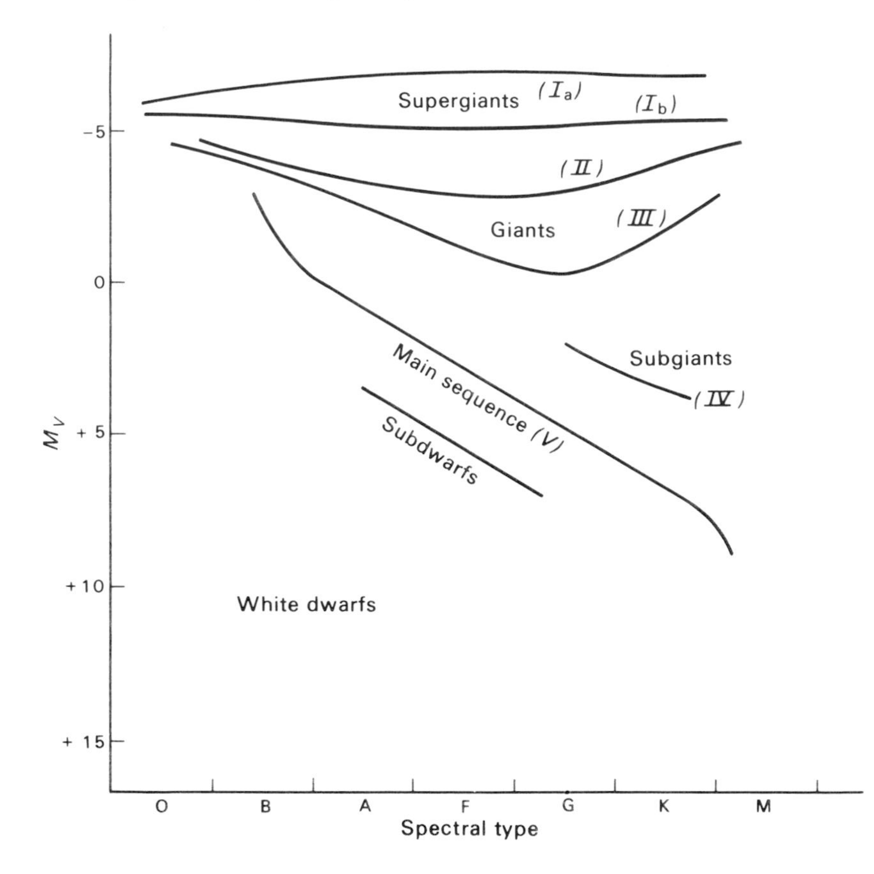

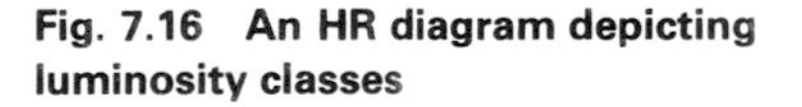
Fig. 7.16 An HR diagram depicting luminosity classes

Star distributions are concentrated predominantly in two lines. The first, or main sequence, runs diagonally across the diagram. Main sequence stars are sometimes known as main sequence dwarfs. The second, or giant branch, has little slope, the luminosities of these stars hardly changing with spectral type.

Above the giant branch are a few stars which are called **supergiants**. A small group of stars is found between the main sequence and the giant branch and they are called the **subgiants**.

Just below the main sequence, there appears to be a group of stars whose change in luminosity with spectral type is similar to that of the main sequence. These stars are called the **subdwarfs**. With a much lower luminosity, there is a small group of stars known as the **white dwarfs**.

It will be seen from Fig. 7.16 that for any particular spectral type there are a few fairly well defined bands in which a star can lie. As a consequence of this a classification based on spectral type and luminosity has been developed at the Yerkes observatory. This system, known as the MKK classification after Morgan, Keenan and Miss Kellman, gives the spectral type which almost always coincides with the Harvard classification and then a Roman numeral, designating the luminosity class (e.g. $B5I_b$). These latter symbols are defined as follows:

I_a – The most luminous supergiants
I_b – Supergiant
II – Luminous giant
III – Giant
IV – Subgiant
V – Main sequence.

Observations show that there is nearly a direct relation between spectral type and colour index $(B-V)$ and therefore in obtaining the HR diagram of a star cluster it is only necessary to plot values of the apparent magnitude (m) or absolute magnitude (M_V) against the colour index $(B-V)$. This is particularly convenient as the time required to obtain spectra of the individual stars, even using an objective prism if this is possible, may be prohibitive if the stars are faint. It is common practice to present HR diagrams as plots of M_V against $(B-V)$.

In the preparation of HR diagrams of clusters of stars, as the stars are all relatively close to each other, no allowance need be made for their individual distances. In fact, apparent magnitudes can be plotted directly against spectral type. By superimposing the main sequence of such a diagram of a cluster onto the general HR diagram with an absolute magnitude scale, the distance modulus of the cluster is obtained immediately.

7.10 The Radius of a Pulsating Star: Practical Exercise

Some variable stars exhibit regular fluctuations of brightness and a **period** can be ascribed to them according to the length of time it takes for the star to go through one cycle of the magnitude variation. A plot exhibiting how the magnitude changes with time is called the **light curve**.

Of the several types of periodic variable, one type also exhibits repetitive changes of colour and radial velocity in periods which match the light curve period. Plots of the change of colour and radial velocity are known as **colour curves** and **radial velocity curves**.

For a star having regular light, colour and radial velocity curves, the observations are interpreted by considering the star to be rhythmically expanding and contracting. Regular oscillations of the radial velocity are indicative of a stellar surface which at one time is approaching the observer and at another receding from him. Variations of colour are indicative of changes of surface temperature as the stellar surface expands and then contracts.

By making observations of brightness, colour and radial velocity of such pulsating stars it is possible to determine their radii. The method of analysis was first devised by Wesselink.

Consider how the method works.

By assuming that a star radiates as a black body, the **luminosity** of a star, or the amount of energy liberated per unit time, depends on the star's surface area and the fourth power of the surface temperature. Thus we may write:

$$L = 4\pi R^2 \sigma T^4$$

where L is the luminosity, R the radius and T the temperature of the star: σ is the Stefan–Boltzmann constant.

Now, the observed magnitude is related to a star's luminosity. Any changes in luminosity will be reflected by there being corresponding changes in the measured magnitude.

The colour of a star is expressed as a **colour index**, the value being obtained by determining magnitudes, as seen through two different colour filters in turn, and forming the difference. The colour index of a star is related to its temperature.

Suppose that at a particular time, t_1, during the star's cycle, the luminosity, radius and temperature are L_1, R_1, T_1 respectively, and at a later time, t_2, the corresponding values are L_2, R_2 and T_2. Using the expression above for luminosity we have two equations:

$$L_1 = 4\pi R_1^2 \sigma T_1^4$$

and

$$L_2 = 4\pi R_2^2 \sigma T_2^4.$$

Dividing the first equation by the second, we have

$$L_1/L_2 = R_1^2 T_1^4 / R_2^2 T_2^4.$$

7.10.1 Use of the Colour Index Curve The colour index of a pulsating star is observed to undergo an oscillation within each period and consequently there are two times within the period when the colour index has a particular value. As colour index is related to temperature, obviously there are two times within its period when the surface temperature of the star has a particular value.

Thus by choosing values of t_1 and t_2 corresponding to times when the colour index is identical we have also chosen times when the temperatures are the same, i.e. $T_1 = T_2$. Hence the ratio derived above reduces to

$$L_1/L_2 = R_1^2/R_2^2.$$

7.10.2 Use of the Light Curve Suppose that at t_1 and t_2, the observed magnitudes are m_1 and m_2 respectively. Now a change of magnitude can be related by Pogson's equation to a change of luminosity. Hence

$$L_1/L_2 = 10^{0\cdot4(m_2 - m_1)}.$$

Suppose that the right-hand side of the equation is evaluated to be equal to N. Then we have that

$$L_1/L_2 = N$$

and, therefore,

$$R_1^2/R_2^2 = N$$

giving

$$R_1/R_2 = \sqrt{N}. \qquad (7.34)$$

By using the colour curve and light curve we have been able to deduce the relative sizes of the star at two particular times during the period.

7.10.3 Use of the Radial Velocity Curve By considering the radial velocity curve, it is possible to provide a second equation relating R_1 and R_2 and with the aid of the relationship (Equation (7.34)) developed in the previous section, solutions for R_1 and R_2 may then be obtained.

Now it must be remembered that when we are looking at the star, we are looking at a hemisphere all of which is expanding or contracting at the particular time. Only the central part of the star is moving along the line of sight. The limb of the star is expanding at right angles to the observer and consequently the light from that part of the star does not contribute to the measured radial velocity. The light from points between the centre and the limb shows a radial velocity which is obviously below the velocity of the stellar surface. Thus the radial velocity that is measured is some kind of average, being below the actual velocity of the stellar surface.

It can be shown that the true velocity of the stellar surface is $\frac{3}{2}$ times the measured radial velocity.

Suppose that at a particular time, the star's surface has a velocity V. Now the velocity is given by the rate of change of the radius. Hence we may write:

$$V = \mathrm{d}R/\mathrm{d}t = \tfrac{3}{2}V_r$$

where V_r is the measured radial velocity. Therefore

$$\mathrm{d}R = \tfrac{3}{2}V_r\,\mathrm{d}t.$$

The change of radius from R_1 to R_2 may therefore be expressed as:

$$R_2 - R_1 = \tfrac{3}{2}\int_{t_1}^{t_2} V_r\,\mathrm{d}t - M, \tag{7.35}$$

the integral being the area under the velocity curve between the limits of time at t_1 and t_2. Using Equations (7.34) and (7.35) it is easy to show that:

$$R_1 = \frac{M\sqrt{N}}{1-\sqrt{N}} \quad \text{and} \quad R_2 = \frac{M}{1-\sqrt{N}}.$$

Thus, by using the three sets of data—light curve, colour curve and velocity curve—the radius of a pulsating star may be determined at two or more particular times.

The three types of curve for η Aquilae are depicted in Fig. 7.17 and a more detailed velocity curve is presented in Fig. 7.18.

(i) By looking at the colour curve choose two times when the colour index has the same value.

(ii) From the luminosity curve, measure the magnitude of the star at the chosen times and evaluate the change in magnitude.

(iii) Determine (N) the ratio of the luminosities at the chosen times.

(iv) Determine (M) the difference of radii at the chosen times by measuring the area under the velocity curve between the chosen times.
(N.B. An *area* may be *negative* if V happens to be *negative*.) To a good approximation this area can be determined by subdividing the time scale into many small equal segments and summing up the areas of the rectangles whose bases are the segments and whose heights are the velocities at the middle of the segments.
Taken from the plot, the units of area are (km s^{-1} × days) and this is easily reduced to km by multiplying the measured value by the number of seconds contained in one day.

(v) Evaluate the radius of η Aquilae at the chosen times in terms of the Sun's radius ($R_\odot = 7{\cdot}0 \times 10^5$ km).

Fig. 7.17 Radial velocity, colour and light curves of η Aquilae

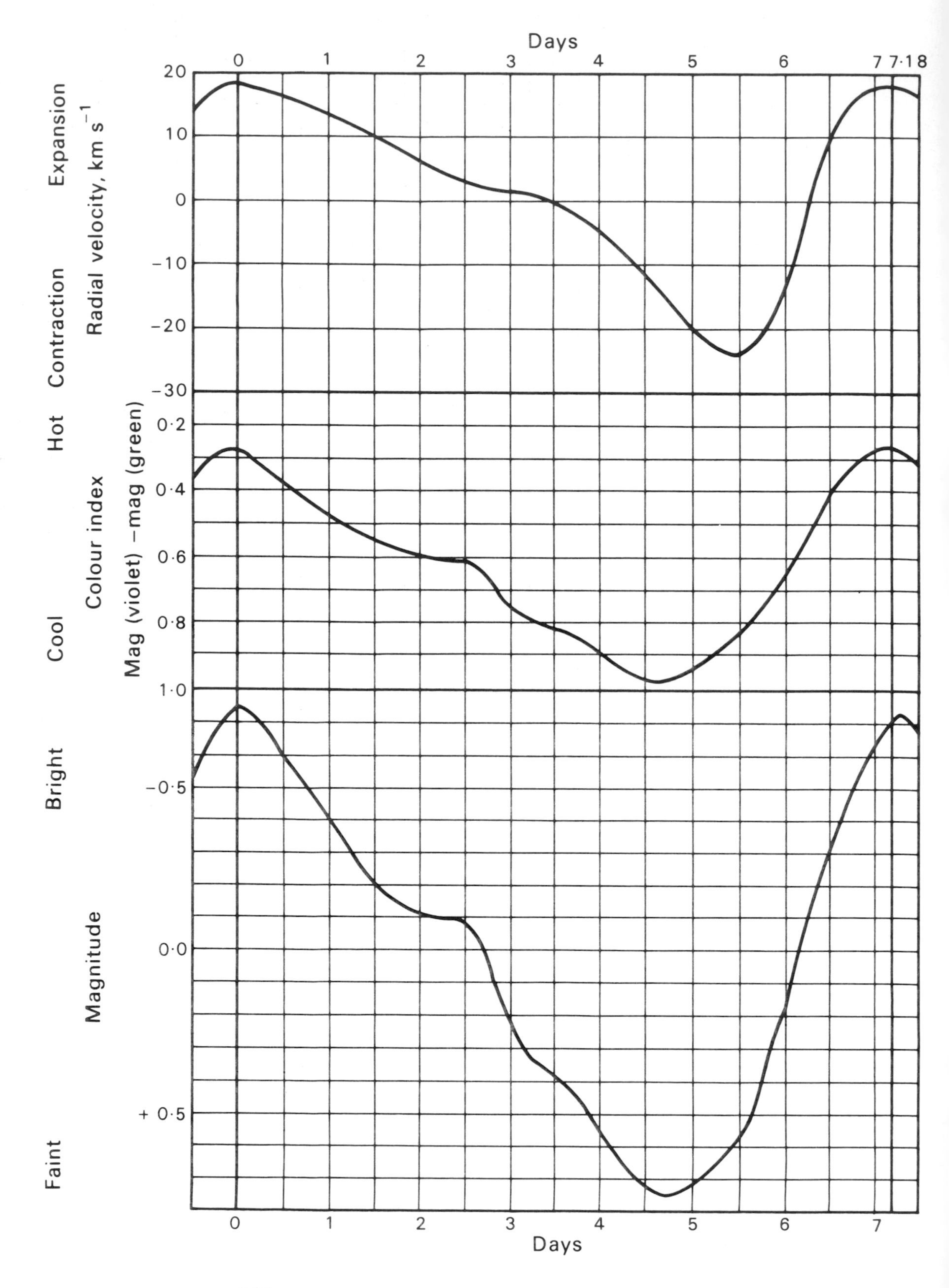

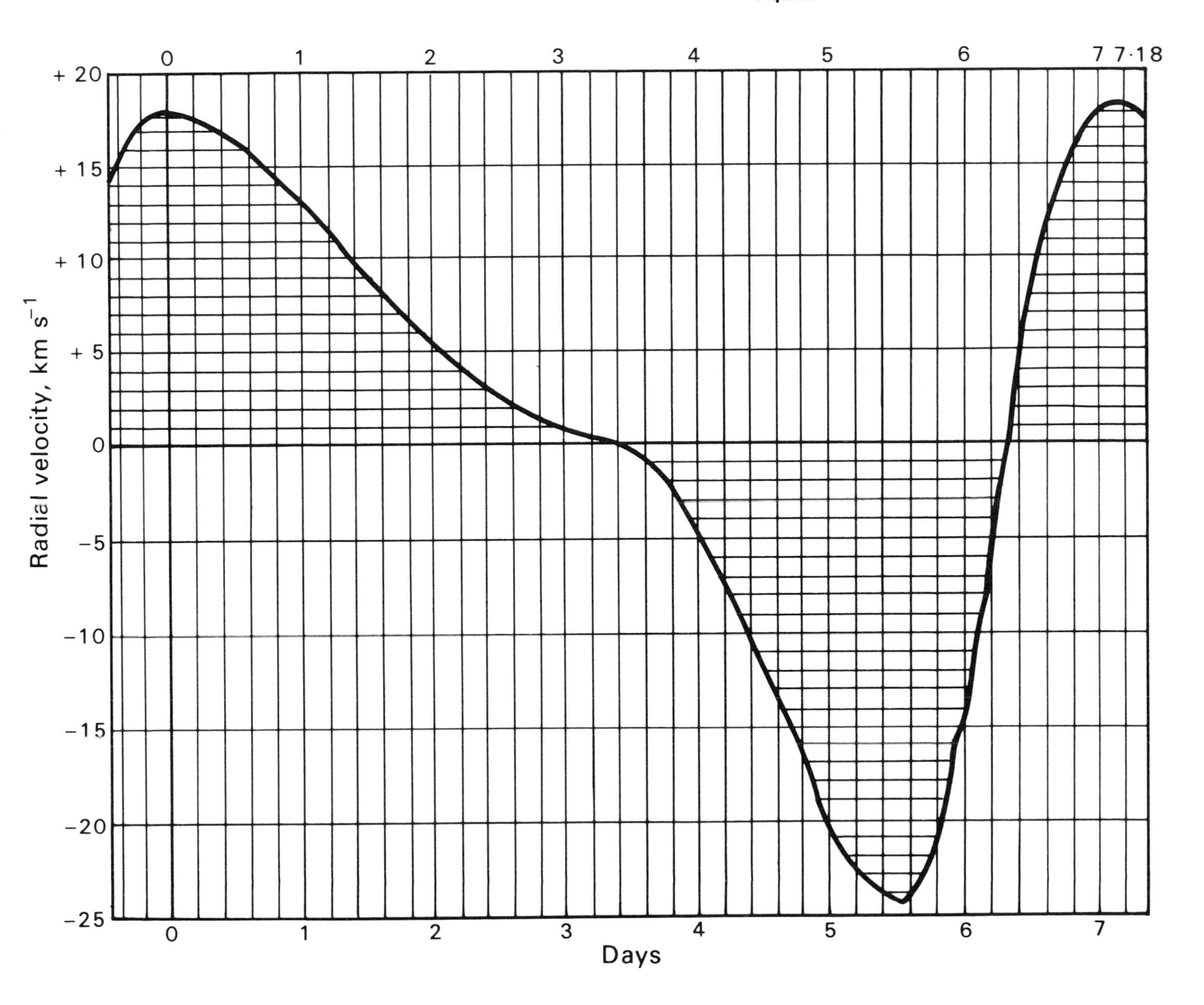

Fig. 7.18 The radial velocity curve of η Aquilae

7.11 Postscript

This chapter has discussed some of the methods by which we may deduce stellar masses, radii and luminosities and in particular the empirical relationships of the HR diagram and the mass–luminosity law. From these basic data we must now turn to their interpretation.

In the following chapter (Chapter 8) the important points concerning the theoretical studies of stellar atmospheres and the reason for the spectral sequence will be discussed.

Chapter 9 investigates stellar interiors, enabling the distribution of the stars on the HR diagram to be understood.

Problems: Chapter 7

1. If the absolute magnitude of a certain type of star is known to be $+0^{m}5$ and such a star is observed with an apparent magnitude of $+16^{m}0$, what is the distance of the star?

2. Determine the absolute magnitudes of the three stars whose apparent magnitudes m and parallaxes P are given:

	m	P
(i)	4·30	0″161
(ii)	5·04	0·011
(iii)	1·70	0·014

3. Determine the ratio of the luminosities (L_A/L_B) of the three pairs of stars whose apparent bolometric magnitudes and parallaxes have been measured:

		m	P		m	P
(i)	Star A	4·82	0″011;	star B	4·10	0″022
(ii)	Star A	4·32	0·108;	star B	2·84	0·027
(iii)	Star A	4·42	0·003;	star B	6·24	0·013

4. The Sun has an absolute bolometric magnitude of 4·75. Taking its luminosity to be unity, determine the luminosities of the three stars listed in Problem 2 above, assuming that the magnitudes given are on the bolometric scale.

5. Assuming that the stars listed in Problem 2 are on the bolometric scale and have the same effective temperature as the Sun whose absolute bolometric magnitude is 4·75 and whose radius is 696 000 km, calculate the radii of the three stars.

6. Evaluate the radii of the three stars below in terms of the solar radius assuming that the effective temperature of the Sun is 6000 K and its absolute bolometric magnitude is 4·75.

	Effective temperature (K)	Absolute bolometric magnitude
(i)	4750	4·00
(ii)	6500	1·25
(iii)	5500	−0·05

7. Two stars, A and B, are found to be brightest at wavelengths 6500 Å and 4000 Å respectively. Given that the radius of A is twice that of B, find the difference in their absolute magnitudes using Wien's law and Stefan's law.

8. Find the masses of two stars in a binary system whose parallax is 0″075 and in which the apparent orbits of the stars are circular, with radii 4″5 and 1″5 relative to the centre of gravity of the system and are described with uniform motion with a period of 300 years.

9. Calculate a first approximation to the parallax of the binary system 70 Ophiuchi for which the angular semi-major axis is 4″50 and the period is 87·7 years.

10. The binary star *Capella* has a total magnitude of $0^{m}21$ and the two components differ in magnitude by $0^{m}5$. The parallax of *Capella* is 0″063; calculate the absolute magnitudes of the two components.

11. An eclipsing binary consists of a red giant and a main sequence star of earlier spectral type with apparent magnitudes $7^{m}1$ and $7^{m}8$ respectively. Assuming that the radius of the giant is 8 times that of the main sequence star, calculate the apparent magnitude of the entire system if observed unresolved when:
    (*a*) both stars are fully visible,
    (*b*) the main sequence star is totally eclipsed,
    (*c*) the main sequence star lies between the observer and the giant.

12. The two components of a binary star are of approximately equal brightness. Their maximum separation is 1″3 and the period is 50·2 years. The composite spectrum shows double lines with a maximum separation of 0·18 Å at 5000 Å. Assuming that the plane of the orbit contains the line of sight, calculate (i) the total mass of the system in terms of the solar mass, (ii) the parallax of the system.

13. In a visual binary, the difference in the apparent magnitudes of the components is 2 magnitudes, the hotter component being the brighter. If the surface temperatures are 12 000 K and 6000 K, calculate the ratio of the hotter star's radius to that of the cooler star.

14. The distance of Jupiter from the Sun is 5·2 AU and its apparent magnitude at opposition is $-2^{m}3$. Calculate its apparent magnitude as seen at its greatest separation from the Sun by an observer near the star α Centauri (parallax $0''758$). (Neglect the brightness variation with scattering angle.)

15. An eclipsing binary has a constant apparent magnitude $4^{m}35$ between minima and apparent magnitude $6^{m}82$ at primary minimum. Assuming that the eclipse is total at primary minimum, calculate the magnitudes and the relative brightness of the components.

16. A triple star consists of three components A, B and C; A is three times brighter than B and B four times brighter than C. If the apparent bolometric magnitude of A is $3^{m}45$, find the magnitudes of B and C and of the system as a whole.
    If the effective temperatures of A, B and C are 3000 K, 4500 K and 6000 K respectively, find the radii of A and B in terms of the Sun's radius as unit, given that the radius of C is equal to the Sun's radius.

8 Stellar Atmospheres

8.1 Introduction

The nature of the physical structure of the stars, and of the Sun in particular, constituted a problem that fascinated mankind from the dawn of astronomy. With Kirchhoff's discoveries about spectra, a great step forward in understanding was achieved. It became known that a star could be regarded as a very hot and incandescent gas or solid under a cooler gaseous atmosphere. The substance of the outer parts of the stars at least was made up of the elements already familiar to terrestrial physicists and chemists.

We have seen in the previous chapter that stellar spectra exhibit a wide variety of forms. Some stars exhibit relatively simple spectra with only a few chemical elements apparently controlling the pattern. Other spectra display very complex patterns with many chemical elements effecting their contribution. We have seen that it is the range of stellar temperatures which influences, at least to some major degree, the appearance and the strengths of the absorption lines; the spectral classification scheme reflects the range of stellar surface temperatures.

Questions that might be asked are:

(i) How does the surface temperature of a star influence the strengths of the spectral absorption features in the way that it does?
(ii) How do the strengths of the absorption features reflect the abundance in the stellar atmosphere of the various chemical elements?

To answer these questions, we need to set up models of stellar atmospheres so that we can understand the processes going on and their interplay controlling both the overall appearance of the spectrum and the individual shapes of the absorption lines.

The three main quantities to be considered in any theory of stellar atmospheres are temperature, chemical composition and density. We consider these in turn.

8.2 Temperature

The black-body radiation law was formulated by Planck (see Roy & Clarke, *op. cit.*, Chapter 14). His law allows prediction of spectral energy distribution curves to be made for black bodies of various absolute temperatures. An important property of such curves was Wien's law stating that the wavelength where the maximum energy is radiated is given by

$$\lambda_{\max} T = \text{constant} = 2{\cdot}90 \times 10^{-3} \text{ m K}.$$

It was also noted that stellar spectral distributions corresponded closely to black-body radiation, owing to the high absorbing power of stellar material. Stellar atmospheres would therefore absorb all radiation falling on them and hence act like black bodies. For example, the high opacity in the case of the Sun is evident from the sharpness of its disk, the depth of optical penetration into its interior being of the order of 100 km, in comparison with the solar radius which is of the order of 700 000 km.

The discussion of the practical determination of the surface temperature of a star (see Roy & Clarke *op. cit.*, Chapter 21) pointed out that difficulties arose owing to the fact that for the higher temperature stars, λ_{max} was outside the range of wavelengths available to ground-based astronomers. By using two-colour photometry to derive the colour index, application of the Planck formula leads to an absolute temperature for the surface of the star and therefore the temperature of its atmosphere.

8.3 Chemical Composition

It will be recalled that it was in 1859 that Kirchhoff and Bunsen first identified the Fraunhofer D lines in a spectrum as being due to sodium. Five years later Huggins began the first systematic study of the lines found in stellar spectra.

The stellar spectral sequence for classifying stars described in the previous chapter follows a progression according to temperature. The stars of lowest surface temperature have spectra showing bands due to molecular absorption. As the temperature increases along the sequence, hydrogen lines become more prominent, becoming strongest at a temperature of 10 000 K.

Recapitulating Kirchhoff's work, he showed that

(I) A hot solid (or dense gas) gives a continuous bright spectrum.
(II) A hot tenuous gas gives a spectrum of bright lines with no continuous background.
(III) Light from a hot solid (or dense gas) that has passed through a cooler tenuous gas gives a spectrum with dark lines on a continuous bright background, the pattern of dark lines being the same pattern of bright lines emitted from the tenuous gas when it is observed alone.

The model of a star inferred from the above laws is that of a sphere of hot dense gas or a hot solid sphere surrounded by a cooler tenuous atmosphere. Kirchhoff and Bunsen showed that the lines found in the solar spectrum coincided with those due to familiar elements such as oxygen, hydrogen, sodium, calcium, and so on.

Further progress was made on the theoretical side when Bohr put forward his theory of the atom. Making use of quantum mechanics, he was able to show that spectral lines, both in emission and absorption spectra, could be explained by supposing that the electrons in atoms could exist only in discrete energy levels or orbits of certain size. Only when an electron passed from one level to another did it emit or absorb electromagnetic radiation. The frequency ν of the radiation of energy ΔE emitted or absorbed in such a transition was given by

$$\Delta E = h\nu,$$

where h was Planck's constant.

Thus Bohr's theory explained absorption or emission in discrete frequencies or wavelengths. The amount of energy ΔE was the difference in energy between the energy levels occupied by the electron before and after the transition. It will be recalled that the lowest energy level (the orbit nearest the atomic nucleus) is referred to as the **ground state**; all others are **excited states**.

If we refer this now to stellar atmospheres, it is found that absorption can occur in three ways:

(*a*) Low density atmospheres with atoms in the ground state. Absorption of energy occurs, putting the absorbing atoms into the first excited state with re-emission of the radiation at the same wavelength, but in all directions, when they return to ground state. This is known as **pure scattering** (see Fig. 8.1(*a*)).
(*b*) Low density atmospheres with atoms in excited states. Absorption is followed by re-emission partly at a different wavelength (see Fig. 8.1(*b*)).

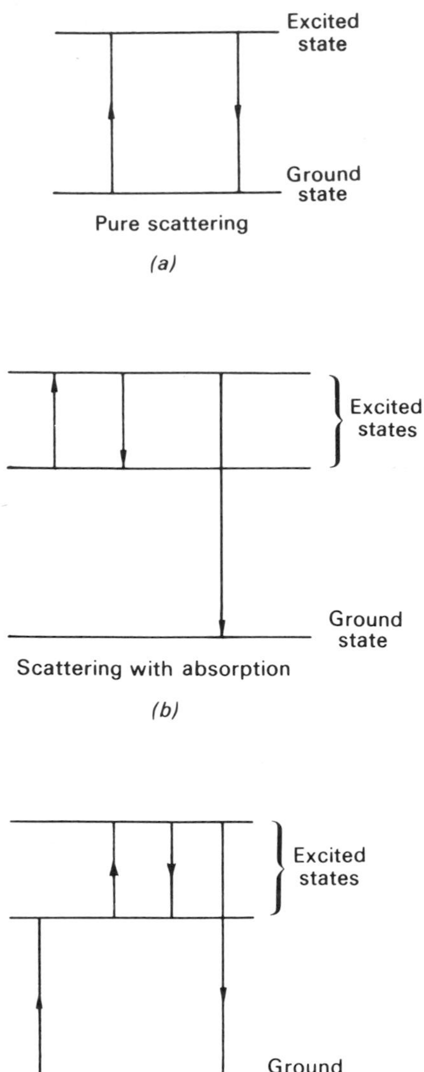

Fig. 8.1 Different modes of absorption in stellar atmospheres

(*c*) High density atmospheres in which collisions are likely to occur between atoms before they have had a chance to re-emit radiation by downward transitions. In this case absorption is followed by further excitation to higher energy levels by collisions. Re-emission occurs at other wavelengths entirely. This is known as **pure absorption** (see Fig. 8.1(*c*)).

Although all three processes are important in stellar atmospheres, we will restrict ourselves to a consideration of pure absorption only.

In a stellar atmosphere, the distribution of atoms and ions in the various energy levels is determined by the temperature and density. Before proceeding further, we have to consider various concepts advanced in the early part of the 20th century.

8.4 The Photo-Absorption Cross-Section of an Atom

If radiation is made to pass through cool gaseous material, the amount of absorption will depend on the ability of the atoms to absorb and on the number of atoms through which the radiation passes. Let us first consider the way in which the absorption ability of an atom is expressed.

In Fig. 8.2, suppose A is an atom placed in a beam of radiation strength F J s^{-1} m^{-2} in the wavelength of the spectral line produced by a gas made up of such atoms.

Let R be the rate of absorption by the atom in J s^{-1}. Now

$$R \propto F,$$

so that

$$R = \sigma F,$$

where σ is a constant with the dimensions of area. The actual value of σ will depend upon the structure of the atom. We may say that from the point of view of the atom's power to absorb radiation, it may be looked upon as an opaque sphere of cross-section σ. Such a sphere would intercept radiation at the same rate as the atom, namely at a rate σF. Hence σ is referred to as the **photo-absorption cross-section** of the atom.

In the case of the hydrogen atom, σ can be found from a simple formula. For helium the calculations are not exact but can be supplemented from experiment. Values can also be found for other elements from laboratory experiments.

The size of σ, which is known for any element, is therefore a factor in producing an absorption line in the spectrum of the stellar atmosphere.

It will be recalled that the second factor is the number of atoms of that element present, per cubic metre, in the atmosphere.

8.5 Model Stellar Atmospheres

Let us now consider the absorption spectrum sketched in Fig. 8.3(*a*). The B_λ axis shows the brightness of the spectrum at various wavelengths λ. The continuous spectrum has dips in it, these dips representing the absorption lines. Some, such as L_1 and L_3 are strong; others such as L_2 and L_4 are weak. The shape of a line is called the **line profile**. Let us take any particular line, say L_5, and consider the spectral brightness in its vicinity. A more detailed shape of the line profile of L_5 is given in Fig. 8.3(*b*).

Absorption lines appear in the continuum when the various atoms in the cooler atmosphere of the star absorb radiation by orbital transitions of electrons within the atoms. It is obvious that the fraction of the energy which is absorbed from the continuum by an atmosphere by a particular transition is equal to the area of the line profile; the broader and deeper the profile, the more energy has been absorbed.

Now the actual recorded shape of the line which is presented in a spectrum depends to some extent on the quality of the spectrometer (i.e. on its instrumental profile). One with good resolving power gives line profiles which are deep and sharp; one with lower resolving power presents the same lines but with shallower and broader profiles. However, the area of any line in the light from a given star is hardly dependent on the particular spectrometer. Because of this, the strength of a spectral line is noted *not* by its recorded depth, but by its area.

For convenience, the term **equivalent width** is used to describe the strength of an absorption line. This is defined as the width that a rectangular spectral line would have if it had the same area as the observed line, but with a depth complete from the continuum to zero brightness (see Fig. 8.3(b)). Any measured equivalent width of a particular line in the light from a given star is practically independent of the spectrometer that is used.

Now the equivalent width of any line can be measured directly from any spectrum. We know also that in theory its value depends upon the absorbing power of the individual atoms, that is, on the photo-absorption cross-section σ of the atoms involved, and on the number of atoms per cubic metre of the element producing the line. If the line, for example, is due to hydrogen, then in principle it is possible from the measured value of the equivalent width and a knowledge of the photo-absorption cross-section σ for hydrogen to obtain from theory a value for the number of absorbing hydrogen atoms per cubic metre present in the stellar atmosphere.

Idealized model atmospheres have been studied, taking advantage of these facts. For example, the model atmosphere introduced by Schuster in 1905 consisted of a plane transparent gaseous atmosphere at uniform temperature on top of a solid black-body surface at the same temperature. Radiant energy enters the atmosphere from the bottom, is selectively absorbed by the atoms within it, and leaves at the top. If the emerging radiation is then spectrometrically examined, an absorption spectrum is observed. It is instructive to consider the Schuster model in a little detail.

Suppose that the plane transparent atmosphere has a height H, that it has no temperature change through its depth and that it is superincumbent upon a solid black-body surface at the same temperature.

Let F_0 be the upward flow of energy into the atmosphere per unit area of the black-body surface in a wavelength range whose boundaries just extend beyond the limits of the line profile. (F_0 is in units of J s^{-1} m^{-2}.)

Similarly let $F(h)$ be the flow of energy at a height h and F_H the emergent energy flow. Suppose that at a height h, the number of absorbing atoms per cubic metre is $n(h)$.

Consider the energy flow through a slab of unit area at a height h and with thickness dh. The energy flowing into the slab across the lower face is $F(h)$ and the energy leaving the slab from across the upper face is $F(h+\mathrm{d}h)$. The rate of loss of energy from the beam is $F(h)-F(h+\mathrm{d}h)$.

Now the rate of absorption in the slab is equal to the energy flow multiplied by the total projected photo-absorption cross-section provided by the absorbing atoms, i.e. it equals $\sigma F(h)n(h)\,\mathrm{d}h$. Then

$$F(h)-F(h+\mathrm{d}h)=\sigma F(h)n(h)\,\mathrm{d}h$$

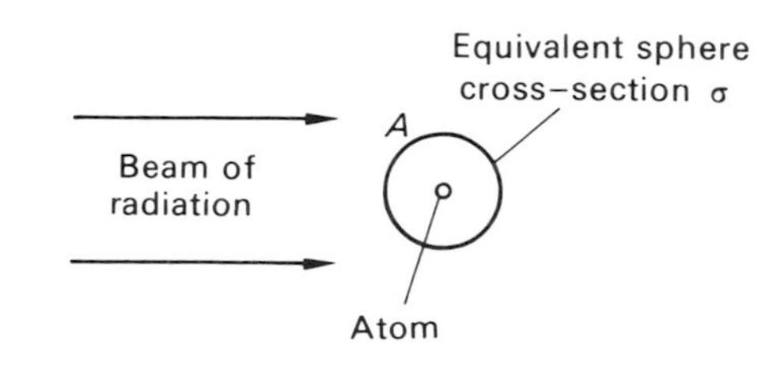

Fig. 8.2 The photo-absorption cross section of an atom

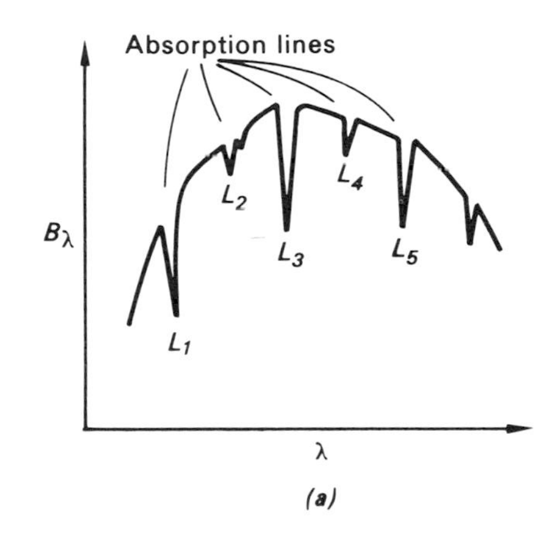

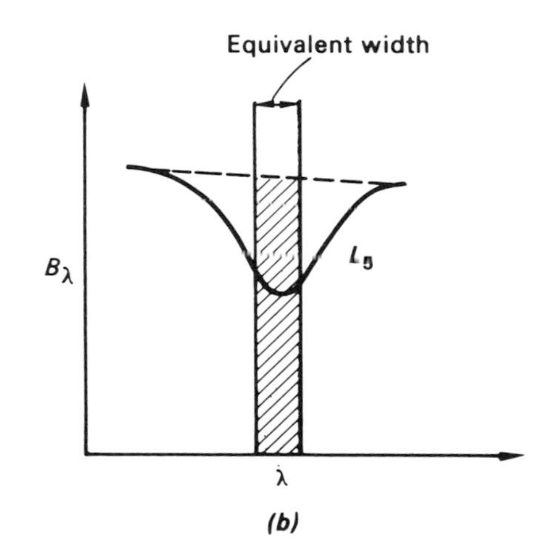

Fig. 8.3 (*a*) A star's energy envelope exhibiting absorption lines, (*b*) An exploded view of the line profile of L_5 demonstrating its equivalent width

therefore

$$\mathrm{d}F/\mathrm{d}h = -\sigma F(h) n(h)$$

thus

$$\frac{1}{F(h)} \frac{\mathrm{d}F}{\mathrm{d}h} = -\sigma n(h).$$

By integrating this equation with respect to h, from zero to H, i.e. from the bottom of the atmosphere to the top, we have that

$$[\log_e F(h)]_0^H = -\sigma \int_0^H n(h)\,\mathrm{d}h.$$

Now $F(0) = F_0$ and $F(h) = F_H$ and by letting W be the ratio of the equivalent width of the line to the wavelength range used to describe F_0,

$$F_H = F_0(1 - W).$$

Hence,

$$\log_e (1 - W) = -\sigma N_0$$

where N_0 is the total number of absorbing atoms contained in an atmospheric column of unit area cross-section, i.e.

$$N_0 = \int_0^H n(h)\,\mathrm{d}h.$$

Thus

$$N_0 = \frac{1}{\sigma} \log_e \frac{1}{1 - W}. \tag{8.1}$$

To a first order, measurement of the equivalent width of a line and knowledge of the photo-absorption cross-section allows the determination of the number of absorbing atoms in a column of a stellar atmosphere.

In order to convert this value into the actual number, N_T, of atoms of a given element in the stellar atmosphere, we need to know what fraction of N_T are in the required state for absorbing energy appropriate to the generation of the spectral line. The measured line results from a group of neutral atoms, or maybe ionized atoms, having an electron in a particular energy level; the number of atoms in this state is obviously lower than N_T. To proceed requires the application of two laws, the first established by Saha and the second by Boltzmann.

8.6 Saha's Equation

For the sake of argument, let us consider an absorption line which results from the neutral atom of a given element. Because of the temperature of the gas in the atmosphere, some of the atoms will be ionized and will therefore be unavailable to take part in the production of the given line. The derivation of the fraction of atoms that is ionized is based on the laws of chemical equilibrium and is known as **Saha's equation**. In an atmosphere which is considered to be in a steady state, the average rate at which atoms are ionized equals the average rate at which recombination takes place. The process may be written in a similar way to that of a chemical equation:

$$A \rightleftarrows A^+ + e,$$

where A and A^+ represent the atom and its ionized form respectively and e represents an electron.

Suppose that x represents the fraction of the total number of atoms that are ionized and n^+ the number of ionized atoms per unit volume (m^{-3}). We may write this as

$$n^+ = xn_T.$$

The number of atoms which are not ionized obviously equals $(1-x)n_T$.

Now the number of atoms ionized per second is proportional to the number available for ionization and is also related to the temperature. Thus the number ionized per second may be written as

$$(1-x)n_T f(T).$$

The number of ions recombining is proportional to both the number of ions and the number of electrons and is again also related to the temperature. Hence the number of recombinations per second may be written as $n^+ n_e g(T)$, where n_e is the electron density.

By equating the rate at which atoms are ionized and the rate of recombination we have that

$$(1-x)n_T f(T) = n^+ n_e g(T) = xn_T n_e g(T)$$

therefore

$$\frac{xn_e}{1-x} = \frac{f(T)}{g(T)} = F(T).$$

This formula, allowing x to be determined from knowledge of n_e and $F(T)$, which for the Schuster model is assumed to be constant with atmospheric depth, is known as Saha's equation. The function $F(T)$ depends on the atom under consideration and in the case of the hydrogen atom, the function is

$$F(T) = \frac{(2\pi mkT)^{3/2}}{h^3} \exp\left(-\frac{V}{kT}\right) \tag{8.2}$$

where k is Boltzmann's constant, h is Planck's constant, m is the mass of the electron and V is the minimum energy required to detach from the atom an electron initially in the ground state (13·6 electron volts for the hydrogen atom).

There are similar formulae for other atoms, with different values of V. Some examples are given in Table 8.1.

Table 8.1

Element	V (electron volts)
Sodium	5·1
Calcium	6·1
Iron	7·9
Hydrogen	13·6
Helium	24·6

Fig. 8.4 shows the behaviour of x for calcium, iron and hydrogen at various temperatures for a value of the number of free electrons typically found in stellar atmospheres.

Thus as the temperature rises, the proportion of ionized atoms grows from zero until by a certain temperature few if any atoms are un-ionized, the value of x reaching unity.

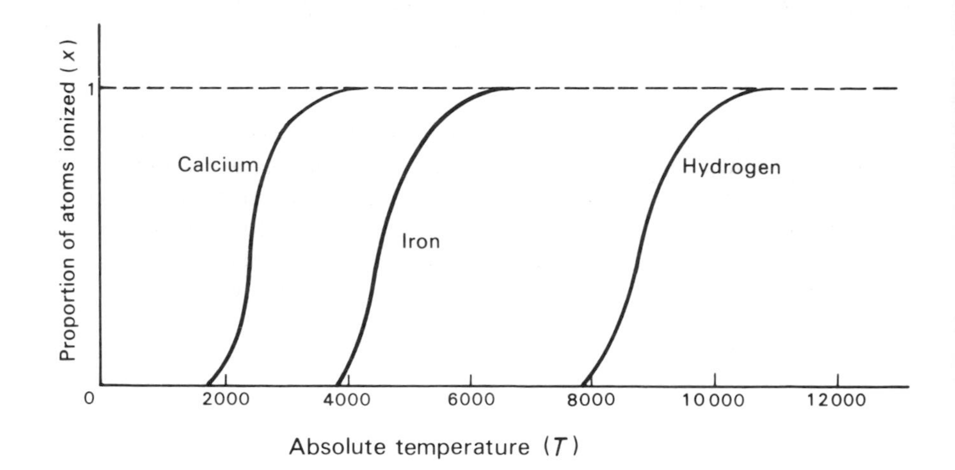

Fig. 8.4 The variation of the proportion of atoms ionized with temperature in typical stellar atmospheres for calcium, iron and hydrogen

Saha's ideas can be made more familiar to us by considering a closed container with some liquid in it. The liquid begins to evaporate, more and more molecules entering the space above the liquid surface until the vapour pressure above the surface becomes so high that just as many molecules are returning to the liquid as are leaving it. The system is then said to be in equilibrium or to have reached a steady state. If, however, the temperature of the liquid is raised, evaporation begins again; the vapour pressure will rise before a new steady state is reached. Again, it is well known that different kinds of liquid, such as water and alcohol, have different degrees of volatility so that the evaporation rate will differ for two liquids, even if they are at the same temperature.

If we now look at the ionization of atoms of different substances in a stellar atmosphere we see that the rate of detachment of electrons from the atoms will depend upon (i) the temperature, (ii) the electron gas pressure and therefore the availability of electrons to recombine with ionized atoms, and (iii) the ease with which electrons can be detached from atoms of a given element.

Application of Saha's equation is not, however, the full procedure required to calculate the total number of atoms of a given kind. Of the atoms that are not ionized, there is a range of excitation and only a fraction of their number will be at an energy level which by further absorption of energy gives rise to the particular absorption line.

Suppose that the number of atoms in an energy level, a, is n_a; this value is a fraction of the total available number $(1-x)n_T$ and may be expressed as

$$n_a = q_a(1-x)n_T.$$

Obviously, the range of values that q takes depends on the temperature in the atmosphere. For the lowest energy level where $a = 1$, q_1 decreases with increasing temperature, while the values of q for the higher energy levels increase.

Where there is a condition of thermal equilibrium, i.e. the average number of atoms in any energy level remains constant with time, we may apply Boltzmann's equation (from statistical mechanics) giving

$$\frac{q_a}{q_b} = \frac{g_a}{g_b}\exp\left(\frac{E_b - E_a}{kT}\right),$$

where $(E_b - E_a)$ is the difference of energy between the two levels and g_a and g_b are the statistical weights of the two energy levels. The need to introduce these last two factors comes about because of the role angular momentum plays on the energy levels. Any atom may take an energy level in a different number of ways. This multiplicity of the level is equivalent to there being more atoms of a given kind which take a share of the available energy. Hence, we may write

$$n_T = \frac{n_a}{q_a(1-x)}.$$

Summarizing the reduction, an equivalent width is measured to determine the number of absorbing atoms n_a and then corrections are applied using Saha's equation and Boltzmann's equation to evaluate the total number of atoms, n_T, of a given kind. The scheme may be applied to every absorption line and, by an extension of the argument, to lines generated by ionized atoms.

The Schuster model has now been replaced by more realistic and more complicated models for which the radiation is considered to be generated at a range of levels in the atmosphere which in turn is assumed to have a temperature gradient.

8.7 Measured Abundances of Normal Stellar Atmospheres

When the abundance determinations are carried out for normal stars in the galactic plane in the solar neighbourhood, the results tabulated in Table 8.2 are obtained, where N is the number of atoms of a particular element present per square metre of atmosphere.

Table 8.2 Abundance of elements in stellar atmosphere

Element	$\log_{10}N$	Approximate percentage by mass
Hydrogen	28·2	70
Helium	27·2	27
Carbon, Nitrogen, Oxygen, Neon	25·3	2
Sodium, Magnesium, Aluminium, Silicon, Calcium, Iron	24·4	1
Remainder		1

In certain high temperature stars, a lower hydrogen abundance is found. In globular cluster stars (discussed later) there is in general a lower abundance of the heavier elements.

8.8 Densities in Stellar Atmospheres

These can be obtained by considering the pressure within the atmosphere. Suppose P is the total gas pressure at the foot of the atmosphere, measured in newton m^{-2}. The upwards force due to this pressure acting vertically upwards over an area of one square metre is P newtons. This must balance the weight of the column of gas, of mass M kg on top of this area.

If g is the acceleration due to gravity within the stellar atmosphere (taken to be constant throughout since the thickness of the atmosphere is small compared to the stellar radius), we may write

$$P = Mg.$$

It is possible to calculate M from a knowledge of the numbers of atoms of the various elements forming the atmosphere and the preponderance of hydrogen. In the Sun's atmosphere, for example, it is found that M is about 37 kg m^{-2}, g is $2{\cdot}74 \times 10^2$ m s^{-2}, giving a value for P of about 10^4 newton m^{-2}.

Also, by the perfect gas law,

$$P = nkT,$$

where n is the total number of particles per cubic metre. Hence

$$n = P/kT = Mg/kT.$$

The density of the individual elements (that is, the number of atoms of each present per cubic metre) is then obtained from n by applying the known relative abundances by number (Table 8.2).

It is to be noted that for a given temperature T,

$$n \propto g.$$

Hence, assuming a constant relative abundance of free electrons, we can also say that

$$n_e \propto g,$$

where n_e is the electron density.

8.9 Spectrum Classification

Consider an assembly of stars, all with the same chemical composition. The line strengths in their absorption spectra depend therefore on two atmospheric parameters, namely the absolute temperature, T, and the acceleration due to gravity, g. The second parameter is required because of the appearance of n_e in Saha's equation.

The dependence of the line strengths on T and g is best demonstrated by considering the effect of varying T and g independently:

(*a*) Let g and therefore n_e be constant throughout the assembly of stars but let each stellar atmosphere have a different absolute temperature T.

Consider the strength of a given line in the different stars. It has been shown that

$$n_a = n_T q_a (1 - x) \tag{8.3}$$

where

(i) n_a is the number of absorbing atoms in energy level a existing in unit volume,

(ii) n_T is the total number of particles of the element (neutral atoms + ions) in that volume,

(iii) $q_a(1-x)$ is the proportion factor relating these two numbers.

In this last product, q_a is the proportion of atoms in energy level a, while $(1-x)$ is the proportion of particles that are un-ionized—that is, neutral.

Now n_a is derived from the measured line strength, q_a is calculated from quantum mechanical theory, while $(1-x)$ can be found in Saha's equation since n_e is a constant throughout the sequence of stars.

But n_T is also constant in the sequence by hypothesis, so that

$$n_a \propto q_a(1-x).$$

If we take as an example the lines in the Balmer series of hydrogen for $a=2$, we find that $q_2(1-x)$ increases very quickly from 6000 K to 10 000 K. Thereafter it diminishes as the temperature grows higher. The consequence of this is that although hydrogen predominates in abundance, it only produces its absorption line patterns at their maximum intensity in the A-type stars where the temperatures are of the order of 10 000 K.

For atoms of other elements, the maximum occurs at different temperatures, namely at lower values of T for lower values of V, the ionization potential. In Table 8.3, some examples are given.

Table 8.3

Atom	Temperature at which lines are at maximum strength (K)	Spectral type
Singly-ionized helium	35 000	O
Helium	14 000	B
Hydrogen	10 000	A
Iron	7 000	F
Singly-ionized calcium	5 500	G
Sodium	4 000	K
(Molecules)	3 000	M, R, N, S

(*b*) We now consider the effect of decreasing the electron density n_e in our assembly of stars while keeping the atmospheric temperature T constant. Decreasing n_e is, as we have seen, a consequence of decreasing g from star to star (T being constant).

Now by Saha's equation,

$$xn_e/(1-x)=F(T).$$

But T being constant, the right-hand side is constant. It is easy to see, therefore, that under such conditions a decrease in value of n_e must result in an increase in x. This in turn means that the ionization curve of any given element is shifted to a lower temperature throughout. The temperature at which the spectral lines due to that element are at their strongest is consequently lowered.

This feature has been used as a means of discriminating between giant and dwarf stars (see Section 7.8) where there are differences in surface gravity between the two classes.

In the light of the ideas presented in this chapter, the student will find it profitable to re-read the section on the spectral sequence, viz. Section 7.7.

9
Stellar Interiors

9.1 Introduction

The problem of investigating the internal structure of a star can be illustrated by the analogy of the "black box". Suppose that a black box containing machinery of some kind is presented to a scientist. He is not allowed to open it to inspect the machinery but he is allowed to examine any output from the box. The output may consist of rods that move mechanically in and out of apertures in the box sides or power that can be tapped from terminals inserted in the box. From his examination, the scientist must construct a plausible theory concerning the machinery in the box. He can never be certain that his deductions are correct but if his theoretical black box behaves exactly like the real one, he may be confident that he is on the right track. If indeed his model predicts certain effects that he had not hitherto discovered in the real black box output and he subsequently confirms that they exist, his confidence is justifiably increased.

A star can be looked upon as an example of a black box problem. In general we know certain facts about it. From these facts, its internal structure must be deduced. These data have already been presented in Chapter 7.

9.2 Stellar Data

For many stars the mass M, radius R and luminosity L are known. We have seen that the masses can be derived from a study of binary systems. Radii may sometimes be deduced from eclipsing binary systems. The luminosity L (the total energy radiated per second from the star's surface) can be found from the deduced absolute bolometric magnitude M_{bol}.

Two important properties giving an insight into the structure and evolution of stars can be derived from the above quantities.

The mean density of the star, $\bar{\rho}$, is given by

$$\bar{\rho} = \frac{M}{\frac{4}{3}\pi R^3}.$$

The mean energy generation rate per unit mass, $\bar{\varepsilon}$, is defined by

$$\bar{\varepsilon} = L/M.$$

For example, for the Sun

$$M_{\odot} = 2 \times 10^{30}\ \text{kg},$$

$$R_{\odot} = 7 \times 10^{8}\ \text{m},$$

$$L_{\odot} = 4 \times 10^{26}\ \text{W},$$

$$\bar{\rho} = 1{\cdot}4 \times 10^{3}\ \text{kg m}^{-3},$$

$$\bar{\varepsilon} = 2 \times 10^{-4}\ \text{W kg}^{-1}.$$

It should be noted that for the Sun we have one additional piece of information. Geological records show that complicated forms of life have existed on the Earth for around 10^9 years. It is unlikely that such life-forms could have survived if the Sun's energy output had changed substantially within that period of time. Thus, although the Sun is not in equilibrium, as there is an enormous outward flow of energy, the long constancy of its output allows us to assume that a state of *quasi-equilibrium* exists. States of quasi-equilibrium can be considered for most of the normal stars. Only by making this assumption are we able to determine the internal conditions of stars.

Assembling the available data, and expressing the mass M, radius R and luminosity L in solar units $M_\odot$, $R_\odot$ and $L_\odot$, it is found that the physical properties of most stars are included in the ranges given by

$$10^{-1} \leq M \leq 50,$$

$$5 \times 10^{-3} \leq R \leq 3 \times 10^{2},$$

$$10^{-3} \leq L \leq 2 \times 10^{5}.$$

It can be seen that the range in the mass is of the order of 10^3. On the other hand, stars can be as small as the Earth in size or so large that the orbit of Mars could lie within them—a range of 10^5. The largest range of all is found with respect to luminosity where some stars are of the order of 10^8 times as bright as others.

Examining the derived quantities $\bar{\rho}$ and $\bar{\varepsilon}$, it is found that

$$10^{-3} \text{ kg m}^{-3} \leq \bar{\rho} \leq 10^{11} \text{ kg m}^{-3},$$

while

$$2 \times 10^{-6} \text{ W kg}^{-1} \leq \bar{\varepsilon} \leq 2 \text{ W kg}^{-1}.$$

The lower end of the $\bar{\rho}$ range is about the density of a normal stellar atmosphere; the upper is such that a matchbox full of the stellar material would have a mass of 1000 tonnes.

The large range in mean energy generation per unit mass indicates that whatever the energy source is, it must be sensitive to physical conditions within the stars. It is also found that $\bar{\varepsilon}$ decreases along the main sequence of the HR diagram (see Section 7.9) from upper left to lower right.

In fact the HR diagram together with the mass–luminosity relationship (see Section 7.6) play key roles in solving the problem of stellar structure.

9.3 Stellar Models

Some of the questions that require answers in problems of stellar structure are:

(*a*) What is the significance of the groupings of stars in the HR diagram?
(*b*) Why should there be a mass–luminosity relation?
(*c*) What is the physical state of a stellar interior? Is it solid, liquid or gas?
(*d*) What is the mode of energy generation in a star?
(*e*) What was the mode of origin of the chemical elements?
(*f*) How does a star form?
(*g*) How old are the stars?
(*h*) What will be the subsequent evolution of the stars?

Consider a mass of gas isolated in space. It is a perfect gas obeying the perfect gas law. At first it will shrink under the gravitational attraction of the inner parts of the gas volume and by symmetry will take a spherical form. As its volume decreases, the gravitational energy is released in the form of heat and the temperature of the material rises. The gas pressure and density will also increase. Finally, as a state of quasi-equilibrium is reached, no further shrinking will occur because at any level within the gas sphere, the weight of gas above that level will now be supported by the upwards pressure at that level. The gas sphere is now said to be in a state of hydrostatic equilibrium.

Consider a star which is in equilibrium and consider a thin spherical shell of thickness dr at a distance r from the centre. In equating the forces acting on a slab of unit area of the shell, its curvature can be neglected (Fig. 9.1). Now

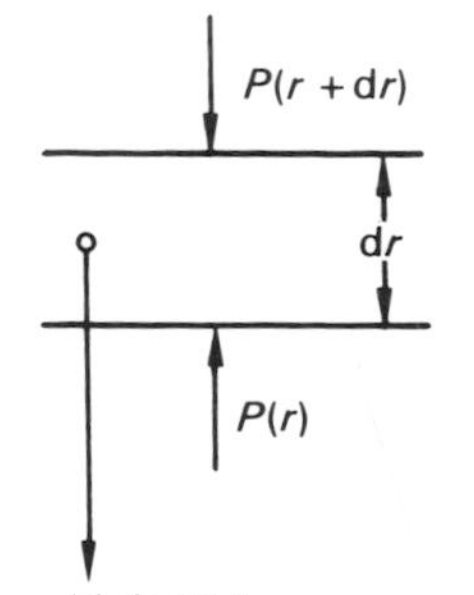

$$\begin{aligned}&\text{the force acting radially inward}\\&\quad=\text{the weight of the slab}+\text{the pressure on the upper face}\\&\quad=\rho(r)\,\mathrm{d}r\,g(r)+P(r+\mathrm{d}r)\end{aligned}$$

where $\rho(r)$ is the density of the material at the radius r, and $g(r)$ is the local acceleration due to gravity.

$$\begin{aligned}&\text{The force acting radially outward}\\&\quad=\text{the pressure on the lower face}\\&\quad=P(r).\end{aligned}$$

For equilibrium the inward force must be balanced by the outward force. Hence

$$\rho(r)\,\mathrm{d}r\,g(r)+P(r+\mathrm{d}r)=P(r)$$

therefore

$$\mathrm{d}P/\mathrm{d}r=-\rho(r)g(r). \tag{9.1}$$

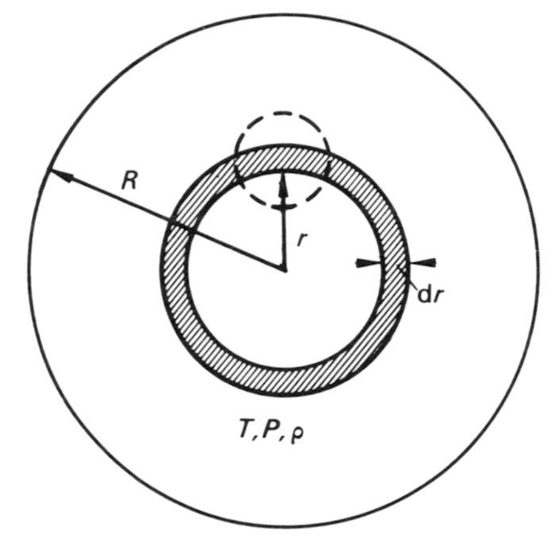

Fig. 9.1 A spherical shell of thickness d*r* inside a star

This relationship is known as the **equation of hydrostatic equilibrium** and describes the rate of change of pressure within a star with increasing distance from the star's centre in terms of the density and acceleration due to gravity.

Now, at any point within the sphere,

$$g(r)=GM_r/r^2,$$

where G is the universal constant of gravitation and M_r is the mass of gas within the volume of radius r. Hence,

$$\mathrm{d}P/\mathrm{d}r=-\rho(r)GM_r/r^2. \tag{9.2}$$

The mass of gas dM_r contained in a spherical shell of infinitesimal thickness dr is given by

$$\mathrm{d}M_r=4\pi r^2\times\mathrm{d}r\times\rho(r),$$

where $4\pi r^2$ is the area of the shell, dr is its thickness and $\rho(r)$ is its density. Hence

$$\mathrm{d}M_r/\mathrm{d}r=4\pi r^2\rho(r). \tag{9.3}$$

Equations (9.2) and (9.3) are obviously not sufficient to describe the stellar interior. We can, however, get some idea of the values of the parameters involved by making some very crude approximations. Consider Equation (9.2) at a point within the Sun midway between the centre and its surface, putting

$$\begin{aligned}r&=\text{half solar radius}=\tfrac{1}{2}R_\odot,\\\rho(\tfrac{1}{2}R_\odot)&=\text{determined mean solar density},\\M_{\frac{1}{2}R_\odot}&=\text{half solar mass}=\tfrac{1}{2}M_\odot,\\\mathrm{d}P&=P_{\text{centre}}-P_{\text{surface}}\approx P_{\text{centre}}\end{aligned}$$

and

$$\mathrm{d}r=\text{radius of Sun}=R_\odot.$$

We obtain

$$\begin{aligned}P_{\text{centre}}&\approx 2\bar{\rho}_\odot GM_\odot/R_\odot\\&\approx 6\times10^{14}\ \text{newton m}^{-2}\\&\approx 6\times10^{9}\ \text{atmospheres}.\end{aligned}$$

Now the highest pressures achieved in the laboratory—and these only for very short times—are of the order of 10^6 atmospheres. We are therefore dealing with matter in a state which cannot be investigated within the laboratory.

Further manipulation of Equations (9.2) and (9.3) allows derivation of a minimum value for the central pressure of a star whose mass and radius are known. What is perhaps remarkable is that such pressures may be evaluated without knowledge of the materials which constitute the stars.

By dividing Equation (9.2) by (9.3) we have

$$\frac{\mathrm{d}P}{\mathrm{d}r}\bigg/\frac{\mathrm{d}M_r}{\mathrm{d}r}=\frac{\mathrm{d}P}{\mathrm{d}M_r}=-\frac{GM_r}{4\pi r^4}.$$

If this equation is now integrated with respect to M between the centre of the star and its surface

$$P_{\text{centre}}-P_{\text{surface}}=\int_0^{M_{\text{surface}}}\frac{GM_r}{4\pi r^4}\,\mathrm{d}M_r\geq\int_0^{M_{\text{surface}}}\frac{GM_r}{4\pi r_{\text{surface}}^4}\,\mathrm{d}M_r$$

thus

$$P_{\text{centre}}=P_{\text{surface}}+\frac{GM_{\text{surface}}^2}{8\pi r_{\text{surface}}^4}.$$

The central pressure of a star is the sum of the pressure at its surface and the term derived from its mass and its radius. If the surface pressure is assumed to be zero, a slightly lower central pressure will be calculated. Thus we may write

$$P_{\text{centre}}>GM^2/8\pi R^4.$$

Substitution of the values for the Sun gives

$$P_{\odot\text{centre}}>4{\cdot}5\times10^{13}\ \text{newton m}^{-2}$$

or

$$P_{\odot\text{centre}}>4{\cdot}5\times10^{8}\ \text{atmospheres}.$$

If we assume—and surprisingly it turns out to be true—that the ideal gas law holds, we can relate the pressure in terms of density and temperature as follows.

Now the equation of state of a perfect gas may be written as

$$P=n_{\text{total}}kT$$

where P is the pressure, n_{total} is the total number of gas atoms or molecules per unit volume, k is Boltzmann's constant and T the temperature.

The total number of molecules is given by

$$n_{\text{total}}=\rho/\mu m_{\text{H}}$$

where ρ is the density of the gas and μ the molecular weight (=mass/mass of hydrogen atom, m_{H}).

Hence the equation of state may be written

$$P=\frac{\rho}{\mu m_{\text{H}}}kT$$

and since $k/m_{\text{H}}=\mathcal{R}_{(\text{SI})}$ (the gas constant) $=8{\cdot}32\times10^3\ \text{J K}^{-1}\ \text{kg}^{-1}$ it may be written as

$$P=\frac{\mathcal{R}_{(\text{SI})}}{\mu}\rho T.$$

For any point from the star's centre we may write, therefore,

$$P(r)=\frac{\mathscr{R}_{(\mathrm{SI})}}{\mu}\rho(r)T(r) \tag{9.4}$$

and hence

$$T(r)=\mu P(r)/\mathscr{R}_{(\mathrm{SI})}\rho(r).$$

The value for the mean molecular weight is given by

$$\mu=\frac{\sum_{\text{all particles}}(\text{abundance})\times(\text{mass of particle})}{\text{total number of particles}}.$$

As we shall see, the temperatures in stars are extremely high and since hydrogen dominates and is ionized into protons and electrons in equal numbers μ may be approximated to half the proton mass. (The mass of a proton is very much greater than that of the electron.)

By using the same approximation as before for the point $r=\frac{1}{2}R_{\odot}$ and putting $P(R_{\odot}/2)=\frac{1}{2}P_{\odot\text{centre}}$, we obtain

$$T_{\odot}\approx 3\times 10^{7}\ \text{K}.$$

Under these conditions within the Sun, with pressures of a thousand million atmospheres and temperatures of tens of millions of degrees, molecules cannot exist, the lighter elements will be completely ionized and the heavier ones stripped almost bare of their electrons. The "gas" which we are considering is composed of a mixture of particles—protons, helium nuclei (α particles), electrons and ions—each with dimensions of the order of nuclei or less. The distances between them are, therefore, great compared with their size and, in fact, the normal stellar interior is a very good approximation to an ideal gas.

A quantitative feel for this may be had by considering the relative abilities of hydrogen atoms and protons to "pack" in a given volume, as follows.

The volume of a hydrogen atom $\sim\frac{4}{3}\pi r^{3}$ where r is the radius of the first Bohr orbit (see Roy & Clarke, *op. cit.*, Equation (14.18)). Hence the volume of a hydrogen atom $\sim 5\times 10^{-31}\ \text{m}^{3}$. The maximum number of hydrogen atoms that can be contained in unit volume (m^{-3})

$$\sim 1/(5\times 10^{-31})\sim 2\times 10^{30}.$$

Therefore

$$\begin{aligned}\text{maximum density} &\sim m_{\text{H}}\times 2\times 10^{30}\ \text{kg m}^{-3}\\ &\sim 1{\cdot}67\times 10^{-27}\times 2\times 10^{30}\ \text{kg m}^{-3}\\ &\sim 3{\cdot}34\times 10^{3}\ \text{kg m}^{-3}.\end{aligned}$$

But we know that stellar densities range much higher than this and on this basis the perfect gas laws could not apply.

If, however, we consider the gas as being highly ionized, it is then comprised of free protons and electrons. Now the "classical" volume of a proton $\sim\frac{4}{3}\pi a^{3}$, where $a\sim 3\times 10^{-15}$ m. In this case

$$\begin{aligned}\text{maximum density} &\sim 1{\cdot}67\times 10^{-27}\times 8{\cdot}85\times 10^{42}\\ &\sim 1{\cdot}48\times 10^{15}\ \text{kg m}^{-3}.\end{aligned}$$

This figure greatly exceeds the densities found in the vast majority of stars.

It must be pointed out, too, that "volume" at such high densities has no precise meaning. At these densities classical concepts no longer hold and particles must be described by quantum mechanics. In fact under stellar conditions, gases begin to deviate from the perfect gas laws at densities $\sim 10^{9}\ \text{kg m}^{-3}$.

9.4 The Effect of Temperature on the Mean Molecular Weight

By a crude approximation we have seen that internal temperatures of stars like the Sun are high, of the order of tens of millions of degrees. The notion of such high temperatures can be presented more rigorously by considering the equation of hydrostatic equilibrium and the equation of state.

Consider determining a mean value, $\bar{T}$, for the temperature of a star, the mean being averaged with respect to mass. This value may be expressed as

$$\bar{T}=\frac{\int T\rho \, \mathrm{d}\tau}{\int \rho \, \mathrm{d}\tau}$$

where $\mathrm{d}\tau$ is an elemental volume and $\rho \, \mathrm{d}\tau$ the mass in that volume, the integral being evaluated over the volume of the star. Obviously the integral in the denominator gives the total mass of the star. If we choose the elemental volume $\mathrm{d}\tau$ to be an elemental spherical shell of radius r and thickness $\mathrm{d}r$, then

$$\mathrm{d}\tau = 4\pi r^2 \, \mathrm{d}r.$$

The mean temperature of the star may now be written as

$$\bar{T}=\frac{1}{M}\int_{r=0}^{R} T\rho 4\pi r^2 \, \mathrm{d}r.$$

But from the equation of state we have that

$$T\rho = \mu P/\mathcal{R}_{(\mathrm{SI})}.$$

Therefore,

$$\bar{T}=\frac{4\pi\mu}{\mathcal{R}_{(\mathrm{SI})}M}\int_{0}^{R} P(r)r^2 \, \mathrm{d}r.$$

Integrating this expression by parts gives

$$\bar{T}=\frac{4\pi\mu}{\mathcal{R}_{(\mathrm{SI})}M}\left[\frac{1}{3}Pr^3\right]_0^R-\frac{4\pi\mu}{3\mathcal{R}_{(\mathrm{SI})}M}\int_0^R \frac{\mathrm{d}P}{\mathrm{d}r}r^3 \, \mathrm{d}r.$$

The first part of the above expression reduces to zero since at $r=R$, the pressure is approximately zero and P is finite at $r=0$. By substituting for $\mathrm{d}P/\mathrm{d}r$ from Equation (9.2), we have

$$\bar{T}=\frac{4\pi\mu G}{3\mathcal{R}_{(\mathrm{SI})}M}\int_0^R \rho(r)M_r r \, \mathrm{d}r.$$

Changing the variable of integration from r to M_r and noting that

$$\mathrm{d}M_r = 4\pi\rho r^2 \, \mathrm{d}r,$$

$$M_r=\begin{cases}0 & \text{at } r=0,\\ M & \text{at } r=R,\end{cases}$$

we have

$$\bar{T}=\frac{\mu G}{3\mathcal{R}_{(\mathrm{SI})}M}\int_0^M \frac{M_r \, \mathrm{d}M_r}{r}.$$

Now $(1/r)\geq(1/R)$ everywhere in the integral and, therefore, we may write

$$\bar{T}\geq\frac{\mu G}{3\mathcal{R}_{(\mathrm{SI})}M}\int_0^M \frac{M_r \, \mathrm{d}M_r}{R}\geq\frac{\mu G}{3\mathcal{R}_{(\mathrm{SI})}RM}\int_0^M M_r \, \mathrm{d}M_r$$

$$\geq\frac{\mu G}{3\mathcal{R}_{(\mathrm{SI})}RM}\left(\tfrac{1}{2}M^2\right).$$

Hence

$$\bar{T} \geq \mu GM/6\mathcal{R}_{(SI)}R.$$

In order to have a lower limit for the mean value, the same limit must apply to at least one point within the star.

In the case of the Sun with $\mu \sim \frac{1}{2}$,

$$T \geq GM/12\mathcal{R}_{(SI)}R \geq 1{\cdot}91 \times 10^6 \text{ K}.$$

The importance of the theorem is that at such temperatures, hydrogen is completely ionized, even at high electron densities. In a real star, it is expected that T exceeds this limit throughout most of the star. Because of this, all of the material of the star is almost completely ionized.

This result from the early stellar models went far in explaining certain puzzling features with respect to the mass–luminosity relation.

This relation, found from the measured masses and luminosities of real stars, was adhered to by stars as dense as or denser than the Sun. When Eddington constructed mathematically his stellar models, he discovered that the mass–luminosity relation was a consequence of these five relations. But the model stars he dealt with were gaseous throughout and it was also known that at laboratory temperatures a gas, when compressed, ceased to be a perfect gas long before stellar densities were approached. Yet the Sun and much denser stars fitted the mass–luminosity curve.

It was realized, however, that the material within a star would be almost completely ionized. Instead of intact atoms being present, a mixture of atomic nuclei and free electrons would make up the star's bulk. The perfect gas law is thereby enabled to operate at densities far greater than those at which a gas composed of neutral atoms and molecules ceases to obey it. The agreement of the Sun and denser stars with the theoretical mass–luminosity relation was therefore to be expected.

The mean molecular weight μ would also be kept down in value because of ionization of the atoms. An atom of hydrogen would become 2 particles when ionized; instead of unit "molecular weight", the weight would be $\frac{1}{2}$ or 0·5. Ionized helium and its electrons give an average weight of $\frac{4}{3}$ or 1·33. An iron atom, of atomic weight 55·8 and possessing 26 electrons, would become 27 gas particles when completely ionized, of average weight 55·8/27 or of the order of 2. All the elements, in fact, apart from hydrogen and helium, average out at 2 when completely ionized. The structure of the star being dependent on the mean molecular weight as well as on the mass, it is then only necessary to specify the proportions of hydrogen, helium and *the rest.*

9.5 The Stability of Hydrostatic Equilibrium

Let us see how closely the equation of hydrostatic equilibrium is obeyed. Suppose that it is possible to alter suddenly the gravitational force by a fraction f of its original value. At that instant, the pressure does not balance the new gravitational force. To come into equilibrium again, the radius of the star must change, the instantaneous acceleration being given by

$$\mathrm{d}^2r/\mathrm{d}t^2 = -fGM_r/r^2.$$

We now use the well-known formula connecting the distance s through which a body moves from rest in time t under a constant acceleration g, viz. $s = \frac{1}{2}gt^2$.

If we assume that the acceleration remains constant and the initial velocity is zero, then according to this formula, a change in radius of $\mathrm{d}r$ takes place in a time dt given by

$$dr = -\tfrac{1}{2}[f(GM_r/r^2)](dt)^2,$$

giving

$$(dt)^2 = \frac{-2\,dr}{f(GM_r/r^2)}.$$

Keeping in mind the approximate nature of the calculation, it is still instructive to consider how long it takes the radius to respond to the sudden change in the gravitational force. Let us see what time it takes for the radius itself to change by the same fraction f, when the gravitational force is altered by this fraction. Using the same values for the mid-point between surface and centre of the Sun as before and putting $dr = -fr$, we obtain

$$dt = \left(\frac{2R_{\odot}^3}{GM_{\odot}}\right)^{1/2} \approx 19 \text{ min}.$$

Thus the response to any change in the equilibrium is very swift. Since we have good reason to believe that normal stars remain substantially unchanged for periods $\approx 10^9$ years, Equation (9.2) must be very closely obeyed.

9.6 The Method of Energy Transport

As we have seen, the stellar interior behaves as an ideal gas and at every point within the sphere of gas, P, ρ and T are therefore specified by Equation (9.4). Because of spherical symmetry, all points distant r from the sphere's centre will have the same values of P, ρ and T. At any point within the star there is a net outward flow of energy and it must be possible to calculate this flow with any satisfactory model. We therefore have to give values to five variables at any distance r from the sphere's centre, the five variables (which are functions of r) being the pressure, P, the density, ρ, the temperature, T, the mass of gas, M_r, within the sphere of radius r and the flow of energy, L_r, from the sphere.

The three Equations (9.2), (9.3) and (9.4) are therefore insufficient to provide information about temperature, pressure, density, mass distributions and energy flow within the star. Two other equations of structure have to be found.

To obtain these, recourse is had to the method by which energy is transported from the central regions, where it is presumably produced, to the surface of the gas sphere, where it is radiated away. It will be remembered that there are three methods of energy transport, viz. by radiation, by convection and by conduction. The relative part played by each mode will depend upon the physical conditions found in the star and in practice, one usually dominates at a particular level within the star.

The required energy transfer equation must therefore involve the method of energy transport. When radiative transfer is considered the equation can be written as

$$\frac{dT}{dr} = -\frac{3}{4ac}\frac{\chi\rho}{T^3}\frac{L_r}{4\pi r^2} \tag{9.5}$$

where a is the radiation density constant, c the velocity of light, χ the optical absorption coefficient per unit mass and L_r the "luminosity" of the star at radius r. It is found that this last function can be expressed in a form similar to Equation (9.3), i.e.

$$dL_r/dr = 4\pi r^2 \varepsilon(r)\rho(r) \tag{9.6}$$

where $\varepsilon(r)$ is the energy released per kilogram per second.

We now have five equations (Equations (9.2) to (9.6)) which can be used to explore and evaluate models of stellar interiors. To use them in an attempt to understand the structure of a star, values of the parameters involved must be given. In particular, the total mass of gas M in the gas-sphere must be stated. A value for the mean molecular weight is also postulated. For example, we might suppose that the sphere consists of 10^{27} kg of pure hydrogen. Whatever we start with, the five relations can be used (that is integrated numerically) to give values not only of the pressure, density and temperature at any point within the gas-sphere but to assign a radius R and luminosity L to the star. For it turns out that where masses not too far removed from that of the Sun are concerned, the central temperatures are so high that the gas-sphere shines. It is, in fact, a star.

The important conclusion that may be drawn from such studies is that a star's structure is entirely determined by its chemical composition and mass.

With the aid of Equations (9.2), (9.4) and (9.5) we are in a position to make a rough comparison between theory and observation of the sort already carried out in obtaining approximate temperatures and pressures for the centre of the Sun. We can say that ρ will be proportional to M/R^3. Introducing this into Equation (9.2) and replacing the differentials by the differences, we obtain

$$P \propto M^2/R^4.$$

Substituting for ρ and P into the ideal gas law (Equation (9.4)) gives

$$T \propto M/R. \tag{9.7}$$

Again replacing differentials by differences, we can introduce these values for ρ and T into Equation (9.5) and obtain

$$L \propto T^3 . \frac{R^3}{M} . R^2 . \frac{T}{R}$$

thus

$$L \propto R^4 T^4/M.$$

By eliminating T, using Equation (9.7), we have

$$L \propto \frac{R^4}{M}\frac{M^4}{R^4}$$

therefore

$$L \propto M^3. \tag{9.8}$$

Thus the crude approximations made in the equations describing stellar interiors indicate that the form of the empirical mass–luminosity law (Equation (7.33)) is predictable from theory. Equation (9.8) also indicates that for stars on the main sequence, the transfer of energy from the stellar interior is effected by the radiative process.

In some special circumstances (white dwarfs, for example) convective transfer of energy must be considered and Equation (9.5) takes on a different form.

9.7 The Main Sequence

When a set of stellar models is calculated, with the chemical composition of the Sun but possessing masses that range from about ten solar masses to one solar mass, their luminosities and effective temperatures place them satisfactorily along the main sequence of the HR diagram with the most massive stars lying at the upper left as they should. This agreement is supporting evidence that our ideas on stellar structure are not too far removed from the truth. In the case of the Sun itself, it transpires that the chemical composition of the solar interior is probably very similar to that in the chromosphere. Hydrogen, helium and the other elements are present by mass in the proportions of about 75%, 25% and 1% respectively. The central temperature is about 14×10^6 K and the central density is of the order of 10^6 kg m^{-3}. It may also be shown that the central temperature is surprisingly constant along the main sequence, varying at most by a factor of two.

That this is so follows at once from Equation (9.6), namely that if T_c is the central temperature,

$$T_c \propto M/R.$$

This result, obtained for a star by means of a crude approximation, is in fact unaltered when a rigorous analysis is used.

Now at the lower end of the main sequence, M and R, in solar units, are both of the order of 10^{-1}. At the top of the main sequence they have values of the order of 40 and 20 respectively.

Substituting these pairs of values in the relation for T_c, it is found that the ratio of the values of T_c at the top and lower ends of the main sequence is 2.

9.8 Early Stages in the Life of a Star

It is now accepted that stars are formed through instability arising in an interstellar gas cloud. Parts of the gas cloud will separate and contract under their own internal gravitational forces. These protostars may have a radius of 1 parsec. Consider the central temperature T_c of a protostar. Initially it is of very low value.

Now it has been seen that

$$T_c \propto M/R.$$

Since the mass of the protostar does not change, the central temperature must rise as the radius shrinks. Calculations show that this contractional stage in the life of a star like the Sun lasts of the order of 10^7 years. By the end of this time the central temperature is measured in millions of degrees. During this time, the behaviour of a star on the HR diagram is sketched in Fig. 9.2, where the dotted lines show the paths of stars of different masses during this phase in the stars' lives.

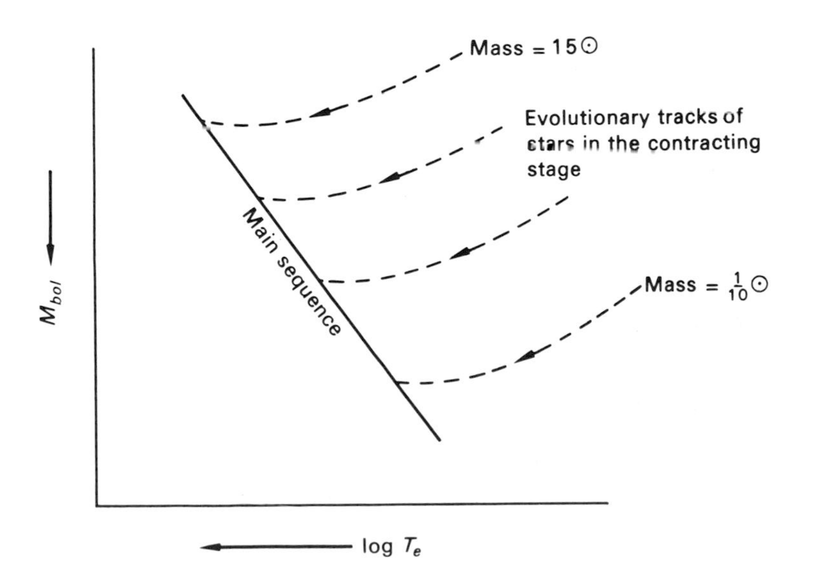

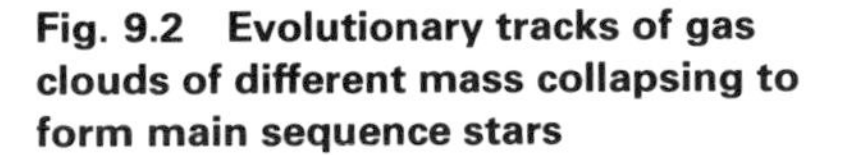
Fig. 9.2 Evolutionary tracks of gas clouds of different mass collapsing to form main sequence stars

As far back as the middle of the 19th century von Helmholtz showed that in the case of the Sun, a contraction of 50 m per year would supply enough energy by the transformation of gravitational potential energy into heat to keep the Sun shining at its present brightness. Let us see for how long the Sun could shine, considering the amount of gravitational energy that is available.

Now it is easy to show that the total thermal energy, E_T, of a star is given by

$$E_T = \int_0^R \left(\frac{3}{2}\frac{\mathscr{R}_{(SI)}}{\mu}T\right)4\pi\rho r^2\,\mathrm{d}r \tag{9.9}$$

where $(3\mathscr{R}_{(SI)}T/2\mu)$ is the energy of one kilogram of ideal gas with mean molecular weight μ. The total gravitational energy, E_G, is given by the expression

$$E_G = \int_0^R [-G(M_r/r)]4\pi\rho r^2\,\mathrm{d}r. \tag{9.10}$$

By multiplying Equation (9.2) by $4\pi r^3$, we have

$$4\pi r^3\frac{\mathrm{d}P}{\mathrm{d}r} = [-G(M_r/r^2)]4\pi\rho r^3 = [-G(M_r/r)]4\pi\rho r^2.$$

Integrating through the star gives,

$$\int_0^R 4\pi r^3\frac{\mathrm{d}P}{\mathrm{d}r}\mathrm{d}r = \int_0^R\left(-G\frac{M_r}{r}\right)4\pi\rho r^2\,\mathrm{d}r.$$

It will be seen immediately that the right-hand side of this last equation is equal to E_G. Thus

$$E_G = \int_0^R 4\pi r^3\frac{\mathrm{d}P}{\mathrm{d}r}\mathrm{d}r.$$

By using integration by parts and remembering that when $r=0$, $P=P_{\mathrm{centre}}$ and when $r=R$, $P\approx$ zero we obtain

$$E_G = -\int_0^R 3P\,.\,4\pi r^2\,\mathrm{d}r.$$

Substituting for P by using Equation (9.3), we have

$$E_G = -\int_0^R 3\frac{\mathscr{R}_{(SI)}}{\mu}\rho T\,.\,4\pi r^2\,\mathrm{d}r$$

$$= -2\int_0^R\left(\frac{3}{2}\frac{\mathscr{R}_{(SI)}}{\mu}T\right)4\pi\rho r^2\,\mathrm{d}r.$$

Comparison of this last equation with Equation (9.9) allows us to write

$$E_G = -2E_T. \tag{9.11}$$

Thus as a star contracts from the original gas cloud, its gravitational energy decreases. Equation (9.11) tells us that half of the energy released is converted into thermal energy, the other half being liberated as radiation. The net energy store available for radiation is therefore equal to the thermal energy.

By using the crude approximation method of taking differences in place of differentials and applying it to Equation (9.9) for the case of the Sun, we find that

$$E_{T\odot} \approx 5\times10^{41}\ \mathrm{J}.$$

Now measurements show that the output of the Sun ($L_\odot$) is of the order of $3{\cdot}8\times10^{26}\ \mathrm{J\,s^{-1}}$. If then the Sun relies on its source of energy being gravitational collapse, this source would be used up in a time given by

$$E_{T\odot}/L_{\odot} \approx 10^{15} \text{ seconds} \approx 3 \times 10^{7} \text{ years.}$$

This value is less than 1% of the Sun's estimated life so far, based on geological evidence.

We must conclude that gravitational energy release is only important in the contractional phase of a star's life. There must therefore be other sources of stellar energy that in some way are "switched on" once the stars have reached the main sequence.

9.9 The Source of Stellar Energy

Even if we know nothing in detail about the method by which the stars produce energy, we can make three statements about it.

(*a*) In Section 9.2 we saw that $\bar{\varepsilon}$, the mean rate of energy generation per kilogram, had a range of 10^6, taking it from 2×10^{-6} to $2 \text{ J s}^{-1} \text{ kg}^{-1}$ (or W kg^{-1}) although, as pointed out in Section 9.5, the central temperature of main sequence stars varies by a factor of 2 only, from one end of the main sequence to the other.

Hence the source of energy has a generation rate very sensitive to temperature change.

(*b*) Whatever the method of production, it must become available at temperatures around 10^7 K.

(*c*) In the case of the Sun, at least, it must produce energy for 5×10^9 years at a minimum.

Chemical reactions are inadequate by several orders of magnitude to produce such enormous quantities of energy. It has been shown by Gamow that if the Sun had been made of the best coal, set alight at the time of the first Pharaohs 5000 years ago and supplied with oxygen to keep it burning, it would be dead ashes by the present century. Chemical burning is a process resulting in the formation of chemical compounds and we know in fact that the Sun is too hot to allow complicated compounds and molecules to exist. We have also seen that the amount of energy required for the long life of the Sun cannot be achieved by gravitational collapse. Some entirely different energy source must be found.

9.10 Einstein's Mass–Energy Relation

Einstein showed that mass and energy are related by the following now-famous formula

$$E = mc^2, \tag{9.12}$$

where E is energy, m is mass and c is the velocity of light. Thus the conversion of one kilogram of matter into energy would provide almost 10^{17} J of energy.

The Sun's mean energy rate of generation is $2 \times 10^{-4} \text{ J s}^{-1} \text{ kg}^{-1}$ and its age is of the order of 5×10^9 years. The total energy generated to date is therefore at least 3×10^{13} joules per kilogram. We see then that in the conversion of mass into energy, an adequate energy source is available.

The process by which matter is converted into energy is by **nuclear reaction**. In estimating the masses of various atomic nuclei, one might consider the total number of protons and neutrons that any particular nucleus has. When this is done it is found that the expected mass, according to the number of protons and neutrons, does not tally with the observed nuclear mass. The **mass deficiency** is related via Einstein's equation to the amount of energy that is available if one type of nucleus is transformed into another.

The greatest mass defect that could possibly be achieved would be by the conversion of hydrogen into iron. For every one kilogram of hydrogen converted in this way, 8 g would be released in the form of energy. The first step in this long sequence of reactions is the most important one; the conversion of hydrogen to helium has a mass defect of 0·007. By assuming that the Sun is composed entirely of hydrogen, the amount of energy which would be released if it were to be converted to helium is

$$E_{N\odot} \approx 0{\cdot}007 M_{\odot} c^2 \approx 0{\cdot}007 \times 2 \times 10^{30} \times 9 \times 10^{16} \approx 1{\cdot}3 \times 10^{45}\ \mathrm{J}.$$

By remembering that the Sun's luminosity is $3{\cdot}8 \times 10^{26}\ \mathrm{J\ s^{-1}}$, we can say that the Sun could shine at its present rate for

$$\frac{1{\cdot}3 \times 10^{45}}{3{\cdot}8 \times 10^{26}}\ \text{seconds} \approx 3 \times 10^{18}\ \text{seconds} \approx 10^{11}\ \text{years}.$$

Thus, nuclear processes are capable of providing sufficient energy for the Sun to shine over a period longer than the lifetime of the Earth. The same nuclear reactions must also be going on within the other stars. It is the high temperatures within the stars providing high thermal energies to the particles which allows them to interact and fuse together.

9.11 The Fusion of Hydrogen into Helium

Because stars are mostly hydrogen and helium, early workers concentrated on finding nuclear reactions by which hydrogen could be fused into helium, the mass that disappeared in the process being the source of the energy released. Two processes are accepted that produce enough energy in stars in this way. One is the **proton–proton reaction**. The other is the **carbon–nitrogen cycle**.

In main sequence stars the following particles are involved:

Protons, denoted by the symbol p
Neutrons, denoted by the symbol n
Electrons, denoted by the symbol β_-
Positrons, denoted by the symbol β_+
Neutrinos, denoted by the symbol ν
Compound nuclei (protons + neutrons) such as α-particles. (An α-particle is a completely-ionized helium atom.)

Table 9.1 lists the elements involved. An isotope of an element has the same atomic number as the element (that is, the same number of electrons) and so behaves chemically like the element but has a different atomic weight because its nucleus is different.

The proton–proton reaction is described by the following reactions:

$$p + p \rightarrow \mathrm{D} + \beta_+ + \nu,$$
$$\mathrm{D} + p \rightarrow {}^3\mathrm{He},$$
$${}^3\mathrm{He} + {}^3\mathrm{He} \rightarrow {}^4\mathrm{He} + p + p.$$

Thus the net result of this reaction is to turn four hydrogen nuclei into a nucleus of helium. In the process isotopes of hydrogen and helium have been created and destroyed. In addition, radiation has been released.

Table 9.1

Atomic number	Atomic weight	Element	Isotope
		Hydrogen	
1	1	$^{1}H(p)$	
1	2		$^{2}D(p+n)$
		Helium	
2	4	$^{4}He(2p+2n)$	
2	3		$^{3}He(2p+n)$
		Carbon	
6	12	$^{12}C(6p+6n)$	
6	13		$^{13}C(6p+7n)$
		Nitrogen	
7	14	$^{14}N(7p+7n)$	
7	15		$^{15}N(7p+8n)$

The carbon–nitrogen cycle passes through four stages given below.

$$^{12}C+p\rightarrow {}^{13}C+\beta_{+}+\nu,$$

$$^{13}C+p\rightarrow {}^{14}N,$$

$$^{14}N+p\rightarrow {}^{15}N+\beta_{+}+\nu,$$

$$^{15}N+p\rightarrow {}^{12}C+\alpha.$$

Again, the net result has been to fuse four hydrogen nuclei into one helium nucleus. The carbon and nitrogen involved in this reaction are in a sense catalysts. Without them the reaction could not take place, but their abundance within the star is unaltered. The original production of these heavier elements is considered later in Sections 9.13 and 18.4.

The energy released in both the proton–proton reaction and the carbon–nitrogen cycle comes from the slight mass discrepancy between the mass of the helium nucleus (4·00392 mass units) and that of the 4 hydrogen nuclei ($4\times 1{\cdot}00828=4{\cdot}03312$ mass units) from which it is formed. This lost mass is transformed into energy according to Einstein's relation, viz. Equation (9.12). When the relation is applied to the fractional loss of mass, namely 0·029/4·004, it shows that every kilogram of hydrogen transformed into helium would release some $6{\cdot}6\times 10^{14}$ J. This is more than enough to keep the Sun shining at its present rate for 5×10^{9} years, if the rate of release of energy fits. Calculation shows that with the temperatures found in the Sun and stars of lesser mass on the main sequence, the proton–proton reaction is the major source of energy and operates at the required rate. For more massive stars further up the main sequence, where the central temperatures are higher, the carbon–nitrogen cycle takes over from the proton–proton reaction as the main energy producer.

It is here, in fact, that we come up against the neutrino problem already mentioned in Section 2.6. From the above equations for stellar energy it should be possible to calculate the neutrino flux escaping from the Sun and consequently the number of solar neutrinos crossing any given area at the Earth's surface. Unfortunately the number detected by experiment is about one order of magnitude lower than expected, a timely warning that we should not be too complacent in our belief that we have complete understanding of the energy-liberating processes in main sequence stars.

9.12 Stellar Evolution after Arrival on the Main Sequence

At the high temperatures within the central regions of a star, hydrogen is transformed to helium, producing a core enriched with helium. For more massive stars in the upper parts of the main sequence, calculation shows that the core will be convective and will be surrounded by an envelope in radiative equilibrium. Lower main sequence stars, on the other hand, have radiative cores surrounded by convective envelopes.

As less and less hydrogen "fuel" remains to be "burned", energy generation dies away within the core. The process of transforming hydrogen into helium then occurs in an expanding shell surrounding the core. Detailed calculation of the changing structure shows that little alteration in position of the star in the HR diagram takes place until some 20% of the original hydrogen has been transformed into helium.

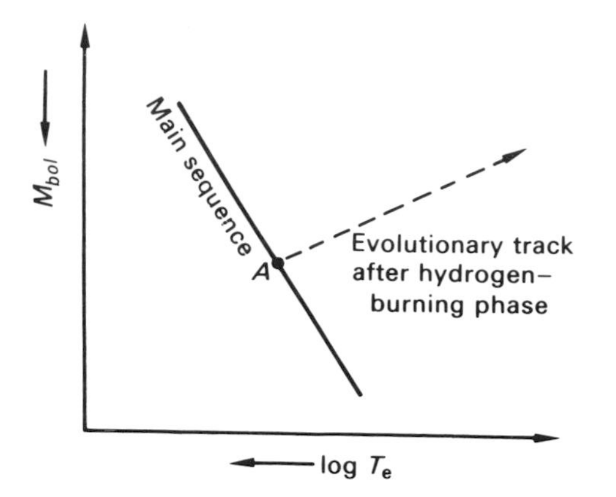

Fig. 9.3 Evolutionary track of a star after the hydrogen-burning phase

At this stage the star is unable to maintain a temperature in the hydrogen-burning shell high enough for the described nuclear reaction. The helium-core then contracts at a rate partially determined by the residual hydrogen-burning in the shell. This is faster than the contraction time scale for the complete star. The total energy of the star is therefore substantially unaffected. But the energy of the contracting helium-core is decreasing, hence the energy of the hydrogen-envelope surrounding it must increase. This it does by expanding. The star therefore expands and becomes a giant. This expansion process, transforming the main sequence star into a giant, occurs in a shorter time than it spent on the main sequence. Its path on the HR diagram during this phase of its life is sketched in Fig. 9.3.

The interpretation of the main sequence on the HR diagram is that it is a region within which most stars spend most of their lives. Forming out of the interstellar cloud, their fast shrinking sends them rapidly leftwards on the HR diagram onto the main sequence. Hydrogen-burning keeps them there for a much longer time before they embark on their giant phase, moving rapidly off the main sequence to the upper right.

9.13 The Ages of Star Clusters

Suppose that out of an interstellar gas cloud a number of stars of different mass are formed at the same epoch making up a cluster of stars. Each star contracts, its position on the HR diagram shifting rapidly on to the main sequence (see Fig. 9.2).

Now we have seen that $\bar{\varepsilon}$, the mean rate of energy generation per kilogram, is least at the bottom right-hand end of the main sequence and greatest at the top left-hand end. The range in $\bar{\varepsilon}$ is very large, of the order of 10^6. Stars at the top of the main sequence are therefore the first to consume some 20% of their hydrogen and leave the sequence to become giants. The main sequence of this star cluster is therefore destroyed from the top producing, at any subsequent time, an HR diagram similar to the one sketched in Fig. 9.3, with the sequence above A missing. The point A is referred to as the **turn-off point** on the HR diagram.

At point A, consequently, a star has just burned away 20% of its initial hydrogen. Let us suppose, for simplicity, that the star was initially pure hydrogen. Then, neglecting the relatively short time during which gravitational contraction from the protostar took place, we may calculate the age of the star cluster as follows:

Age of cluster = time taken to transform 20% of total mass of star into helium

$$= \frac{\text{total energy released in nuclear transformation}}{\text{luminosity of star at turn-off point}}$$

$$= 0{\cdot}2 \times M \times 6{\cdot}6 \times 10^{11}/L \text{ seconds}$$

$$= 1{\cdot}3 \times 10^{11}/\bar{\varepsilon} \text{ seconds.}$$

Example 1. If turn-off point is near the top of the main sequence,

$$\bar{\varepsilon} \approx 2\ \mathrm{J\ s^{-1}\ kg^{-1}}.$$

Hence cluster age is 2×10^6 years.

Example 2. If turn-off point is near the Sun's position,

$$\bar{\varepsilon} \approx 2 \times 10^{-4}\ \mathrm{J\ s^{-1}\ kg^{-1}}.$$

Hence cluster age is 2×10^{10} years.

9.14 Application to Real Clusters of Stars

Among the star fields there are localized condensations or clusters of stars. More than 500 such **galactic** or **open clusters** are known. Some contain only a few score stars; others have as many as 1000 members. Perhaps the best-known open cluster is the Pleiades, visible to the naked eye on a clear night. Apart from the seven bright stars (the Seven Sisters) known from antiquity, there are over 100 fainter members of the cluster, all within one degree of *Alcyone*, the brightest member. Another well-known cluster, the Hyades, is rather more wide-spread, its stars (under 100 in number) occupying an area roughly 20° across. The double cluster, *h* and χ Persei, is also well known (see Fig. 9.4).

Fig. 9.4 **The double cluster, *h* and χ Persei, an example of a galactic cluster. (By courtesy of the Mt. Wilson and Palomar Observatories)**

The members of an open cluster are so close directionally that it is probable they are close to each other in space, relatively speaking. For clusters not too far from the Sun, their members show similar radial velocities and proper motions. If the stars forming a cluster are in fact approximately equi-distant from the Sun, their space velocities must be almost identical and the cluster is an entity, at least semi-permanent, moving through space.

It is at least reasonable to suppose that the stars in a cluster were formed at the same time from the same gas cloud. They therefore provide a test of the ideas presented above.

A second type of cluster found in space is called a **globular cluster** because of its shape. Over one hundred of these objects are known to belong to our own Galaxy. Very few are visible to the naked eye, a telescope being required to detect their structure. Each globular cluster is a system of anything up to 500 000 stars, the average distances between the stars in a cluster being much smaller than those between stars in the general distribution of stars. The cluster shape is usually spherical with the stars forming the cluster being more and more closely packed together the nearer they are to the centre of the cluster. It should be noted, however, that even at the centre the distances separating the stars are vast compared with the radii of the stars themselves. It is only the great distances of the clusters that make them look compact (see Fig. 9.5 for example).

Fig. 9.5 NGC 6205 (M13), the famous globular cluster in Hercules photographed by the 200-inch Mt. Palomar telescope. (By courtesy of the Mt. Wilson and Palomar Observatories)

For both types of cluster, a colour–magnitude diagram can be plotted. To draw an HR diagram for a cluster, the spectra of the stars forming it must be obtained and it is only for the nearer clusters that this is possible. But colour, as we have seen, is closely related to spectral type so that we may look upon the diagram obtained as essentially an HR diagram. Also, since the stars in a cluster may be assumed to be all at the same distance from the observer, differences in apparent magnitude can be taken to be differences in absolute magnitude. If, by some means, the cluster distance can be found, as can be done with many of them, the magnitude scale will be an absolute one.

If now the HR diagrams for a number of clusters are drawn and superimposed, we can study the resultant figure (Fig. 9.6) to examine what differences in age there might exist between clusters.

The open or galactic clusters show turn-off points at various positions along the main sequence. There is also evidence of a gap that occurs between the main sequence and the giant branch. This gap is known as the **Hertzsprung gap**.

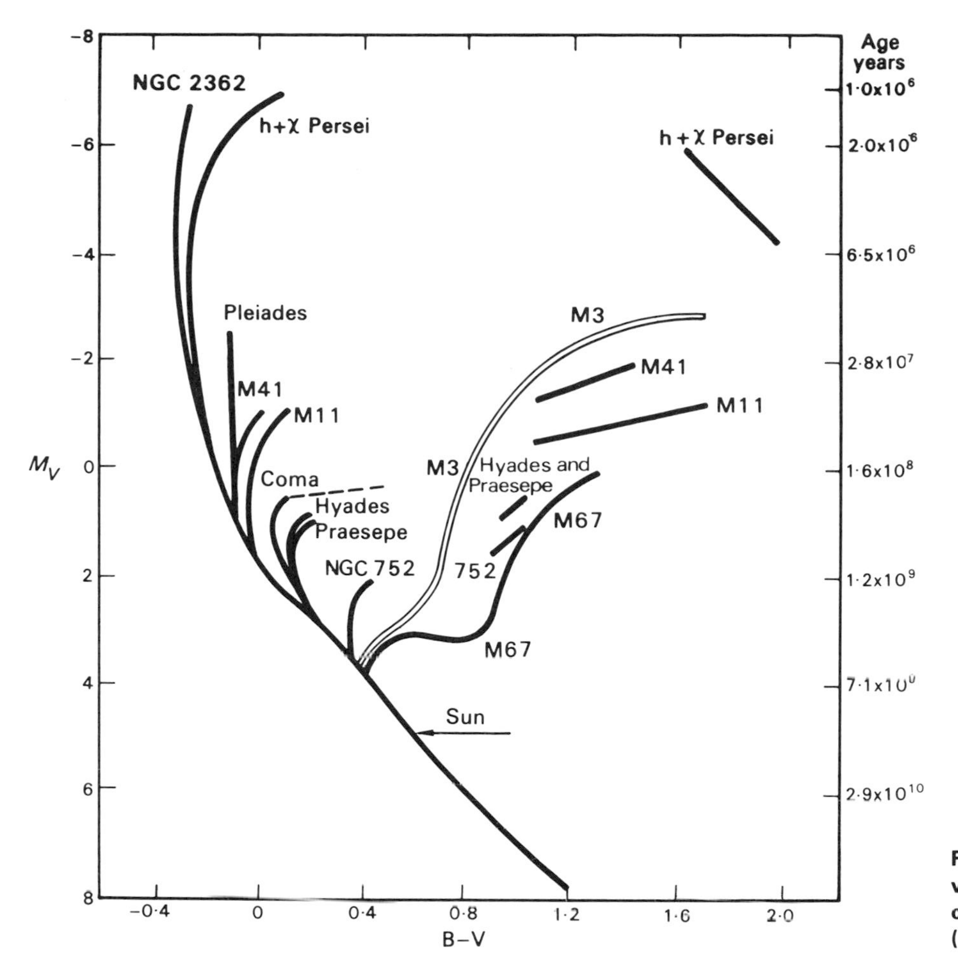

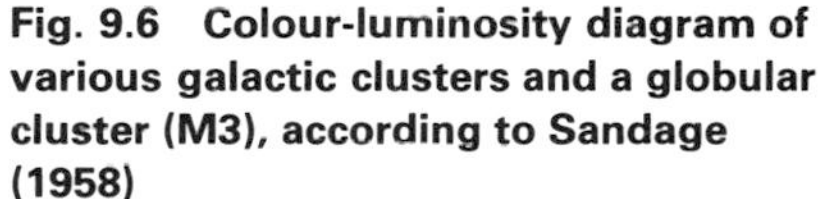
Fig. 9.6 Colour-luminosity diagram of various galactic clusters and a globular cluster (M3), according to Sandage (1958)

By using the method of Section 9.13 we can calculate ages for these clusters. It is found, for example, that the double cluster h and χ Persei is about 5×10^6 years old; the Pleiades is some 2×10^7 years of age, the Hyades is 4×10^8 years old. All of these, it may be noted, are much younger than the Earth. In fact the range in age for open clusters is from 2×10^6 to 6×10^9 years.

It may be noted that for very young clusters, some of their stars may be observed in the contracting stage—that is, they have not yet reached the main sequence.

In the case of the globular clusters, it is found that they too have a turn-off point on the main sequence and a giant branch attached to the main sequence at this point. They also have a horizontal branch running across the HR diagram (Fig. 9.7) from the giant branch to the main sequence, meeting it above the turn-off point. In this horizontal branch there is often a region inhabited by RR Lyrae stars—blue giant stars of variable luminosity similar to the well-known star RR Lyra.

The ages calculated for the globular clusters are all close to 6×10^9 years. This difference between the ages of globular clusters and open clusters has a bearing on the structure and evolution of the Galaxy as a whole and will be discussed later.

One important conclusion may be drawn from the ages found for the youngest open clusters. A time-scale of 2×10^6 years is small compared with the age of the oldest stars, of the order of 6×10^9 years, and indicates that star-formation may still be in progress. There are certain groups of stars, often O and B type main sequence stars of high luminosity, that have velocities

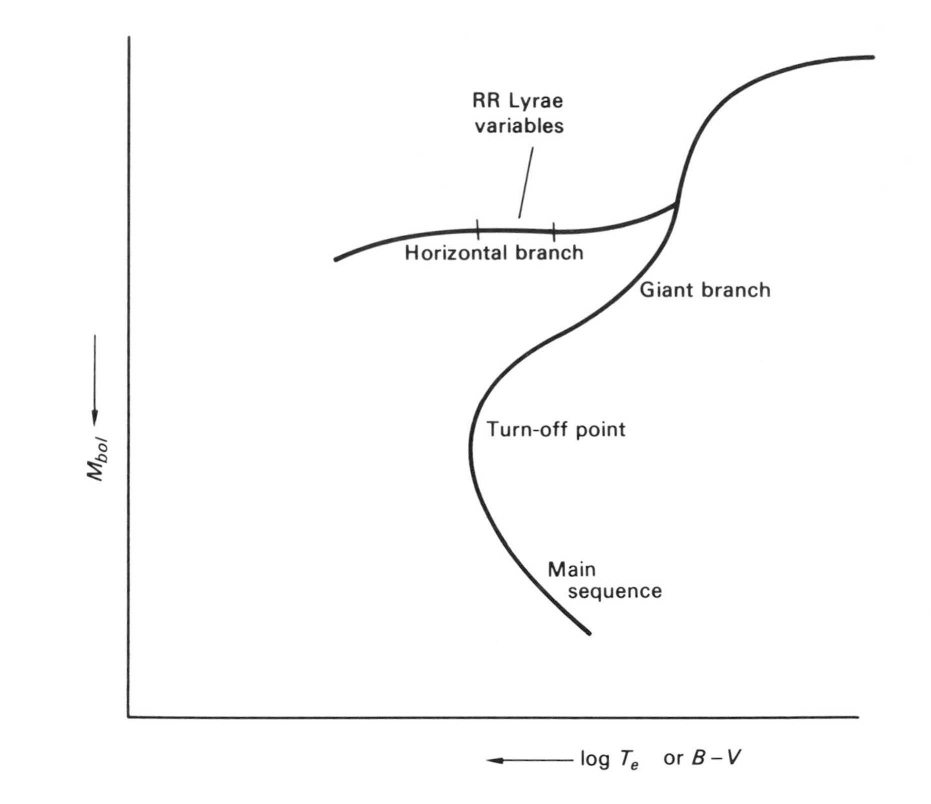

Fig. 9.7 The HR diagram of a globular cluster

indicative of a common origin. The stars in any group would have been close together just over 10^6 years ago and it is believed that at such a time the stars were formed from an interstellar cloud. An example of such **stellar associations** is the ζ Persei group of 17 stars.

9.15 The Later Stages of Stellar Evolution

After the giant stage of stellar evolution other thermonuclear reactions may take place. The expansion of the star was caused by the contraction of the core so that the central temperature rose until the onset of the next reaction involving heavier atoms. This occurs when the temperature is of the order of 10^8 K. It involves three α-particles or helium nuclei. The first step is the building of beryllium 8, viz.

$$^{4}\text{He} + {}^{4}\text{He} \rightarrow {}^{8}\text{Be}.$$

The second step uses the beryllium 8, thus:

$$^{4}\text{He} + {}^{8}\text{Be} \rightarrow {}^{12}\text{C}.$$

Although ^{8}Be is unstable, it is formed at such a speed at a temperature of 10^8 K that there exist sufficient of these nuclei to fuse with the helium nuclei. The product, carbon 12, is stable. In the process, energy is released.

The contraction of the core is now halted; the star expands and moves rapidly to the right on the HR diagram. There is a deepening convection zone in the outer part of the star, eventually mixing hydrogen into the inner helium region. The tendency is therefore to render the star more uniform in chemical composition with a resulting movement back towards the main sequence. Calculation bears this out and accounts for the horizontal branch in the globular cluster HR diagram.

The final stages in stellar evolution are far less certainly understood. The stages reached by particular stars depend upon their mass. Without going into detail, it is possible to show that elements more massive than carbon are synthesized as the central temperature increases still further. Some stars may reach a stage where central densities become so high that the material enters a fourth state of matter which is not gas, liquid or solid. This is said to be **quantum mechanically degenerate**. Special mathematical techniques have to be used to describe the behaviour of stars in this state. White dwarfs are thought to contain degenerate matter. They form about 3% of all the stars in our galaxy. Densities as high as 10^5 times that of water have been calculated to exist within them. Other stars in their post-hydrogen burning phase may proceed through the red giant phase to an unstable state where they explode, becoming **supernovae**. Yet others may become members of the different classes of variable stars discussed in Chapter 12.

Problems: Chapter 9

1. The components of a visual binary system are identical main sequence stars with apparent bolometric magnitudes of $+5^{m}\!.0$ and effective surface temperatures of 11 000 K. The semi-major axis of the relative orbit is $0''\!.1$ and the orbital period is 14 years. Assuming that the central temperature of the components is 1·2 times the Sun's central temperature, calculate the masses and radii of the components in solar units and the parallax of the system. (Sun's absolute bolometric magnitude $= 4^{m}\!.75$ and its effective surface temperature = 5800 K).

2. Calculate from the following data, the age of a binary star in which the components have equal masses and are both known to have consumed 20% of their initial hydrogen at constant luminosity.

   Period of orbit = 20 years
   Angular separation of the two stars is constant and $= 0''\!.75$
   Plane of the orbit is at right angles to the line of sight
   Annual parallax $= 0''\!.1$
   Apparent bolometric magnitude of each component $= +0^{m}\!.2$
   Sun's absolute bolometric magnitude $= +4^{m}\!.75$
   Sun's luminosity $= 4\times10^{26}$ J s^{-1}
   Sun's mass $= 2\times10^{30}$ kg
   Energy produced in the conversion of hydrogen to helium $= 6{\cdot}6\times10^{14}$ J kg^{-1}

3. The main sequence of a star cluster contains stars up to $0^{m}\!.0$ brightness in absolute bolometric magnitude, but none brighter. Assuming that the central temperature of stars is constant along the main sequence, calculate the age of the cluster from the following data, assuming that stars turn off the main sequence after consuming 20% of their hydrogen.

   Sun's mass $= 2\times10^{30}$ kg
   Sun's effective temperature = 5800 K
   Sun's luminosity $= 4\times10^{26}$ J s^{-1}
   Sun's absolute bolometric magnitude $= +4^{m}\!.75$
   Stars of $0^{m}\!.0$ bolometric magnitude have an effective temperature of 11 000 K
   Energy liberated from conversion of hydrogen to helium $= 6{\cdot}6\times10^{14}$ J kg^{-1}

10 Stellar Motions

10.1 Introduction

Halley's discovery in 1718 that the positions of three stars had changed relative to the background of faint stars gave impetus to attempts to measure stellar motion and parallax. According to Halley, comparison of the then current positions of *Sirius*, *Aldebaran* and *Arcturus* with those given by Hipparchus almost two millenia before showed definite discrepancies. Halley concluded that these stars were moving in space relative to the Sun and that in all probability every star was in motion. It was only the enormous distances to the stars that prevented all but the fastest motions being detected.

The observed changes in a star's position due to aberration and parallax that we have already discussed would be superimposed on any change in its direction due to its own intrinsic velocity relative to the Sun, and in this chapter we will assume that the effects of aberration and parallax caused by the Earth's motion about the Sun are already removed. Then what is left will be due to the star's own velocity with respect to the Sun. It goes without saying that the effects of precession and nutation are also eliminated by using a fixed mean equator and equinox for reference.

We therefore define the **proper motion** of a star to be the annual angular change in its **heliocentric** direction on the celestial sphere due to its space velocity relative to the Sun.

The Sun being a star, we would expect it also to have a space velocity measured with respect to some fixed set of axes so that the velocity of a star with respect to the Sun is the relative velocity of star and Sun. For the sake of clarity, we will first of all consider only the stars' velocities relative to the Sun, forgetting the Sun's own space velocity for the moment. Later we will see how it can be taken into consideration, defined and indeed found.

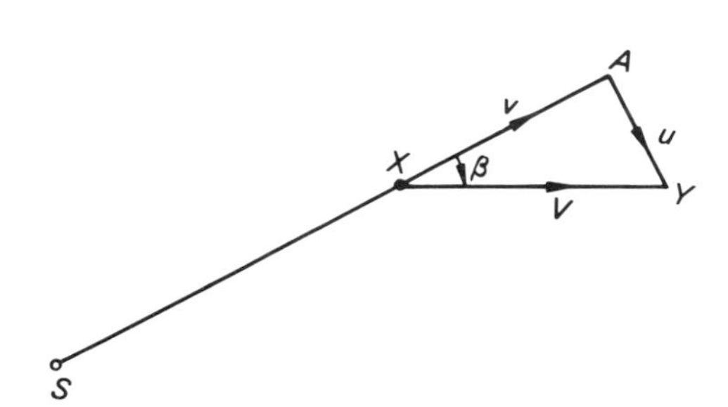

Fig. 10.1 A star's velocity resolved into transverse and radial components

10.2 The Transverse and Radial Velocities of a Star

In Fig. 10.1, let S be the Sun and X a star with a velocity V relative to the Sun. Let XY be a vector representing V and let it make an angle β with SX produced. Then we may form the triangle of velocities XAY, where XA and AY are the components of the star's velocity along the line of sight from S to X and perpendicular to the line of sight. Let v and u be these components respectively so that

$$v = XA, \qquad u = AY.$$

Then

$$v = V\cos\beta; \qquad u = V\sin\beta. \tag{10.1}$$

The component v is the **radial velocity** of the star. Since it lies along the line of sight from the Sun it cannot effect any change in the star's heliocentric direction. By convention, the radial velocity is taken to be positive if the star is receding from the Sun. If it is approaching, the radial velocity is negative. Usually such velocities are expressed in kilometres per second.

The component u, across the line of sight, is the **transverse velocity** of the star. It produces a change in the star's heliocentric direction and so must be related, with the star's distance, to the proper motion. It is worth noting that a star may have a large space velocity V relative to the Sun but, unless the star is comparatively near and has a large transverse velocity, its proper motion may be very difficult, if not impossible, to detect.

10.3 The Components of Proper Motion

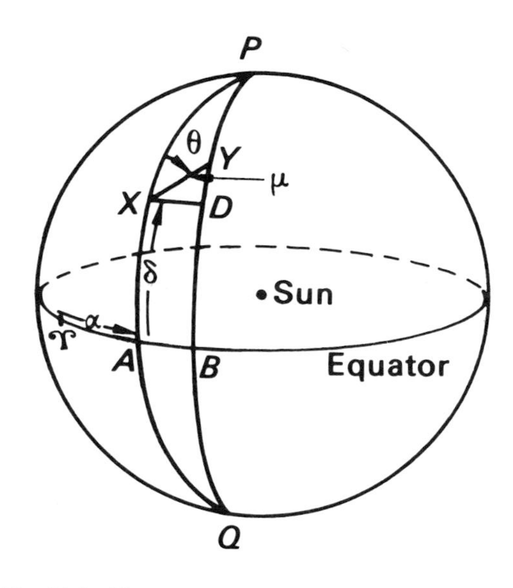

Fig. 10.2 The proper motion of a star on the celestial sphere

The proper motion of a star will be shown as a shift in the star's right ascension α and declination δ. In general, the shift in one year will be small and so measurements are made at epochs a number of years apart, photographic plates being taken of the area including the star. The shifts in right ascension and declination are measured. Corrections are made for stellar parallax, aberration, precession and nutation, the residual shifts then being due to the star's velocity relative to the Sun. These shifts are divided by the number of years between the epochs, thus obtaining the components of the star's proper motion in right ascension and declination.

We now obtain the relationships between these components in right ascension and declination, μ_α and μ_δ, respectively, the proper motion μ, and the so-called position angle θ. In Fig. 10.2, a star X of right ascension α and declination δ moves in one year to Y on the heliocentric celestial sphere. The great circle arc XY is therefore the star's annual proper motion μ; $\angle PXY$, where P is the north celestial pole, is the **position angle** θ. Draw two great circles from P through X and through Y to meet the equator in the points A and B respectively. Let a small circle arc XD be drawn through X parallel to the equator to meet the great circle through Y in D. Then

$$DY = \mu_\delta; \qquad AB = \mu_\alpha.$$

Hence

$$XD = \mu_\alpha \cos \delta,$$

where δ is the star's declination.

Triangle XYD can be taken to be a small plane triangle since μ is a small quantity. Hence

$$XD = \mu \sin \theta; \qquad YD = \mu \cos \theta.$$

We therefore have

$$\begin{aligned} \mu_\alpha \cos \delta &= \mu \sin \theta, \\ \mu_\delta &= \mu \cos \theta. \end{aligned} \qquad (10.2)$$

In practice, μ_α is usually expressed in seconds of time per annum while μ_δ is expressed in seconds of arc per annum.

10.4 Measurement of the Transverse and Radial Speeds of a Star

Let a star's transverse linear speed be u km s^{-1}. If n is the number of seconds of time in one year, the total distance moved by the star in that time in a direction at right angles to the line of sight will be nu km. Let this distance be XU in Fig. 10.3. The star's distance from the Sun is SX, equal to d km, say. Then $\angle XSU$ must be the star's proper motion μ. Hence

$$\tan \mu = nu/d,$$

or, since μ is a small angle,

$$\mu = 206265\, nu/d, \qquad (10.3)$$

where μ is now expressed in seconds of arc.

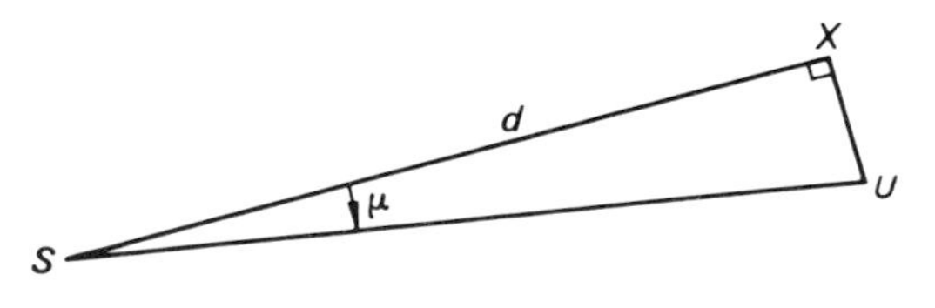

Fig. 10.3 A star's proper motion and its distance

Let P be the star's parallax also in seconds of arc. Then, if a is the Earth's orbital semi-major axis, in kilometres, we have

$$P = 206265\, a/d. \tag{10.4}$$

Dividing Equation (10.3) by Equation (10.4), we obtain

$$\mu/P = nu/a$$

or

$$u = \frac{a}{n}\frac{\mu}{P}.$$

Now $a = 149{\cdot}6 \times 10^6$ km; $n = 31{\cdot}56 \times 10^6$ s, so that

$$u = 4{\cdot}74\, \mu/P. \tag{10.5}$$

Measurement of μ and P therefore enables u to be calculated.

The star's radial velocity can be found by spectrometric means, using the Doppler formula, viz.

$$\Delta\lambda/\lambda = v'/c,$$

where $\Delta\lambda$ is the observed shift in wavelength of a line of wavelength λ, c is the velocity of light and v' is the star's velocity of approach or recession relative to the observer. Since the observer is on the Earth, which has its own orbital velocity about the Sun, a correction has to be made to v' to obtain v, the heliocentric radial velocity of the star. Indeed, if his equipment is sufficiently sensitive, a correction for the observer's velocity due to the Earth's diurnal rotation should be made.

10.5 The Velocity of a Star Relative to the Sun

Let us suppose that the star's transverse linear speed u and radial velocity v have now been found. Then if V is the star's speed relative to the Sun, in a direction making an angle β with the line of sight, we see immediately from Fig. 10.1 that Equation (10.1), viz.

$$V \cos \beta = v,$$

$$V \sin \beta = u,$$

gives

$$V^2 = u^2 + v^2 \quad \text{or} \quad V = (u^2 + v^2)^{1/2}, \tag{10.6}$$

and

$$\tan \beta = u/v. \tag{10.7}$$

Example 10.1 For a star the following data are available: apparent magnitude 0·14, parallax $0''.124$, radial velocity (referred to Sun) is -14 km s^{-1}, declination 38° 43′ N. The components of its proper motion are: in right ascension $0^s.0164$, in declination $0''.284$. Find

(i) the total proper motion and the position angle,
(ii) the transverse linear speed,
(iii) the space velocity relative to the Sun,
(iv) the parallax at closest approach to the Sun,
(v) the epoch of closest approach,
(vi) the apparent magnitude at closest approach, assuming the absolute magnitude to be constant.

(i) By equations (10.2),

$$\tan \theta = \mu_\alpha \cos \delta / \mu_\delta.$$

Now $\mu_\alpha = 0^s\!.0164 = 0''\!.246$. Hence

$$\tan \theta = 0{\cdot}246 \times \cos 38^\circ 43' / 0{\cdot}284,$$

giving

$$\theta = 34^\circ 03'.$$

Now

$$\mu = \frac{\mu_\delta}{\cos \theta} = \frac{0{\cdot}284}{\cos 34^\circ 03'} = 0''\!.343.$$

(ii) By Equation (10.5),

$$u = 4{\cdot}74 \times 0{\cdot}343 / 0{\cdot}124 = 13{\cdot}1 \text{ km s}^{-1}.$$

(iii) By Equation (10.6), using the given radial velocity

$$V^2 = 14^2 + 13{\cdot}1^2 = 367{\cdot}7,$$

giving

$$V = 19{\cdot}2 \text{ km s}^{-1}.$$

Also

$$\sin \beta = u / V,$$

giving

$$\beta = 43^\circ 05'.$$

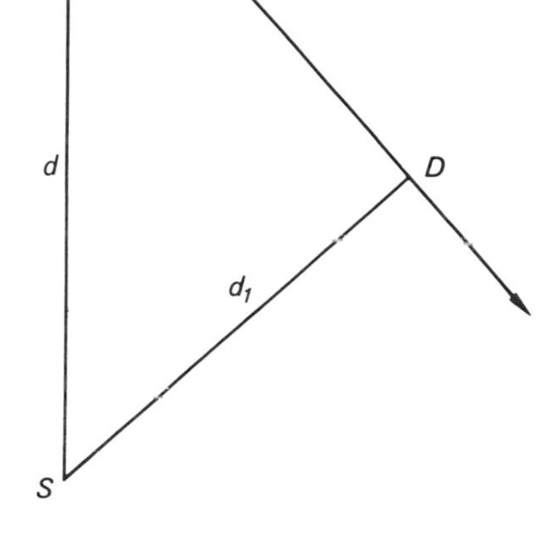

Fig. 10.4 Calculation of the closest approach of a star

(iv) Since the radial velocity is negative, the star is approaching. In Fig. 10.4, its present position is A, its velocity is V, given by AE, and it will be at its nearest to the Sun when it is at D, with $\angle ADS = 90^\circ$.

Let its present and future distances be d and d_1, as shown. Let P and P_1 be the corresponding parallaxes.

Then if the parallaxes are expressed in seconds of arc and the distances in parsecs, we have $P = 1/d$, $P_1 = 1/d_1$, and we have

$$P_1 = P(d/d_1).$$

Now

$$\sin \beta = d_1 / d,$$

so that

$$P_1 = P / \sin \beta.$$

But in $\triangle ABE$,

$$\sin \beta = u / V,$$

so that we may write

$$P_1 = PV / u,$$

giving

$$P_1 = 0{\cdot}124 \times 19{\cdot}2 / 13{\cdot}1 = 0''\!.182.$$

(v) The time taken for the star to reach D is t, where $t = AD / V$. Now

$$AD / AS = \cos \beta = v / V,$$

so that

$$AD = d \times v/V.$$

Hence

$$t = \frac{dv}{V^2} = \frac{1}{P} \times \frac{v}{V^2} \times \frac{206265 \times 149{\cdot}6 \times 10^6}{31{\cdot}56 \times 10^6} \text{ years,}$$

giving

$$t = 300\ 100 \text{ years.}$$

The epoch of closest approach therefore occurs 300 100 years hence.

(vi) Let m_1 be the apparent magnitude at the time of closest approach. Then by Equation (7.4),

$$M = m + 5 + 5 \log_{10} P,$$

$$M = m_1 + 5 + 5 \log_{10} P_1,$$

where M is the absolute magnitude of the star.

Hence, by subtraction,

$$m_1 = m + 5 \log_{10} (P/P_1) = 0{\cdot}14 + 5 \log_{10} (0{\cdot}124/0{\cdot}182)$$

$$= 0{\cdot}14 - 0{\cdot}83$$

or

$$m_1 = -0{\cdot}69.$$

10.6 The Convergent Point of a Galactic or Open Cluster

We have already seen that much valuable information concerning the ages and evolution of the stars can be obtained by a study of the HR diagrams of star clusters. In this section, however, we confine our attention to open or galactic clusters such as the Hyades or Pleiades. The members of a cluster of this type show similar radial velocities and proper motions and we have looked upon this feature as indicative of the common origin of the stars forming a group of this sort.

It is found that there is a tendency for the proper motion vectors, if produced, to converge to a point on the celestial sphere. This feature is an effect of perspective. Although all the members of the cluster have the same space velocity relative to the Sun, it is not in general at right angles to the line joining Sun to cluster. The stars in the cluster are all therefore moving either away from or towards the Sun as if on parallel rails. And just as railway tracks, although parallel to each other, appear to converge in the distance, so do the proper motions of the cluster stars. We now show how the existence of a convergent point for an open cluster may be used to obtain the parallaxes of the cluster members.

In Fig. 10.5 let X_1 and X_2 be two members of an open cluster. Their directions from the Sun S are SX_1 and SX_2. Because they are both members of the same cluster their space velocity V will be the same. Let it make angles β_1 and β_2 with the stars' directions. The stars' velocities V can be resolved in the usual way into components u_1, v_1 and u_2, v_2 respectively and it is seen that we can write

$$\begin{aligned} u_1 &= V \sin \beta_1; \qquad v_1 = V \cos \beta_1; \\ u_2 &= V \sin \beta_2; \qquad v_2 = V \cos \beta_2. \end{aligned} \tag{10.8}$$

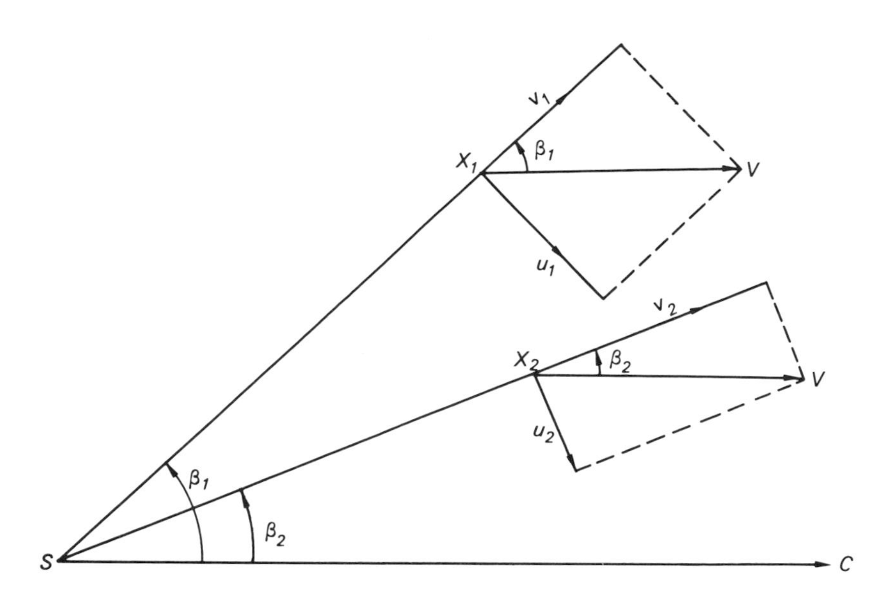

Fig. 10.5 Motions of a cluster of stars

The transverse velocities u_1 and u_2 will give rise to proper motions μ_1 and μ_2 for the stars X_1 and X_2, with position angles θ_1 and θ_2 in turn. In Fig. 10.6 we suppose these to be plotted on the heliocentric celestial sphere. If the great circles along which these proper motions lie are now produced, they will meet at a point C which is the **convergent point**. It is easy to see from Figs 10.5 and 10.6 that X_1C is β_1 and X_2C is β_2.

Because of observational inaccuracies and, for some clusters, the closeness of the members in the sky, it is not possible to obtain the convergent's position to high accuracy from only two stars in a cluster. As many as possible are used so that the position is as well-determined as possible. It cannot be said either that the method has very general application; it can only be used for a few clusters where the proper motions of the individual stars are really significant in relation to the uncertainties of their measurement.

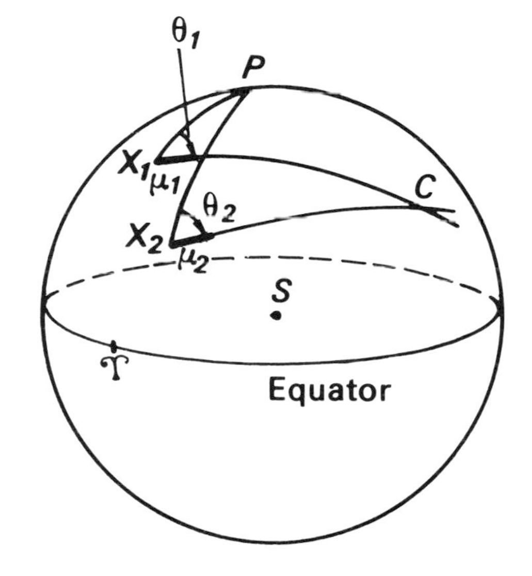

Fig. 10.6 The convergent point of a cluster

10.7 The Parallaxes of the Stars in an Open Cluster

Let us suppose that the radial velocity v_1 of a cluster star X_1 has been measured. Then by Equation (10.8),

$$v_1 = V \cos \beta_1. \tag{10.9}$$

As well as v_1, β_1 is known once the position of the convergent point C has been found, allowing V to be calculated from Equation (10.9). To obtain V as accurately as possible, all the stars in the cluster with known radial velocities are used. The transverse velocity of any star, say X_1, can now be found from the equation

$$u_1 = V \sin \beta_1.$$

Using Equation (10.5), we have

$$u_1 = 4{\cdot}74(\mu_1/P_1),$$

where P_1 is the parallax of X_1. Hence

$$P_1 = 4{\cdot}74\mu_1/V \sin \beta_1, \tag{10.10}$$

from which the parallax P_1 of the cluster member X_1 can be found.

In this way, the parallaxes of all the cluster stars can be determined. It is important to appreciate that distances can be found for clusters well outside the range within which the method of trigonometrical parallax falls, largely because a proper motion is a secular quantity and therefore cumulative over the years, whereas a parallactic shift is periodic.

We shall consider later what other significant information can be derived from a study of clusters, but in the meantime we turn to the question of the Sun's own space velocity.

10.8 The Solar Motion

Consider now the reverse of the problem already dealt with. Let us assume that the stars are fixed and that it is the Sun which moves. We can define its velocity V, in kilometres per second, with respect to a set of rectangular axes OA, OB, OC defined by the system of the fixed stars. Let the Sun move so that in the course of a year it travels from S_1 to S_2, as shown in Fig. 10.7.

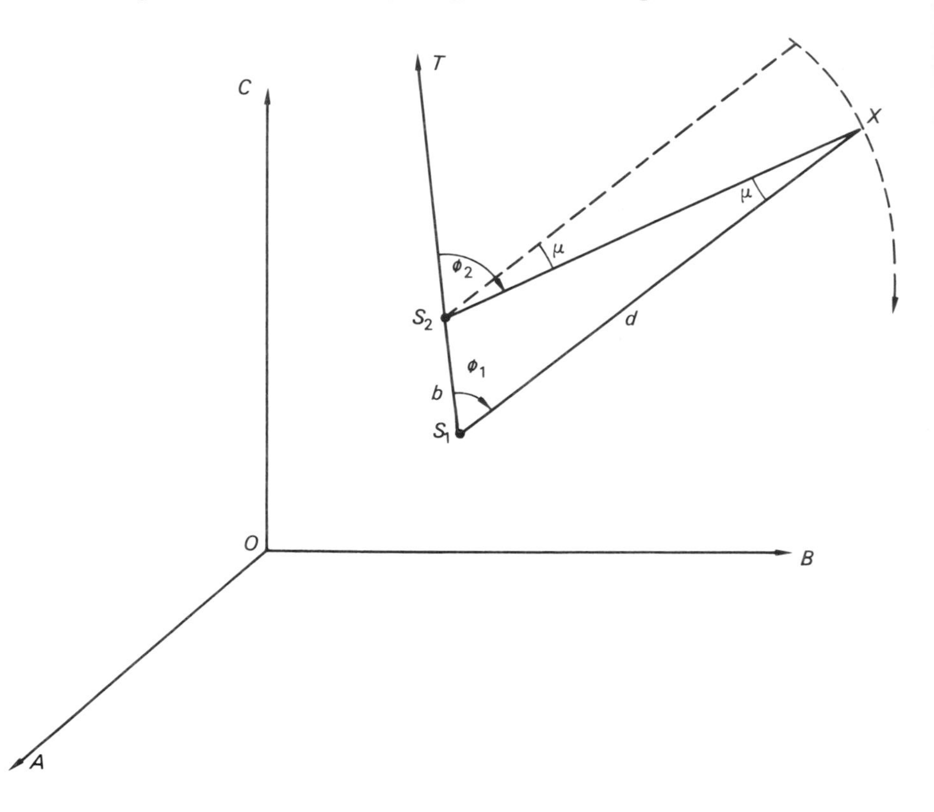

Fig. 10.7 The parallactic proper motion produced by the solar velocity

A star, which was seen to lie in the heliocentric direction S_1X at the beginning of the year, lies in the direction S_2X at the end of the year. It will therefore have exhibited a proper motion μ, given by $\angle S_2XS_1$.

Let T indicate the direction in which the Sun is travelling. This direction makes angles ϕ_1 and ϕ_2 with S_1X and S_2X. If $b = S_1S_2$, and $d = S_1X$, then we may write

$$\sin \mu/b = \sin(180-\phi_2)/d$$

or

$$\sin \mu = \frac{b}{d} \sin \phi_2. \qquad (10.11)$$

Now μ is a small angle so that the suffix "2" can be omitted, since either ϕ_1 or ϕ_2 may be used in the equation. Hence

$$\sin \mu = \frac{b}{d} \sin \phi. \qquad (10.12)$$

Also

$$b = nV \tag{10.13}$$

where n is the number of seconds in a year.

Introducing the star's parallax P, we have

$$\sin P = a/d, \tag{10.14}$$

in which a is the Earth's orbital semi-major axis.

Hence, by Equations (10.12), (10.13) and (10.14), we obtain

$$\frac{\sin \mu}{\sin P} = \frac{nV}{a} \sin \phi,$$

or

$$\mu = \frac{VP}{4{\cdot}74} \sin \phi, \tag{10.15}$$

putting in values for n and a.

This formula shows that if the Sun had a space velocity V, not only would the stars have proper motions even if they were at rest, but that a star's proper motion would be proportional to its parallax and also proportional to the sine of the angle its direction makes with the direction in which the Sun is moving.

The proper motion of a star due to the Sun's own motion is called a **parallactic proper motion** because it is an apparent shift due to the observer's own movement.

Let us consider what kind of pattern the parallactic proper motions of the stars should exhibit. In a general way, the effect is familiar to us in everyday life when we travel in a train or car. Objects such as trees or buildings fixed in the landscape appear to have angular velocities. Those ahead of us separate, moving away from the road ahead. Those which we have passed, close in towards the way behind. Objects on either side stream past. It is noticeable, too, that in any given direction, the nearer objects have more rapid parallactic displacements than those farther away. The "landscape" effect is two-dimensional. In the case of stars, however, the effect is extended to three dimensions. It is expressed by Equation (10.15).

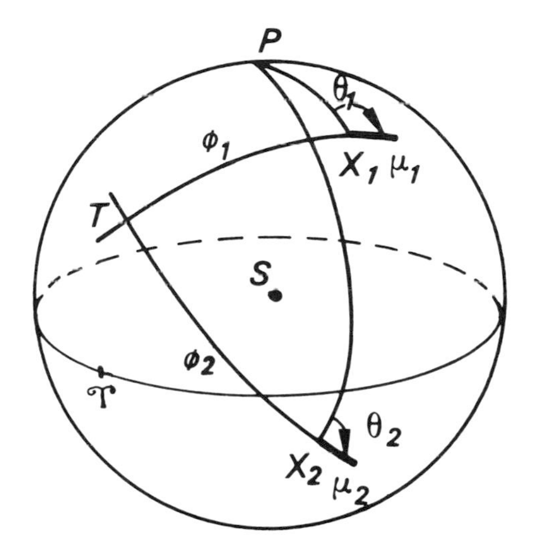

Fig. 10.8 The parallactic proper motion on the celestial sphere

From the point of view of the heliocentric celestial sphere, we can proceed in a manner reminiscent of finding the convergent point of a cluster and its space velocity.

Let X_1 and X_2 be two stars, in heliocentric directions making angles ϕ_1 and ϕ_2 with the direction ST in which the Sun is travelling. Then their parallactic proper motions μ_1 and μ_2 and position angles θ_1 and θ_2 could be plotted on the sphere as shown in Fig. 10.8. If produced *backwards* these proper motions would provide two great circles that would intersect at T, the position on the stellar background to which the Sun is moving. This position is called the **solar apex**. The point opposite it in the sky is called the **solar antapex**.

Once the apex has been found, the values of ϕ_1, ϕ_2, etc. can be determined and a value of V, the solar motion, calculated. The problem is, of course, complicated by the fact that both stars and Sun are moving. This makes it necessary to re-argue the problem and in particular to consider how the set of axes OA, OB and OC can be defined. We have to make some assumption about the distribution of velocities among the stars, including the Sun. This is discussed in the following section.

10.9 The Distribution of Stellar Velocities

Consider a flock of birds hovering in the air. Although the flock does not have a systematic velocity in any direction, the birds themselves will have their own individual velocities within the flock. These velocities can be referred to a set of rectangular axes OA, OB, OC fixed with respect to the flock. In Fig. 10.9 the birds $B_1, B_2, \ldots B_r, \ldots$ and so on, are shown, the arrows in length and direction denoting the birds' own velocities with respect to the axes. If the flock of birds did have a systematic velocity then we could simply apply that velocity to every bird and to O, the origin of the axes.

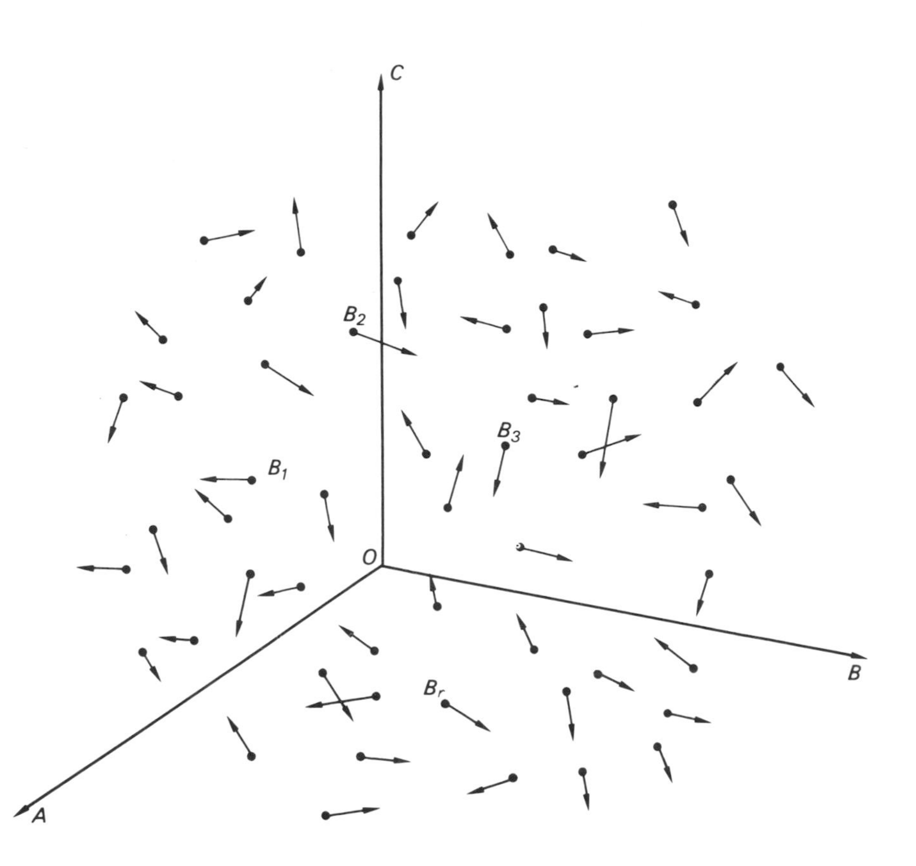

Fig. 10.9 Velocities of a flock of birds

We now assume that there are many birds and that their intrinsic velocities are random in distribution. Then if we sum algebraically the components of their velocities along any given direction, say OA, the sum would be very small indeed because, in all probability, there would be as many positive as negative velocity components of any given size.

The above argument is taken to apply to the tens of thousands of stars in the vicinity of the Sun, *and including the Sun.* It is also taken to apply to any reasonably large sub-group of such stars, numbering a few hundreds, say.

This argument is now used to provide a means of subtracting out from the observed proper motions, the systematic parallactic proper motions imposed on the stars by the solar motion. We want to detect that systematic component and use it to measure (i) the position of the solar apex, (ii) the solar velocity with respect to the "flock" axes OA, OB, OC.

In Fig. 10.10, let all the stars within a small region of the sky be chosen. Let X be the centre of the region, its heliocentric direction S_1X making an angle ϕ with the direction to the solar apex T. For example, we might include all stars within 5° of X.

Suppose there are N stars within the region. Let the observed proper motion in right ascension of the rth star be $\mu_{\alpha r}$. Then we may write

$$\mu_{\alpha r} = \mu'_{\alpha r} + \mu_{\alpha}, \qquad (10.16)$$

where $\mu'_{\alpha r}$ is the proper motion in right ascension produced by the star's own velocity relative to origin O, and μ_α is the parallactic proper motion in right ascension produced by the solar motion. It is seen from Equation (10.15) that the quantity μ_α will be the same for all stars in the region if the values of ϕ for these stars are substantially the same and also their parallaxes. By choosing a small enough region of the sky the former is assured; the latter is more difficult to achieve but, by choosing stars within a narrow magnitude range, some approach towards satisfying this condition is made.

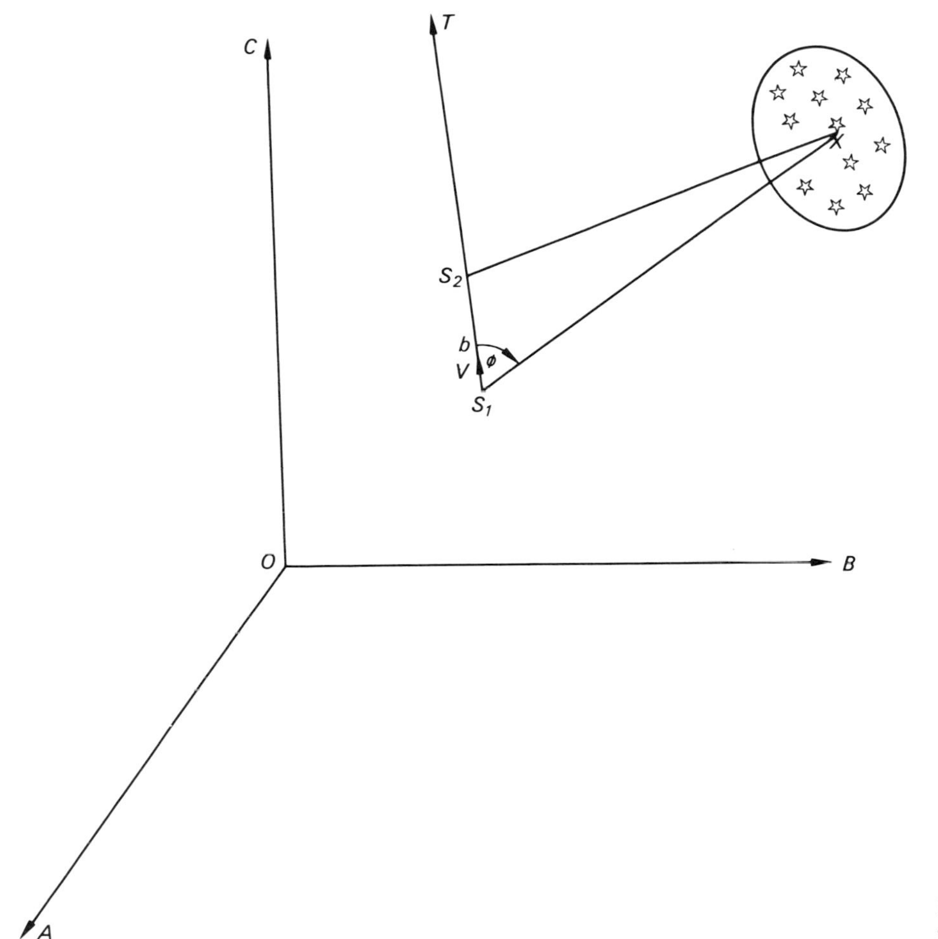

Fig. 10.10 Use of a small group of stars to detect the parallactic proper motion

Then, adding together all the N equations of type (10.16) obtained for all the stars in the region, we may write

$$\mu_{\alpha 1}+\mu_{\alpha 2}+\cdots+\mu_{\alpha r}+\cdots+\mu_{\alpha N}=(\mu'_{\alpha 1}+\mu'_{\alpha 2}+\cdots+\mu'_{\alpha r}+\cdots+\mu'_{\alpha N})+N\times\mu_\alpha.$$

Because of the random nature of the stars' own velocities, the quantity within the bracket will be small in comparison with $N\times\mu_\alpha$ or with the left-hand side which is observed. Hence,

$$\mu_\alpha=\frac{1}{N}(\mu_{\alpha 1}+\mu_{\alpha 2}+\cdots+\mu_{\alpha r}+\cdots+\mu_{\alpha N}).$$

Similarly, if μ_δ is the component of parallactic proper motion in declination for the region stars, we obtain

$$\mu_\delta=\frac{1}{N}(\mu_{\delta 1}+\mu_{\delta 2}+\cdots+\mu_{\delta r}+\cdots+\mu_{\delta N}),$$

with obvious notation.

Thus, for any region whose centre has known co-ordinates α and δ, values are found for μ_α and μ_δ, the components of the parallactic proper motion. By observing a different region of the sky, similar information can be derived enabling the solar apex to be found by using the procedure illustrated in Fig. 10.8.

In practice, many regions are used. The co-ordinates of the solar apex are found to be: RA 18^h, Dec 34° N, approximately the co-ordinates of *Vega*. It is of interest that a value for the position of the solar apex quite close to this was obtained by Sir William Herschel, the first man to tackle this problem, in 1783 from just 13 proper motions.

10.10 The Solar Speed

The solar speed can be found by using the above argument on the observed radial velocities. Again let a small region of the sky centred on X be taken as in Fig. 10.10. The component of the Sun's own velocity V along the heliocentric direction to the region is $V \cos \phi$. This must produce on each star in the region a **parallactic radial velocity** of $-V \cos \phi$.

Now let V_r be the observed radial velocity of the rth star and V'_r the radial velocity caused by its own space velocity. Then

$$V_r = V'_r - V \cos \phi.$$

Summing over all the stars in the region, of number N, say, we obtain

$$V_1 + V_2 + \cdots + V_r + \cdots + V_N = (V'_1 + V'_2 + \cdots + V'_r + \cdots + V'_N) - NV \cos \phi, \tag{10.17}$$

giving

$$V = -\frac{1}{N \cos \phi}(V_1 + V_2 + \cdots + V_r + \cdots + V_N),$$

since, as before, we may assume that the quantities in the bracket in Equation (10.17) will largely cancel out, being randomly distributed in magnitude and direction along S_1X. Again, in practice, a large number of regions are taken.

The value found for V is 19·5 km s^{-1}. It is in fact possible to find the position of the solar apex from radial velocity studies as well. There is good agreement with that found from a study of the proper motions.

It is worth mentioning again that these studies, although involving thousands of stars, are concerned with only a small localized region of the Galaxy surrounding the Sun, a sphere of radius 1000 parsecs (cf. the radius of the Galactic equatorial plane which is 15 000 parsecs).

10.11 Secular Parallaxes

The method of measuring stellar distances by trigonometrical parallax is confined to distances of only a few hundred parsecs. However, use can be made of the Sun's space velocity to obtain a measure of the distances of faint stars.

The **secular parallax** of a star is the angle subtended at the star by the distance through which the Sun moves in one year. Thus

$$\sin \omega = b/d, \tag{10.18}$$

where ω is the star's secular parallax, d its distance and b is the distance moved by the Sun in a year.

But we may write

$$\sin P = a/d,$$

where P is the star's trigonometrical parallax and a is the Earth's orbital semi-major axis.

Then, remembering that both ω and P are small angles, we may write

$$P=\frac{a}{b}\omega. \tag{10.19}$$

But $a=149{\cdot}6\times10^6$ km; also $b=nV$, where $n=31{\cdot}56\times10^6$ s and $V=19{\cdot}5$ km s^{-1}, giving

$$P=0{\cdot}243\ \omega. \tag{10.20}$$

Now the parallactic proper motion μ is given by Equation (10.12), viz.

$$\sin\mu=\frac{b}{d}\sin\phi,$$

so that

$$\sin\mu=\sin\omega\sin\phi, \tag{10.21}$$

or

$$\mu=\omega\sin\phi. \tag{10.22}$$

If stars within a small region of the celestial sphere and within a small magnitude range are chosen, a value of the parallactic proper motion μ can be derived for them. Since ϕ is known for the centre of the region, Equation (10.22) can be used to calculate the secular parallax ω of these stars. Equation (10.20) can then be applied to obtain their trigonometrical parallax P. The process does not yield an accurate value of P for the stars. For one thing, the assumption that the stars within a small magnitude range are all at approximately the same distance is certainly not true. The value obtained for P in this way, however, may be considered to be of the right order of size and can be used in a number of investigations where high accuracies are not required.

10.12 The Convergence of an Open Cluster: Graphical Exercise

Details of some of the stars of the Hyades cluster are presented in Table 10.1. The data have been extracted from *Catalogue of Bright Stars* (Yale University Observatory, editor D. Hoffleit (1964)). For each star the RA and declination are given for 1900, the magnitude listed and the annual proper motions are given in seconds of arc. A scheme for performing the necessary calculations is presented in Table 10.2. This same table lists the radial velocity of each star.

Table 10.1 The Hyades

Star HR	RA h m	Dec.	Magnitude	$\mu_\alpha \cos\delta$	μ_δ
1201	3 47	+17° 02′	5·97	+0″141	−0″028
1279	4 2	+14° 54′	6·01	+0·131	−0·024
1283	4 3	+19° 21′	5·49	+0·103	−0·032
1319	4 10	+15° 09′	6·32	+0·115	−0·029
1373	4 17	+17° 08′	3·76	+0·105	−0·031
1389	4 20	+17° 42′	4·3	+0·107	−0·029
1392	4 21	+22° 35′	4·29	+0·100	−0·047
1394	4 21	+15° 23′	4·48	+0·111	−0·023
1409	4 23	+18° 58′	3·54	+0·106	−0·038
1411	4 23	+15° 44′	3·85	+0·101	−0·028
1427	4 25	+15° 59′	4·78	+0·105	−0·028
1473	4 33	+12° 19′	4·27	+0·099	−0·012
1620	4 57	+21° 27′	4·64	+0·056	−0·043

Table 10.2 Calculations with the Hyades data

Velocity of the cluster

Star (HR)	$\tan\theta$	θ	β	$\sec\beta$	$v(\mathrm{km\,s^{-1}})$	V
1201					+35	
1279					36	
1283					24	
1319					37	
1373					38	
1389					35	
1392					35	
1394					41	
1409					39	
1411					40	
1427					38	
1473					45	
1620					42	

$\Sigma V =$ ......

Velocity of the cluster $(\bar{V}) =$ ......

The position of the convergent point of the Hyades is: $\alpha =$ ...... $\delta =$ ......

Parallax determinations

Star (HR)	$(\mu_\alpha \cos\delta)^2$	μ_δ^2	μ^2	μ	$\sin\beta$	u $(V\sin\beta)$	P
1201							
1279							
1283							

Plot the position of each star. (A convenient scale is 10 mm = 2°; the RA scale should be extended beyond the values given for the tabulated stars, to 7 hours; it is usual to plot increasing RA from right to left.)

Now by Equations (10.2), the position angle of the proper motion of a star is given by

$$\tan\theta = \mu_\alpha \cos\delta/\mu_\delta.$$

After θ has been determined, use a protractor and ruler to draw the apparent direction of the proper motion of each star. For this exercise, the presented data do not have a high accuracy and to a first approximation the curvature of the celestial sphere can be neglected. For clarity, the lines of apparent direction are best drawn in ink or ball-point. Extend these lines across the star plot.

The position of the convergent point of the cluster may be estimated as follows. Place a large pencil dot at each point where any two lines of apparent direction intersect. Draw an oval on the plot so that it includes the part where the pencilled dots have their greatest concentration. Take the position of the centre of this oval as being the position of the convergent point. Read off the RA and declination of this point.

If β is the angular distance of any star from the point of convergence, then the radial velocity v is given by

$$v = V\cos\beta$$

where V is the velocity of the star, this being common to all stars of the cluster. Hence,

$$V = v\sec\beta.$$

For each of the plotted stars, measure the angle β. Using the tabulated values for the radial velocities, determine a value of V for each star and determine a mean value for the velocity of the cluster.

We have also seen by Equation (10.5) that the parallax of a star may be determined if its proper motion and transverse velocity are known, i.e.

$$P = 4{\cdot}74\ \mu/u.$$

Now $u = V \sin \beta$ and $\mu = ((\mu_\alpha \cos \delta)^2 + \mu_\delta^2)^{1/2}$. Hence, by knowing V, β and μ the parallax of any star may be determined. A scheme for evaluating the parallaxes of three stars of the cluster is given in the lower part of Table 10.2.

Problems: Chapter 10

1. A star of declination 55° N has a parallax of $0''130$ and a radial velocity of $+22$ km s^{-1}. Its components of proper motion are

$$\mu_\alpha = 0^{\mathrm{s}}0091; \qquad \mu_\delta = -0''115.$$

Calculate the total proper motion, the transverse speed, and the total space velocity of the star relative to the Sun.

2. In an open cluster a star X, of declination 16° 39′ N, has components of proper motion

$$\mu_\alpha = 0^{\mathrm{s}}010; \qquad \mu_\delta = 0''019,$$

its angular distance from the convergent being 33° 12′. The radial velocity of star Y, another member of the cluster, at angular distance 31° 56′ from the convergent, is $+25{\cdot}6$ km s^{-1}. Calculate the distance, in parsecs, of X.

3. The angular distance of a member of the Taurus cluster from the convergent point is 21° 24′, its radial velocity is 36·9 km s^{-1} and its proper motion is $0''097$. Find (i) the speed of the cluster, (ii) the distance of the star in parsecs.

4. The declination of a star is 42° 57′ N and its proper motion components are

$$\mu_\alpha = -0^{\mathrm{s}}0374; \qquad \mu_\delta = 1''21.$$

Calculate its total proper motion.
If the radial velocity of the star is $-7{\cdot}6$ km s^{-1} and its parallax is $0''376$, calculate its space velocity relative to the Sun, and its total proper motion at the time of closest approach to the Sun.

5. A star with co-ordinates (0^{h}, 0°) has a parallax of $0''100$. The components of its proper motion are

$$\mu_\alpha = 0^{\mathrm{s}}089; \qquad \mu_\delta = -0''325.$$

Spectroscopic radial velocity measurements on June 21st revealed a red shift of 0·513 Å in the line Hβ (4861 Å). Assuming the velocity of light to be 3×10^8 m s^{-1} and the astronomical unit to be of length $149{\cdot}5 \times 10^9$ m, calculate the star's space velocity relative to the Sun.

11 Stellar Populations: Observational Evidence

11.1 Introduction

We are now familiar with the basic measurements and their interpretation concerning the structure of stars and their motions in space. It might be thought that there should be no reason for there to be connections between individual variations in the physical constitutions of stars and their space velocities. However, during the last thirty years a loose empirical connection has been discovered. This relationship is interpreted by considering the stars to be of different ages and not born at the same time. In turn, by now using this established link of age between the associated properties of constitutions and space velocities, either a star's particular physical nature or its space velocity may be used separately to act as an indicator of its age.

Although there is no direct observational method for determining a star's age—with the exception of the Sun, where geophysical evidence gives a minimum age $\approx 4{\cdot}5 \times 10^9$ years—there are many pointers to a relative age scale. We shall see that slight differences in constitution and the differences in space velocities form part of the information which is used in developing ideas about the ages of stars.

In our own galaxy there are basically two kinds of star cluster, the **open** or **galactic cluster** and the **globular cluster**. It is reasonable to assume that the stars in a particular cluster were all formed at about the same time, but that the individual clusters may have formed at different times. We have already seen in Chapter 9 that the HR diagrams of clusters vary from one cluster to the next and that a particular pattern is indicative of the cluster's age. Let us look more closely at the types of stars that constitute stellar clusters.

11.2 Stellar Clusters

By considering the brightest stars in clusters, it is found that in galactic clusters they are blue supergiants, whilst in globular clusters they are red giants. It is also noted that blue supergiant stars are only found in or near clouds of interstellar gas and dust. This suggests that these stars are young and have not, as yet, had sufficient time to dissociate themselves from the material from which they have condensed. Several galactic clusters containing blue supergiants are observed to be expanding. Measurements of their present size and of the expansion rate tell us that their ages are less than 10^8 years which, cosmically speaking, is young.

There is little direct evidence for the ages of the luminous red giants found in globular clusters. However, it can be noted that these systems are devoid of gas and dust, suggesting that sufficient time has elapsed for any such material to have been swept up by the stellar members. Globular clusters are highly stable dynamical systems and are capable of a long life without significant dispersion of their members.

On this simple reasoning, stars have been divided into two groups called **Population I** and **Population II**. Population I contains the blue supergiants and refers to recently formed stars; Population II contains the red giants and refers to old stars. This definition may be summarized in an abbreviated way by:

Population I—Young stars
Population II—Old stars.

That galactic clusters, containing Population I stars, and globular clusters, containing Population II stars, are different in nature is perhaps not unexpected. We shall see in Chapter 15 that their distributions in space within our galaxy are fundamentally different.

Other evidence, presented below, also suggests that the division of stars into two populations—young and old—is reasonable. We find that no distinct line can be drawn between the two populations and that the populations can be subdivided. There is therefore a range in population classification. Blue supergiants represent the extreme of Population I stars and the luminous red giants represent the extreme of the Population II stars. Differences in stellar populations can be clearly seen in other galaxies (see Fig. 11.1, for example).

Fig. 11.1 (*a*) A spiral arm of the Andromeda Nebula (blue light) showing giant and supergiant stars of population I and (*b*) NGC 205 (yellow light) showing stars of population II. 200-in Hale telescope. (By courtesy of the Mt. Wilson and Palomar Observatories)

11.3 Kinematic Behaviour

It has been found that there is a correlation between the stars' astrophysical characteristics and their motions within the Galaxy. When the nearby stars are considered according to their velocities and categorized into two groups of **high velocity stars** and **low velocity stars**, it is found that the HR diagrams for these groups are quite different. A plot of the high velocity stars provides an HR diagram resembling that of a globular cluster. The high velocity stars contain quite a proportion of red giants. On the other hand, the HR diagram of the low velocity stars resembles those of galactic clusters, there being a good representation of O and B stars.

This correlation is strikingly confirmed by considering stellar clusters themselves. Galactic clusters have low space velocities while globular clusters have very high ones.

The highest space velocities are found for RR Lyrae variables (Section 12.2.4) and for the subdwarfs (Section 7.9). RR Lyrae stars are frequently found in globular clusters and so it is no surprise that they have high velocities. The high velocities of the subdwarfs suggest that the apparent main sequence stars of the globular clusters are perhaps more closely related to subdwarfs than to dwarfs.

The most common nearby stars—the ordinary dwarfs on the main sequence—exhibit intermediate velocities between the high and low velocity stars. This suggests that there is a continuous gradation of populations. Thus the range of populations might be represented according to the scheme:

Young I–Intermediate I–Old I–Mild II–Extreme II.

It is also found, as we shall see later in Chapter 15 that if stars are placed in this population scheme according to their velocity, each grouping has a distinctive distribution within the Galaxy.

11.4 Spectral Differences

Detailed investigation of stellar spectra reveals differences between the high and low velocity stars. The spectral differences reveal themselves as abnormal weaknesses of some lines and bands and abnormal strengths of others. They perhaps give means of classifying individual stars into populations rather than applying statistics to groups of stars.

It is also possible to subdivide Population I stars into "weak" and "strong" line groups. Statistical investigations show that the weak-line stars have higher velocities than the strong-line ones. This analysis agrees with the observation that metal lines are weak in globular cluster giants and in subdwarfs. The Sun may therefore be described as a weak-line Population I star.

11.5 Initial Composition

There is no reason to suppose that there are significant initial compositional differences between the centre and surface of a star. However, there are good reasons according to the theory of the internal structure of stars for believing that the surface composition is hardly changed during the major part of a star's lifetime. As the surface composition is responsible for the impressed absorption spectrum of the star, we can say that quantitative analysis of the spectral lines allows the initial composition of a star to be deduced.

It is convenient to consider the constituents as being hydrogen, helium and other elements. Hydrogen is the most abundant of the elements in a star's make-up, followed by helium. The remaining elements are represented in a very small proportion. For example, if we consider the abundances by weight, in the case of the Sun, hydrogen forms 73%, helium 25% and other elements form 2%.

When abundance measurements are applied to the scheme of population classification, it is found that the fraction of "other" elements, Z, (other than hydrogen and helium) changes continuously from one extreme of the classification to the other. Thus for Young I stars, Z is of the order of 4% but for Extreme II stars, Z is of the order of 0·3%. The Sun, with a value of Z equal to 2% fits into the scheme as being an Old I star.

We see that the Old Population II stars have very little of the heavy elements, whereas the Young Population I stars have of the order of ten times more. This suggests a build-up of heavy elements with time in the interstellar material. As stars condense out of this material, they inherit higher heavy element abundances. This growth of heavy elements with time requires explanation. We have seen in Chapter 9 that the very mechanism which provides the energy for stars to shine also provides the source of growth of heavy elements.

11.6 Conclusion

We have now seen that stars can be classified into populations and that the classification scheme is based on independent observational evidence. The scheme is summarized in Table 11.1. Boundaries in the scheme are fairly arbitrary and are not intended as being sharp and distinct. They represent a continuous sequence of stellar ages.

Table 11.1 The scheme of population classification

Population	Typical members	Typical space velocity	Abundance of other elements, Z (%)
Young I	Blue supergiants Galactic clusters	$10\ km\ s^{-1}$	4
Intermediate I	Strong-line } Majority of nearby stars	$20\ km\ s^{-1}$	3
Old I	Weak-line } Majority of nearby stars	$30\ km\ s^{-1}$	2
Mild II	Majority of high-velocity stars	$50\ km\ s^{-1}$	1
Extreme II	Luminous red giants Globular clusters	$180\ km\ s^{-1}$	0·3

It has already been mentioned that the spatial distributions of the groups of stars in this classification are different. This point will be taken up again when the Galaxy is discussed in Chapter 15. At the same time reasons will be suggested as to how such observed differences in the stars, presented in Table 11.1, come about.

12 Stars with Special Properties

12.1 Introduction

So far we have only considered stars or pairs of stars that fit the normal pattern. There are however several other types of star which are found in the universe. The unusual characteristics of the majority of these objects are brought to our attention by changes in brightness. Other stars show variation of spectral features such as the strength and shape of an emission line. Studies of light curves and spectral variations provide means of learning about the interiors and the atmospheres of these stars and their relationships to what might be classed as normal stars. It is found that the brightness changes are due to short term fluctuations in the intrinsic output or luminosity. Some of the spectral changes can tell us about events occurring in the stellar atmosphere; for example, the ejection of material in the form of a shell and the transfer of material from one star to another under the same extended atmosphere, are revealed through changes recorded in spectral lines. As well as providing means of understanding these particular objects, we shall see that their properties have in some cases provided us with more general information about the Universe. There are some stars which exhibit special features which may not be variable. Some of these objects are also described in this chapter.

Fig. 12.1 Pairs of images of stars; note the central pair are of different sizes, illustrating a change of brightness of the star. This is WW Cyg and is recorded here at its maximum and minimum brightness. (By courtesy of the Mt. Wilson and Palomar Observatories)

12.2 Variable Stars

12.2.1 Introduction We have already seen in Section 7.5.4 that some stars exhibit brightness fluctuations which are regular and which can be interpreted by considering the radiating source to be a binary star system exhibiting eclipses. A simple inspection of the light curves of many of the recorded variable stars shows that they cannot be explained by the star being an eclipsing binary. Although in some cases the fluctuations are periodic, an eclipsing binary model cannot be produced to provide a matching shape of light curve to fit the recorded data. Indeed some of the light curves do not exhibit periodicity and are in fact irregular. The brightness fluctuations of the stars discussed below are caused by variations in the stars' luminosities and so they may be said to be **intrinsic variables.**

That some stars show variability has been known for a long time. A sudden appearance of a new star (nova) in 134 B.C. prompted Hipparchus to catalogue the stars and assign brightnesses or magnitudes to them. A few stars are sufficiently bright and exhibit sufficiently marked variations that they can be studied by eye. More stars are added to the list of variables when a telescope is used. Thus the fact that a star like *Mira* (*o* Cet) shows variations in its brightness has been known for a long time. The variation of this star covers a range of more than six magnitudes. When at full brightness it can be seen by the unaided eye; at minimum brightness it falls well below the level of visibility and a telescope is required to observe it. Observations of this star by Fabricius in 1596 can be considered as being the forerunners in what has now developed into a large branch of observational astronomy.

The application of photography to variable star studies provided a revolutionary breakthrough in the method of observation. Not only could the measurements be made on a more objective basis but also many new variable stars were discovered. In 1866, Schoenfeld's list of known variable stars presented 119 such stars. When photography was used, the variable star list quickly grew beyond 1000, many variable stars being discovered on a single plate. The latest edition of the *General Catalogue of Variable Stars* produced by the Russian Academy of Sciences now lists a total of 20 437 variable stars.

Because of the limitation of being able to look at only one star at a time, the advent of photoelectric devices has not significantly increased the known numbers of variable stars. However, the increased sensitivity of photoelectric photometry enables the light curve of a known variable star to be recorded with improved accuracy over the photographic technique.

Some of the brightness variations are exhibited with characteristic patterns and several variable star types have been designated. Each classification of variable is usually labelled according to the name of one famous star within the classification. This star is sometimes known as the prototype of the classification. For example, stars which exhibit a light curve similar to that of δ Cephei are known as Cepheids. As we shall see, there are, in fact, several distinct varieties of Cepheid and each type is labelled accordingly.

Other types of variable are labelled as being T Tauri stars, RV Tauri stars, U Geminorum stars, etc. It would be out of place here to present in detail all the types of variable star which are known. We will confine our discussion to those which provide us with the most information and those which are most interesting or most spectacular.

12.2.2 Cepheid Variables The study of Cepheid variables has played a very important part in the development of our knowledge of the universe. Cepheids are, of course, interesting in their own right but, as we shall see, their chief importance is in providing a means of measuring distances which are beyond the limit which can be measured by trigonometrical parallax. They provide the means of extending our distance scale.

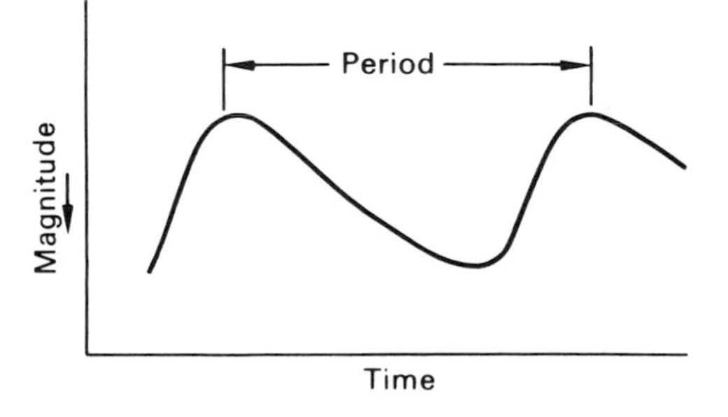

Fig. 12.2 The light curve of a classical Cepheid

The light curves of classical Cepheids are regular in their periodicity and the changes in brightness are smooth. Typical curves (see Fig. 12.2, for example) show that the increase in brightness is rapid and that after maximum brightness, the decrease is at a slower rate. The results from multi-colour photometry show that the brightness variations are more pronounced in the blue than in the red.

Typical changes in magnitude between minimum to maximum are of the order of one magnitude. Cepheids have a range in period from about 1 to 50 days but most of them are close to 5 days. The star δ Cep itself has a range in magnitude from 3·71 to 4·43 with a period of 5·37 days. Another famous star—α UMi; (*Polaris*)—is also a Cepheid with a range in magnitude from 2·08 to 2·17 and a period of 3·97 days.

During each period of the brightness variation, changes in the Cepheid's spectrum can also be observed; the spectrum may change by as much as one classification. Most stars in this group have spectra which at maximum brightness lie in the range F0 to G0 and which change to lie in the range G0 to K5 at minimum brightness. The prototype star, δ Cep, for example, presents an F4 spectrum at maximum brightness and a G6 spectrum when it is at minimum. Classical Cepheids are more luminous than the main sequence stars of the same spectral type. Their absolute magnitudes are typically between -2 and -6. Thus they might be considered as being giant or supergiant yellow stars. Because of their high intrinsic luminosity, they stand out significantly against other stars at the same distance. However, they are not very common. Only about 500 classical Cepheids have been recorded in our galaxy.

Another fascinating feature of the Cepheid is that the spectral lines exhibit Doppler shifts which vary. When the magnitudes of the shifts are plotted against time, it is found that they undergo a periodic cycle. These periods match the light curves exactly. When the star is most luminous, the Doppler shift is a blue shift; when the star is at minimum brightness, the Doppler shift is a red shift. The pattern of change cannot be explained in the same way as the corresponding change in the spectroscopic binary. It can only be interpreted by considering the star to be pulsating in the period given by the light curve. At maximum brightness, the star's surface is approaching the observer and must therefore be expanding, the speed of approach being at a maximum when the luminosity is a maximum. At minimum brightness, the star's surface exhibits its maximum recession from the observer and is in the process of contraction. The velocity of the Cepheid's surface when expanding or contracting is extremely high. Typical speeds are of the order of 20 km s^{-1}.

It would be out of place here to discuss the behaviour of Cepheids in any more detail. One point worth mentioning, however, is that it is possible to show on a theoretical basis that there is a simple relationship between the period, P, of the light curve and the mean density, $\bar{\rho}$, of the Cepheid. If P is measured in days and $\bar{\rho}$ in kg m^{-3}, then the relationship may be expressed in the form

$$P^2\bar{\rho} = 9{\cdot}7. \qquad (12.1)$$

This equation is in good agreement with the observations of P and $\bar{\rho}$ for Cepheids and it is therefore strong confirmation that a pulsating model provides the correct interpretation of the light curve and the periodic changes in the radial velocity.

The pulsating theory also tells us that there should be a connection between the period of a Cepheid and its luminosity. If we consider for simplicity that all Cepheid masses are roughly the same, then, by Equation (12.1), those with the longer periods are less dense and are therefore larger. But, in general, the larger the star the greater is its luminosity. Thus it is expected that the longer the period, the more luminous the Cepheid will be. That this is so was discovered empirically prior to the theoretical work on the pulsating model.

12.2.3 The Period–Luminosity Law In 1912, Miss Leavitt, who was working at Harvard, made an investigation of the star system known as the Small Magellanic Cloud. This system contains numerous Cepheid variables. Now it can be assumed that as the distance of the system of stars is great, each member can be taken to be at the same distance from us. This assumption is reasonable because the area occupied by the system on the celestial sphere is small and it does not show appreciable parallax. Differences in apparent magnitude must therefore reflect differences in absolute magnitude.

From her study of the periods of the Cepheid variables in the cloud Miss Leavitt was able to plot apparent magnitude against the period. In doing so she discovered the *period–luminosity* law. This now-famous curve, normally plotted as magnitude against the logarithm of the period (see Fig. 12.3), showed that for this particular group of Cepheid variables there is a relationship between the apparent magnitude and the period of the light curve. Since there is a direct relation between apparent and absolute magnitude, there must also be a similar relationship between period and *absolute* magnitude.

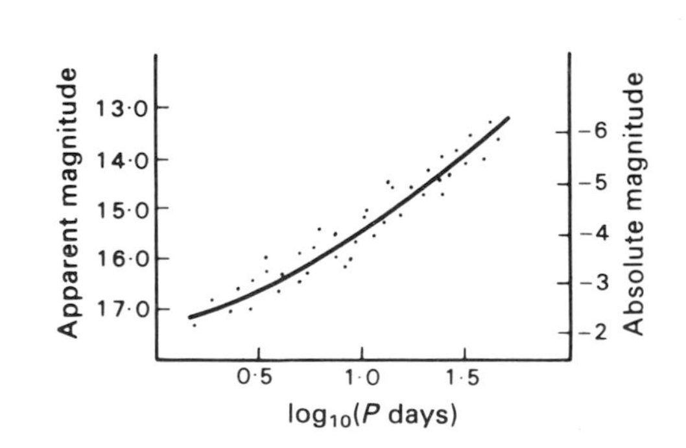

Fig. 12.3 The period–luminosity law. (For convenience, the magnitude of a Cepheid is taken as the mean value given by the maximum and minimum from its light curve)

If a set of Cepheid variables can have absolute magnitudes assigned to them, the relationship between their periods and absolute magnitudes acts as a distance-measuring device. When a Cepheid is discovered, its period can be observed. Knowing the period–luminosity law, an absolute magnitude may be obtained. Comparison of this with its apparent magnitude thus provides its distance modulus. All then that is required for this method to be applicable is to provide a zero point for the period–luminosity relationship or, in other words, investigate the relationship for Cepheids whose distances are accurately known. At the time of the law's discovery, however, the distance of the Small Magellanic Cloud was not known.

The problem of setting a zero point to the law is not easy and is open to systematic errors. Many of the Cepheids are found to lie in dusty regions of the Galaxy and some light is lost due to absorption, making them appear less bright than they would normally be according to their distance. When Shapley set out to determine the zero point, he found that even the nearest Cepheid variables were too far away to provide accurate trigonometric parallaxes and he was therefore unable to obtain reliable values of their absolute magnitudes. He had to resort to the method of secular parallax which, it will be recalled, is essentially a statistical method unable to provide answers of extreme accuracy. Also, he was not aware that there are, in fact, distinct types of Cepheid, belonging to the different Populations I and II, and that these have different values of absolute magnitude.

Accordingly, the simple period–luminosity law found by Miss Leavitt has been modified. We now have period–luminosity laws for each kind of variable which exhibits light curves akin to those of the classical Cepheid. They are depicted schematically in Fig. 12.4. By choosing the correct law appropriate to the type of variable star that is observed, a value is given for its absolute magnitude, M. The distance, r, of a cluster of stars containing the variable can then be determined by measuring the apparent magnitude, m, and applying the distance modulus, given by Equation (7.2), viz.:

$$m - M = 5 \log_{10} r - 5.$$

Before Baade's investigations (see Section 15.9) of the pulsating variables, it was thought that the pulsating stars of short period—RR Lyrae stars—were short-period Cepheids and they they fitted on a single smooth period–luminosity curve. However, after study of the Cepheids in the Andromeda Galaxy, it became apparent that the classical Cepheids are more luminous than had first been thought and that there is a distinct gap in luminosity between the

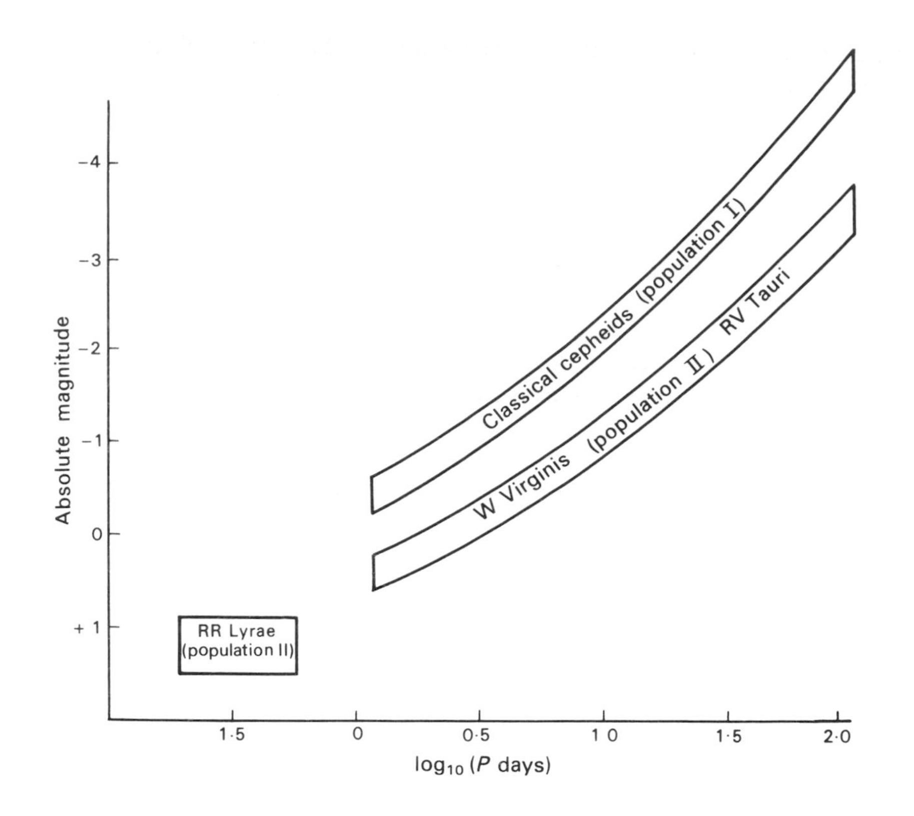

Fig. 12.4 The period–luminosity laws of pulsating stars

short-period classical Cepheids and the long-period RR Lyrae stars. These two types of stars have now been shown to be fundamentally different, the classical Cepheids being Population I stars and the RR Lyrae variables being Population II.

12.2.4 RR Lyrae Stars Although there are some RR Lyrae stars in the local neighbourhood of the Sun, the majority are found in globular clusters and they are sometimes referred to as **cluster-type** variables. They are classified as Population II stars.

The light curves of RR Lyrae stars are quite like those of classical Cepheids. There is a fairly steep rise followed by a slower decline, during which there is sometimes a small secondary rise, before continuing down to the minimum. Their periods, however, are much shorter, being less than a complete day. Typical values range from just over an hour to just over a day. Radial velocity changes are observed to match the period of the light curve, in a similar way to that of classical Cepheids. There are also spectral peculiarities which change during the period. Unlike the Cepheids, though, RR Lyrae stars show no relation between their periods and their mean spectral type. It is noted also that the spectra of the longest period RR Lyrae stars are quite different from those of the shortest period Cepheids. The range in the mean spectra is also different running from A to F. RR Lyrae stars appear to be pulsating stars but they should be not be considered as being short-period Cepheids.

Measurements of RR Lyrae stars can also be used to determine distances according to their period–luminosity law. From observations of stars whose distances are already known, the law suggests that the intrinsic luminosity of RR Lyrae variables is independent of their period. This is illustrated schematically in Fig. 12.4. There remains some uncertainty as to what their absolute magnitudes are, but the most recent data provide a figure of M equal to +0·6. Accordingly, the distance of any group of stars containing RR Lyrae stars whose mean apparent magnitudes can be measured is given by Equation (7.2), in the form

$$m - 0{\cdot}6 = 5 \log_{10} r - 5$$

or

$$m = 5 \log_{10} r - 4{\cdot}4. \tag{12.2}$$

12.2.5 W Virginis and RV Tauri Stars Soon after the discovery that RR Lyrae variables could not be considered as being short-period classical Cepheids, it was also found that some of the longer period pulsating stars were also different from the classical Cepheids. These stars are predominantly found in globular clusters and are therefore classified as belonging to Population II.

They differ from the classical Cepheid in that, after the maximum brightness has been reached, the fall is sometimes bumpy. Bright emission lines of hydrogen are observed or may be enhanced close to the period of maximum brightness. Another interesting feature differentiating these stars from the classical Cepheids is that at the end of the minimum period of brightness, the absorption lines may be seen to be doubled, showing that at that particular moment there are two "waves" of material moving at different speeds within the star.

From observations of these stars in globular clusters, the period–luminosity law obtained is much the same as that for classical Cepheids but with the important distinction that for identical periods the Population II Cepheids are less luminous, typically by 1·5 magnitudes.

Two types of variable which have been classified according to the prototype stars W Virginis and RV Tauri are found to be the Population II Cepheids. Thus it is now accepted that these stars form a continuous range, being basically the same type. The Population II Cepheids have periods of a few days and their continuous range merges into W Virginis stars with periods between 10 and 30 days and these in turn merge into the RV Tauri stars with periods from 30 to 150 days. The period–luminosity law for this sequence of stars is shown schematically in Fig. 12.4.

12.2.6 Long-Period Variables This type of variable is not well defined as the stars in the group show a large range in their characteristics and there are several subdivisions in the group. Periods of this wide class of variable run from just under 100 days to 1000 days. They are all giant stars with spectra which are generally M-type. It will suffice us here to describe the features of just one star from the broad class, this being *Mira* (*o* Cet).

The period of this star is not exactly regular. The time interval between successive maxima in brightness is on average 330 days, but this interval in fact varies from 320 to 370 days. The star's maximum brightness usually lies between 3rd and 4th magnitude, making it possible to be seen by the unaided eye. At minimum brightness it has faded to 9th magnitude. This fluctuation in brightness of six magnitudes therefore corresponds to a change in intrinsic luminosity of a factor of over 100. From maximum to minimum brightness the spectral classification also changes from M6 to M9.

Long-period variables frequently show in their spectra the hydrogen lines in emission and the strength of the emission is seen to vary through the period. Mira again provides a good example of this phenomenon. At maximum brightness of the overall light, the emission lines are strong. When the spectrum is examined at minimum brightness, the line emission is not as prominent.

12.2.7 T Tauri Stars T Tauri stars exhibit a highly irregular light curve with brightness changes which may be as much as three magnitudes. They have spectra which show emission lines which are also variable in strength. Occasionally there are times when the emission lines are not present and the spectrum becomes more like that of an ordinary dwarf star.

These stars are only found in association with gaseous and dusty material. It has been suggested that they are extremely young and of low mass. As we see them, they are still undergoing gravitational collapse prior to settling on the lower part of the main sequence.

12.2.8 Explosive Variables This group of variable stars contains several types of star which exhibit luminosity changes occurring in relatively short periods of time. The range in luminosity change might be as little as a fraction of a magnitude or as large as sixteen magnitudes, taking the extreme case of a supernova. In most cases, the outbursts appear to be quite random in nature. Between outbursts, the star behaves in a stable way and is apparently unaffected by the previous outbursts. However, for the stars which undergo a large luminosity change, typical of a nova or a supernova, the outburst is considered to be catastrophic in that the star does not return to its original quiescent state.

One type of explosive variable comprises the UV Ceti stars. These stars appear to be yellow or red dwarfs. In just a few seconds they may increase in brightness by a few magnitudes and then return to their original brightness over a period of a few tens of minutes. The characteristic pattern of the changes is reminiscent of flare phenomena on the Sun and these variables have been called **flare stars**. It must be pointed out though that the mechanism cannot be exactly the same for, in the case of the Sun, fortunately for us, a sudden flare does not cause an overall luminosity change in the visible light of 10 or more times.

Another type of variable are U Geminorum stars. These stars show sudden increases in brightness over a few hours and then return to their former brightness over a period of a few days. Although the outbursts are irregular, they are usually separated by an interval which does not vary too greatly for any one star. The characteristic interval of quiescence for a star of this type is typically in the range of 20 days to 2 years.

The spectral changes that occur during the outburst of U Geminorum stars resemble those which are observed in novae. For this reason, and also because the light curves resemble those of novae, but on a smaller scale, these stars are sometimes called **dwarf-novae.**

12.2.9 Novae **Novae** is the collective name given to particular stars which increase their magnitudes typically by 12 magnitudes in a relatively short time. The name originates from the early periods in astronomy. It has been recorded in history that new stars (novae) were occasionally observed where previously no star had been seen. An example that can be cited is the nova already mentioned, this being the one noted by Hipparchus which prompted him to make his star catalogue. By careful examination of previously taken photographs we now know that the newly observed novae are in fact already existing stars which have suddenly increased in brightness, perhaps becoming sufficiently bright to be observed by the unaided eye. The name "nova", however, has remained in use to describe this type of event.

After reaching its maximum brightness, the nova begins to fade relatively slowly, taking several months to return to near its pre-outburst brightness. During this period the light level is sometimes seen to oscillate for a few days before continuing to fade. Occasionally some novae are seen to have a second outburst during the period of decay. Without the record of spectra, these observations of the light curve itself would be difficult to interpret.

Spectra of novae taken during their activity reveal that a shell or shells of material are ejected from the original star. Obviously it is not always possible to observe the nova prior to its maximum brightness, for it may not have been discovered by then. Those which have been observed spectrometrically prior to maximum brightness exhibit a large blue Doppler shift revealing that the star is expanding at a great speed. (A radial velocity of -1700 km s^{-1} was observed for Nova Aquilae in 1918.) As soon as the maximum brightness is reached, the spectrum becomes more complicated with broad emission lines beginning to appear.

Nevertheless the general interpretation of the spectrum of a nova is fairly simple. Very briefly, the absorption spectrum of the central star can be seen through the expanding shell. As the shell expands, recombination produces emission lines. These lines are seen to be broad due to the shell's expansion in all directions from the central star—the light that we receive is contributed to from all parts of the shell and therefore suffers a range of Doppler shifts. The light from the near part of the shell is blue-shifted, that from the rear, not obscured by the central star, is red-shifted and that from the edges, having only a transverse velocity, has no shift. The emission lines sometimes show multiplicity, revealing that more than one shell is being blown off the star.

For some novae, the shells may sometimes be photographed directly some months or years after the catastrophic event.

Little is known as to why some stars produce the internal conditions which eventually give rise to these events, or of the types of star which are likely to undergo these dramatic changes. At one time it was suggested that every star must go through the nova stage at some point in its evolution, but consideration of the number of novae which are observed with the total of stars present in the universe and their expected life-spans now rules out this idea.

12.2.10 Supernovae Very occasionally some stars become unstable and exhibit an extremely large outburst, giving a brightness change much larger (by as much as 8 magnitudes) than for a nova. After reaching a maximum value the brightness then decreases over a period of months. These stars are known as **supernovae**.

Supernovae are a relatively rare phenomenon. Only three have been observed within our own galaxy—all prior to the advent of the telescope. These novae occurred in 1054, 1572 (Tycho's nova) and 1604 (Kepler's nova). With the aid of telescopes some fifty supernovae have been recorded in other galaxies.

However, in February 1987, a supernova achieving naked-eye visibility occurred in the Large Magellanic Cloud (LMC). Given the designation 1987A, it has been the focus of intensive study with all the observational armoury available and has provided the stimulus to the advancement of the theories necessary to our understanding of such events. It has been observed across the whole electromagnetic spectrum by satellite and ground-based observatories and, very importantly, provided neutrino detection over an interval of about 10 seconds. The current consensus is that the progenitor was a B3 supergiant star (Sanduleak $-69°$ 202) whose core suddenly collapsed into a 1·3 or 1·4 solar mass neutron star. The trail-off in the arrival of neutrinos rather than a sudden cut-off indicates that a black hole has not been formed.

In the first few weeks following the spectacular outburst, the visual brightness was dominated by energy deposited in the star's outer layers by the explosion's shock waves with the spectra revealing expansion velocities of the order of 17 000 km s^{-1}. Later the light output was supported by energy liberated from the radioactive decays of ^{56}Ni to ^{56}Co to ^{56}Fe, the shape of the light curve being controlled by the associated half-lives. In addition to direct knowledge gained about the cataclysmic processes of such supernova events, the spectra have also provided information about various atomic absorptions in the interstellar medium of the LMC and the path between the Cloud and our own galaxy.

In the following years, supernova 1987A will provide much more observational material for the foundations of our models. Of prime importance will be the search for the pulsar that should eventually be revealed as the debris disperses from the explosion.

From the general behaviour of supernovae, there appear to be two basic types. Group I supernovae achieve absolute magnitudes of the order of -16; their spectra are extremely complicated and it is suggested that they result from accretion by a white dwarf which eventually achieves a mass equal to the Chandrasekhar limit (see Section 14.7). Group II supernovae are two or more magnitudes fainter and are little different from extra bright novae in their display. Their spectra resemble those of ordinary novae but exhibit much higher radial velocities and broader emission lines. The supernova 1987A has been classified as Group I.

Of the three supernovae observed within our own galaxy, the one of 1054 presents the most valuable information. In the constellation of Taurus there is a hazy patch of light or nebula. Telescopes reveal that it has structure and photographs show a complex of filaments emanating from a central blue star. The pattern resembles the outline of a crab and so it was named the Crab Nebula by Lord Rosse in 1844, when he observed it with his 72-inch (1·83 m) telescope.

From the photographs that have been taken of this object over a few tens of years, it is apparent that the structure is expanding in all directions. Measurement of these photographs allows the angular rate of the expansion at right angles to the line of sight to be determined, i.e. the tangential velocity, μ, in seconds of arc per year can be obtained. But, in the case of a star, we have from Equation (10.5) that the proper motion, u, is given by

$$u = 4{\cdot}74\mu/P$$

where P is the star's parallax. Thus we can write the star's parallax as

$$P = 4{\cdot}74\mu/u. \qquad (12.3)$$

Now, with the aid of spectrometry, the radial velocity of the expansion of the Crab Nebula can be measured. By assuming that the transverse expansion is equal to the radial expansion, a value for u is therefore obtained. Insertion of the measured radial velocity into Equation (12.3) has thus allowed the distance of the Crab Nebula to be determined. It is found to be a little greater than 1500 pc.

Knowledge of the rate of expansion also allows us to calculate the time that has elapsed since the explosion for the Crab Nebula to have attained its present size. This figure is about 900 years. That is, some time during the period 1000 to 1100 A.D. we might have expected the event to have been recorded. No such records have been found in the Western civilizations. The evidence can be found, however, in writings both in Chinese and Japanese which state that a new star was apparent during the middle of 1054. This star was recorded as being as bright as Venus. Now that we know the distance of the Crab Nebula we can calculate its absolute magnitude at the time of its discovery. All this information leads us to the conclusion that the Crab Nebula presents to us the remnants of a supernova. There have been numerous attempts to search for the similar debris of Tycho's and Kepler's supernovae but these remained fruitless until very recently.

One of the interesting features concerning the light from the Crab Nebula is that it is strongly polarized. It has been suggested that the high velocity particles are radiating in the presence of a magnetic field by a process known as the **synchrotron mechanism**.

Radiation by this mechanism has been observed in the laboratory during experiments with synchrotrons. These machines are designed to accelerate atomic particles by using magnetic fields. It has been observed that when electrons are orbiting in the magnetic field of the machine, they radiate electromagnetic energy. Synchrotron radiation is observed to be strongly polarized.

Any moving electron when passing through a magnetic field is forced to follow a helical path. The rate at which it is made to execute one turn along the helix is controlled by the strength of the magnetic field. While the electron has this motion, it can be considered to be experiencing a centrifugal force or acceleration. Now it is well known that on the classical theory, any charge radiates when it is accelerated. We therefore expect fast-moving electrons to radiate when passing through a magnetic field. For synchrotron radiation to be generated by the Crab Nebula, a field of the order of 10^{-4} gauss is sufficient to account for the observed brightness. The synchrotron process explains the polarized nature of the radiation from the nebula.

12.3 Wolf–Rayet Stars

These stars are comparatively rare, there being only about 150 listed in the *Henry Draper Catalogue* and its *Extensions*. They are not necessarily variable in any way but are included here because of their special interest.

Their spectra suggest that there are two classes of Wolf–Rayet star which have been labelled WC and WN stars. For the WC stars, emission lines of ionized carbon and oxygen are present in addition to emission lines of hydrogen and helium. For the WN stars, emission lines of nitrogen are present to the exclusion of the carbon and oxygen lines.

The strong division into the two classes depending on whether carbon lines or nitrogen lines are present has aroused speculation that these spectral differences might represent real differences in the abundances of these elements in the two types of star. This idea is very important since it will be remembered that the carbon–nitrogen cycle (see Chapter 9) is one of the possible mechanisms for the production of the energy by which stars shine. Whether the two types of star have different abundances has not yet been settled as the strengths of lines depend on the conditions of the excitation as well as on the abundances of the elements. These conditions are not yet fully understood in the case of Wolf–Rayet stars.

Spectral measurements indicate that Wolf–Rayet stars are extremely hot with surface temperatures in the range 60 000 to100 000 K. It is also concluded that the central bright star is surrounded by an extended atmosphere, thought by some astronomers to be an expanding shell. The measured widths of the emission lines suggest that the shell material is streaming from the central star at speeds of the order of 1000 km s^{-1}. Interpretation of the broad emission lines in this way is not accepted by all astronomers, however.

Finally it may be mentioned that the distribution in space of Wolf–Rayet stars leads us to believe that these stars are extremely young.

12.4 Stars with Extended Atmospheres

Evidence of an expanding shell structure around some stars is provided in the spectra of P Cygni type stars. A typical spectrum and a simple model for its interpretation is shown in Fig. 12.5. Line profiles usually exhibit a shallow

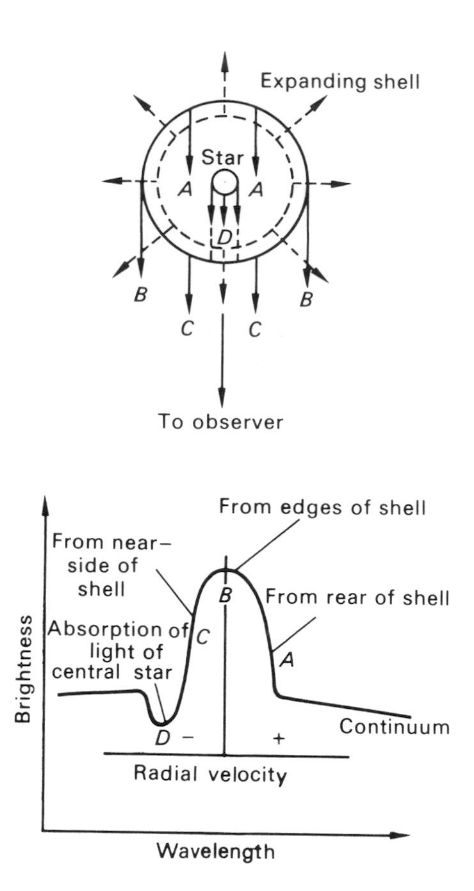

Fig. 12.5 The line profile of a P Cygni type star

blue-shifted absorption feature with a broader emission centred on a wavelength position given by the radial velocity of the star.

The emission line is thought to be generated from material in an expanding shell. The front parts of the shell provide a blue-shifted emission, the rear parts provide a red-shifted emission, while the parts expanding at right angles to the line of sight exhibit no Doppler shifts other than that provided by the motion of the star itself. Light from the central star is absorbed by the part of the shell between the star and the observer and consequently, as this part of the shell provides the greatest radial velocity, the absorption feature exhibits the largest blue-shift in the profile. Expansion speeds are usually smaller than 200 km s^{-1}.

It appears likely that P Cygni stars are related to novae, the prototype star itself having been observed to have a nova-like outburst in the 17th century.

Other stars exhibiting extended atmospheres also display emission lines but the shape of the line profile is different from P Cygni stars. Many of these emission line stars have B type spectra classification and detailed inspection of their absorption line profiles indicates that these stars are rapidly rotating (see next section). Rotational velocities of B class emission stars are extremely high with values of the order of 250 km s^{-1}. Such speeds induce instabilities which lead to a centrifugal force at the stars' equators of the same order of, or greater than, the gravitational force. Consequently some stellar material is sprayed outwards from the equatorial regions and on expansion, an extended atmosphere or shell is formed.

Shell stars of this type appear to be stable for extended periods and then suddenly undergo an active phase. During the activity the emission lines (hydrogen) weaken and may disappear, only to return again after the active phase has passed. A famous star which exhibits this kind of phenomenon is γ Cassiopeiae.

Light which is generated from below the atmosphere of B type emission line stars percolates outwards in all directions and some of it is scattered by the atmosphere from its original direction and towards the observer. Most scattering processes give rise to polarized light. For a spherically symmetric star, we expect that the overall polarization should be zero; the polarization generated by each zone of the atmosphere is cancelled out by that from another zone ninety degrees away on the projected disc of the star. However, if the star has asymmetry, for example, its fast rotation producing an equatorial bulge, the overall polarization may have a measurable value. A study of Be stars has shown that the wavelength dependence of the polarization of their light is different from the curve for stars whose light is made polarized by interstellar material. Thus some of the Be stars may intrinsically produce weakly polarized light. An example of such a star is again γ Cas. The polarization of this star is variable. In addition the polarization drops markedly across the emission lines suggesting that the emission occurs high in the atmosphere and does not suffer the same scattering as does the light which makes up the continuum.

The supergiant and giant stars of spectral classes G, K and M exhibit extended atmospheres and give evidence of an outward flow of material with velocities greater than the velocity of escape. Infra-red observations of the giant M stars indicate the existence of shells of matter in the form of condensed grains. In addition to variations in brightness, some of these stars show variable polarization which again must be intrinsic to the star.

12.5 Fast-Rotating Stars

Observation of the sunspot pattern over a few days reveals that the Sun is rotating slowly about an axis. Our evidence that other stars rotate is presented by broadening of their spectral lines.

If a star is not rotating or its axis of rotation happens to be along the direction of sight then the spectral lines will appear in the ordinary way. For stars which are rotating with their axes of rotation at any angle other than along the line of sight, then the spectral lines appear to be broadened. The broadening results from the differential Doppler shifts in the light from the various parts across the stellar disk.

In order to see how the line is broadened, consider Fig. 12.6(a), which depicts a rotating star with its axis at right angles to the line of sight. Suppose that the star's rotation is in the sense such that the left-hand limb is approaching the observer. The light emitted from the left-hand hemisphere towards the observer will therefore be Doppler-shifted towards the blue, the maximum shift being presented in the light from the stellar limb. By a similar argument, the light from the right-hand hemisphere will be progressively red-shifted in going from the star's centre to the right-hand limb. The total effect on a stellar profile is illustrated in Fig. 12.6(b). Analyses of spectral profiles allows rotational speeds to be determined.

Now the axes of rotation of stars are likely to be distributed at random in space. Any determination of the equatorial velocity of a star provides a value which is the velocity component relative to the observer. Thus if V is the true equatorial velocity, the value that is obtained is equal to $V \sin i$, where i is the inclination of the axis of rotation to the line of sight. However, when stars of the same spectral type are studied and a statistical factor is applied to the probabilities of measuring stars with various values of i, it is found that the mean rotational speeds of such stars vary with spectral type. Stars between B5 and B7 have the maximum rotational speeds ($\approx$200 km s^{-1}). Rotational speeds drop quickly through the A and F spectral types. The Sun, with its small

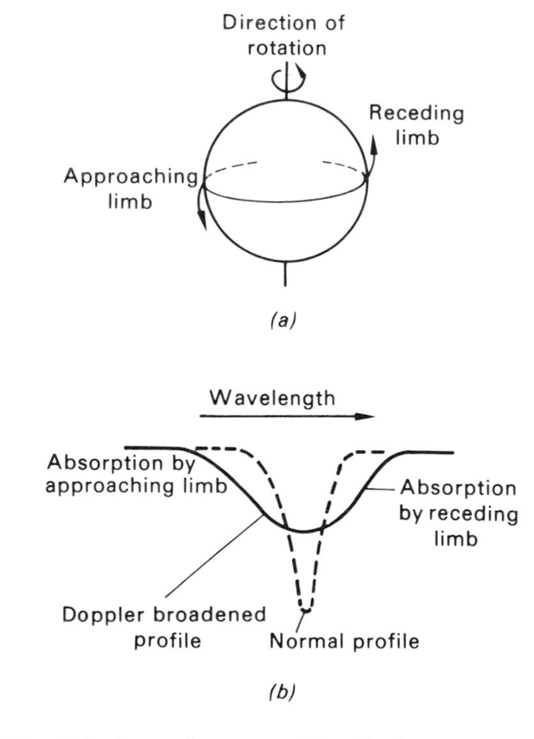

Fig. 12.6 A rotating star and its effect on the spectral line profile

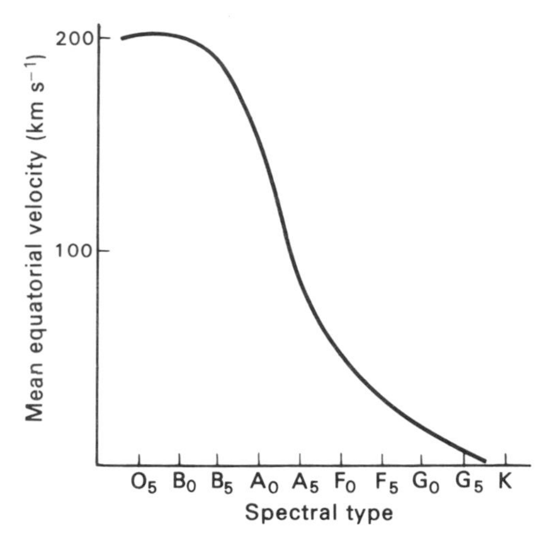

Fig. 12.7 The distribution of rotational speed with spectral type

rotational velocity, is not untypical of a G2 type star. The distribution of the mean equatorial velocities with spectral type is depicted in Fig. 12.7.

12.6 Magnetic Variables

By using polarimetric elements prior to a large dispersion spectrograph, H. W. Babcock was able to measure the Zeeman splitting of lines into two circularly polarized components in the spectra of some stars. His technique allowed him to determine the general magnetic field of these stars. As stellar spectra are difficult to record with a large dispersion, magnetic field strengths must be fairly large to be detected. Babcock's catalogue lists stars with fields of a few hundred to a few thousand gauss. The majority of the magnetic stars have A-type spectra but with peculiarities.

Many of the stars show variability of the field strength which in some cases is periodic. Indeed some stars show that the field reverses its sign periodically. Several of the regular periodic stars have since been shown to exhibit brightness variations with light curves which match the magnetic variation period.

It is interesting to note that for any particular star, the determined strength of the field depends on which Zeeman split line is chosen for measurement. As the absorption lines are effectively produced at different levels above the star's photosphere, measurement of a range of lines allows study of the variation in the strength of the magnetic field with depth through the cooler layers surrounding the star.

Problems: Chapter 12

1. If the mean absolute magnitude of RR Lyrae stars is $+0^m\!.6$ and such a star is observed with an apparent magnitude of $+15^m\!.8$ within a star cluster, what is the distance of the cluster?
2. The shell that surrounds *Nova Aquilae* is observed to be expanding at the rate of 2 arc sec per year. The absorption line of hydrogen at 4861 Å in the spectrum of the centre of the shell is displaced 28 Å towards shorter wavelengths from its position in the spectrum of the star. Deduce the parallax of the nova.
3. In a nearby galaxy a Cepheid is found to be of apparent mean magnitude $+18^m\!.7$. Its period is 22·6 days. Another Cepheid in the Small Magellanic Cloud has a mean apparent magnitude of $+10^m\!.2$ and a period of 22·6 days. Calculate the ratio of the distance of the galaxy to that of the Cloud.
4. In the Andromeda Galaxy a Cepheid is found to be 4·65 magnitudes fainter than a Cepheid of the same period in the Small Magellanic Cloud. Calculate the distance of the Andromeda Galaxy, given that the distance of the Cloud is 53 000 parsecs.
5. Assuming that novae reach a brightness 25 000 times that of the Sun (at the same distance), calculate the distance of the Andromeda Galaxy, given that novae in it reach an apparent magnitude at maximum brightness of +17·8. (The apparent magnitude of the Sun at a distance of 1 AU (=1/206265 pc) is −26·68.)

13 Nebulae

13.1 Introduction

To the unaided eye there are several patches of hazy light which remain at a fixed position relative to the star background. They are commonly known as **nebulae**. The increased light gathering power and angular resolution of the telescope allows the natures of many of them to be discerned. For example, we have already seen that the hazy patch of light in the constellation of Taurus—the Crab Nebula—is the filamentary remnants of a star seen to explode more than nine centuries ago. This type of nebula, as it turns out, is atypical of the types of object that the word "nebula" is used loosely to cover. If the term "nebula" is applied to an object, it is usual to qualify it by another word which describes the nature of the nebulosity.

The number of recorded nebulosities quickly grew with the advent of the telescope. In an effort to avoid confusion in the discovery of comets, Messier, a French observer, published a list of 103 nebulous objects in 1784. The objects in this list are designated by the letter M and followed by a number. It happens that the Crab Nebula is the first object listed in his catalogue and therefore is known as M1.

In the hundred years after Messier, the number of observed nebulosities increased enormously. With the compilation of the *New General Catalogue of Nebulae and Clusters of Stars*, completed by Dreyer in 1888, some 7840 objects were listed. Objects included in this list are designated by the letters NGC and followed by a number. Thus the nebulosity which is just about visible to the unaided eye in the constellation of Andromeda is known as NGC 224. This same object also happens to be listed in the Messier catalogue and it is alternatively known as M31.

Many of the objects listed in the NGC catalogue are known to be agglomerations or clusters of stars, these being the open or galactic clusters and the globular clusters. They have already been described previously and will not be discussed further here. In addition, other members of the Messier and NGC catalogues, the galaxies, are no longer classed as nebulae. Again, M31, or NGC 224, is an example of this, it being the Andromeda Galaxy.

If we omit star clusters and galaxies from these catalogues, we are left with diffuse nebulae (condensations of gas and dust), planetary nebulae and supernova remnants. Their basic natures are described briefly below.

13.2 Diffuse Nebulae

The presence of material in the universe, other than that which has condensed into stars, reveals itself in one of two ways. Either it shines by emitting and/or reflecting light or its absorbs light which has been generated elsewhere.

Examples of bright nebulae are to be found in association with the group of stars known as the Pleiades (a galactic cluster) and the famous Orion Nebula (NGC 1976, M42). Bright nebulae are seen in different forms. They may be fairly compact, exhibiting structure with bands of absorbing material or they may have a delicate filamentary structure with wisps extending over a number of degrees of the sky.

Evidence of dark nebulae is already shown in the structure of some of the bright nebulae. The phenomenon of obscuration and absorption is also obtained by taking photographs of the Milky Way regions; several dark patches are noted and in some cases the edges are illuminated, akin to the appearance of a heavy grey terrestrial cloud when covering the Sun.

The appearance of a diffuse nebula depends upon a number of factors: (i) the density of the gas and dust composing it, (ii) the chemical composition of the nebula, (iii) the existence or not of one or more stars in its neighbourhood, (iv) the spectral types of the stars. In general the nebula will be an **emission** nebula, or a **reflection** nebula or a **dark** nebula. We consider these in turn.

13.2.1 Emission Nebulae These nebulae are always associated with a very hot star, of spectral type B1 or earlier. The star's output of ultra-violet light ionizes the hydrogen atoms in the nebula. Any volume in which the hydrogen is ionized is referred to as an H II region. In addition to hydrogen, other elements, for example helium, oxygen and nitrogen, may be present. A single O-type star can ionize all the hydrogen within a sphere 100 pc in radius, while a BO-type star's ionizing effect has a radius of about 25 pc. The radii of such spheres (called Strömgren spheres after the Danish astronomer who investigated their properties) depend upon the density of the nebula. Such a nebula, as well as exhibiting a stellar spectrum, also exhibits emission lines in its spectrum.

When the first spectra were taken, many of these lines were readily identified. They included lines provided by hydrogen, helium and oxygen, for example. Other emission lines also appeared which for some time defied identification—they were unknown in the laboratory. It was supposed that these lines resulted from atomic transitions of some element which was unknown on the Earth. It was given the name "nebulium".

Nebulium was later to be wiped from the list of elements when, in 1927, Bowen showed that the emission lines corresponded to transitions within atoms such as oxygen and nitrogen. The physical conditions pertaining to the nebulae, being so much different from those which can readily be achieved in the laboratory, allow some atomic transitions to occur which are difficult to demonstrate in the laboratory. Such transitions provide what are known as "forbidden" lines. What is really meant by this term is that these transitions, under normal circumstances, have a small probability of occurring.

From the observed laboratory spectra a set of rules, known as quantum rules, is established which permits us to consider the orbits of the electrons which undergo transitions. The quantum rules allow us to predict whether a particular transition from one orbit to another is an allowed transition, so generating a particular spectral line. Under normal laboratory conditions an electron in an excited orbit very quickly returns to a lower level (i.e. its life-time is of the order 10^{-8} s), thus producing an emission line. There are some excited states where the electron is said to be in a **metastable state**, the probability of falling down towards the ground state being extremely low. The life-time of a metastable orbit may be of the order of minutes. High energies are required to put orbiting electrons into metastable orbits. Although this can be done in the laboratory, the life-times of the orbits are so long that the atoms have sufficient time to suffer many collisions with each other and their energies are off-loaded in this way rather than by the emission of radiation in the form of forbidden lines. Forbidden lines are obviously more likely to be seen if the pressures within the gases are greatly reduced, thus making the atomic collisions less frequent. With the low pressures required, the limited space of the laboratory allows only a small number of atoms to be used in the experiment and there is therefore little chance of the forbidden lines being seen.

In the emission nebulae we have all the requirements for the forbidden lines to be seen. Firstly, the hot stars associated with the nebulosity provide the high energy radiation to excite the atoms in the nebulosity to metastable states. Secondly, the gaseous material is at a sufficiently low pressure for the atoms not to undergo collisions during the life-times of these states and yet the volume of gas involved is large enough for there to be a sufficient number of atoms to be seen performing the downward forbidden transition. Thus emission nebulae provide us with one of the many examples in astronomy where we can make observations of material which happens to be in a condition which is not normally attainable in the laboratory. Measurements of the relative strengths of the forbidden lines in the spectra of emission nebulae allow us to determine the density of the gas or the number of atoms contained in a unit volume of the gaseous material. A typical value would be of the order of 10^8 atoms m^{-3}.

The Orion nebula is the best known example of an emission nebula (see Fig. 13.1). Long-exposure photographs of the region round the middle star in the sword of the constellation Orion show a greenish nebulosity extending across a full degree of the sky. The star itself can be resolved into a group of very hot O-type stars. The spectrum of the nebula is consequently an emission one containing lines originating from hydrogen, ionized helium, oxygen and other elements.

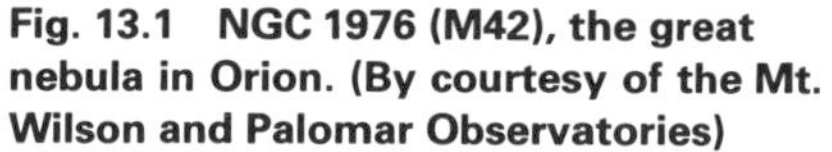

Fig. 13.1 NGC 1976 (M42), the great nebula in Orion. (By courtesy of the Mt. Wilson and Palomar Observatories)

13.2.2 Reflection Nebulae If the stars in the vicinity of interstellar clouds are cooler than spectral type B1, the nebulae may still be bright. The light illuminating them, however, is simply starlight reflected from the material, scattered by the dust particles present in the nebulae. The spectrum exhibited is a continuous one crossed by dark lines and the colours of such nebulae resemble those of the stars embedded in them. Thus photographs (see Fig. 13.2) of the Pleiades cluster in Taurus show reflection nebulosity surrounding each of the six brightest stars in the group.

Fig. 13.2 NGC 1432, the Pleiades cluster, showing the presence of reflection nebulosity. (By courtesy of the Mt. Wilson and Palomar Observatories)

The cause of the scattering can be deduced. Scattering by atoms can be dismissed, for such a process would require far too high densities in a gas cloud to produce the efficient scattering found in these nebulae. Because the colours of a reflection nebula and its star are so similar, it would also appear that the scattering cannot be caused by molecules, which produce a highly wavelength-dependent type of scattering. Particles larger than molecules must produce the observed scattering; such particles are usually termed **dust**.

The composition and shape of the dust particles may even be deduced. The nebula containing them has a high reflectivity. Hydrogen is so abundant in the universe that it seems probable that the particles are made up of solid compounds of hydrogen. Polarization studies suggest further that the dust specks are irregular in shape, being elongated and aligned by a weak interstellar magnetic field.

13.2.3 Dark Nebulae It is well known that the interstellar material absorbs light, making stars appear to be less bright than they would otherwise be according to their distance. This topic is taken up again in the following section. However, the absorption is so severe in some directions that light from the stars is completely extinguished and no stars can be seen in that particular direction. These optically thick clouds of dust and gas are the **dark nebulae**. In addition, if no star bright enough exists in the neighbourhood of a cloud of dust and gas, such a nebula will be dark. Examples of such dark nebulae are the Horsehead

Nebula in Orion, parts of the North American Nebula in Cygnus and M16 in Scutum (see Fig. 13.3).

Fig. 13.3 NGC 6611 (M16); dark nebulae showing in this famous object in Scutum. (By courtesy of the Mt. Wilson and Palomar Observatories)

Information about the distance of a dark nebula can be obtained from star counts. It is always possible to count the number $N(m)$ of stars per square degree between apparent magnitudes m and $m+1$. If a dark nebula exists at a distance of d parsecs, it will have no effect upon the brightnesses (and therefore apparent magnitudes) of stars in the line of sight to it nearer than d parsecs. For stars more distant, however, it will absorb light from them, increasing their apparent magnitude by q magnitudes, say, depending upon its density and thickness. Now, in general, nearby stars are brighter than more distant ones and so it may be argued that a nebula will predominantly affect the brightnesses of stars of higher magnitudes.

Hence, if two graphs are made of $N(m)$ against m for two regions of the sky, one unobscured, the other containing a dark nebula, the second graph should fall below the first for the higher magnitude range. Such graphs are called Wolf diagrams, after M. Wolf whose diagram for the North American Nebula is reproduced in Fig. 13.4. The solid curve is drawn from star counts of the unobscured regions; the dotted line arises from corresponding star counts in the obscured regions. Obviously there is reasonable agreement for the brighter stars. For the fainter and more numerous stars, however, at any given magnitude where there are fewer seen in the obscured regions than in those regions clear of dust. By various methods, such information can be made to give estimates of the nebula distance and thickness.

It has already been mentioned that some bright nebulae are seen to be obscured in part by dark nebulae. Inspection of the dark nebulae in this circumstance reveals that the strongly absorbing material is often found in the form of dark globules of small angular size. It has been suggested by many astronomers that these globules represent the initial phase of the collapse of material prior to the formation of a star.

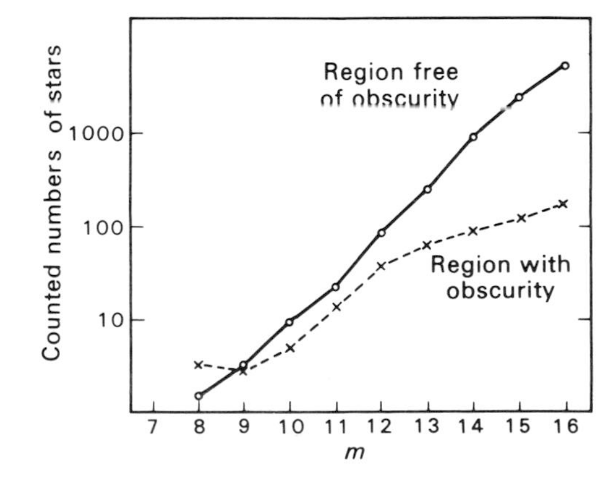

Fig. 13.4 Wolf diagram

According to the proposed scheme, the globules contract from the gaseous dusty material by radiation pressure. The electromagnetic radiation which the material absorbs from all directions exerts a pressure on the globule making it coalesce. After a certain stage the force of gravitation becomes important and the globule collapses, increases its temperature and then initiates the nuclear reactions necessary to allow it to shine as a star. Estimates have been made of the mass of dusty material contained in these globules and there is a satisfactory agreement between these values and the values required to initiate the condensation of a star.

The time for the collapse of a dust cloud is predicted to be very short in relation to the complete life-span of a star, perhaps short enough for changes to be seen by photographic records over a number of years. Certain nebulous objects, known as **Herbig–Haro objects**, have been reported to have changed their shapes over a period of years and become more spherical. Perhaps these objects are the earliest observational evidence of the birth of stars.

It can be pointed out in this context that the star FU Orionis has been seen to go through a remarkable change in a short period of time. This star is found very close to the Orion Nebula (NGC 1976). When it was first photographed in 1936, this star was very faint, being about sixteenth magnitude. In less than one year it increased its brightness to tenth magnitude, since when it has remarkably kept at a steady luminosity. According to Hayashi's theoretical work, it is predicted that the gravitational collapse of a star commences slowly, but during the last phases would speed up dramatically. Thus it has been suggested that FU Orionis is perhaps a star which has been observed to go through the last stages of gravitational collapse. Constant watch is now being kept for similar events in the Orion Nebula and other nebulous regions.

At this stage, it may be useful to present (in highly simplistic form!) a summary of the main distinctions where diffuse nebulae are concerned. Table 13.1 supplies this summary.

Table 13.1 Factors producing bright and dark nebulae

Material in nebula	Bright stars in neighbourhood	Spectral type of stars	Type of nebula	Comment
Dust and gas	Yes	B1 and earlier (i.e. hotter)	Emission	Emission of light from ionized gas (principally hydrogen)
		Later than B1 (i.e. cooler)	Reflection	Starlight reflected by dust particles
	No	—	Dark	Obscuration of more distant objects by dust

13.2.4 Molecular Clouds In a similar way to there being energy states within atoms, the energies of molecules are also quantized. However, in addition to electronic transitions, molecules can hold energies associated with their rotations and also with vibrations of their atomic components, both these modes of motion being quantized with appropriate energy levels. For example, a simple diatomic molecule like CO (carbon monoxide) can rotate about an axis at right angles to the atomic band, the associated energy of rotation being quantized. In jumping from one rotational level to another, radiation will be emitted/absorbed with a particular frequency.

A strong line associated with CO is that at 2·6 mm and observations at this wavelength have been particularly useful in mapping out molecular clouds within the cool regions of the Galaxy. According to the estimated flux from the ultraviolet radiation from stars, the photodissociation of CO within the interstellar medium should be virtually complete. However, dust provides shielding, so boosting the molecular abundance. Molecular clouds and dust are always found in association.

The rotational energy of a molecule depends on the moment of inertia, which in turn depends on the masses of the constituent atoms. The CO molecule may comprise atoms of the common form of carbon—^{12}C—or the rarer isotope ^{13}C. The difference this causes to the rotational energies is considerable and provides distinct separations in the wavelengths of spectral lines, making it relatively easy to determine ratios of isotope abundances. Profiles of the spectral features also allow studies of the velocity distribution within molecular clouds.

Most of the molecular interstellar lines are found in the infra-red and millimetre regions of the spectrum. Well over 50 molecules have been identified including the familiar ones of formaldehyde (H_2CO), methyl alcohol (CH_3OH) and ethyl alcohol (C_2H_5OH).

13.3 Interstellar Absorption

13.3.1 Introduction The fact that the gaseous and dusty material absorbs the light from stars is qualitatively demonstrated by taking photographs of these obscuring patches or dark nebulae. Effects of absorption show up in three other ways. Measurement of the strengths of these effects allows us to interpret the nature of the material causing the absorption and, by taking observations of many stars over a range of directions, the distribution of the absorbing material can be studied. The three interstellar absorption effects which impose themselves on the light emitted by stars appear as spectral features, stellar reddening or as polarization. They are each discussed briefly below.

13.3.2 Interstellar Spectral Lines When stellar spectra are recorded it is noted that, in some cases, absorption bands appear which are not expected from the type of star that is observed. The most common features that are identified correspond to the H and K lines of calcium and the D lines of sodium.

That these lines are produced by interstellar material, rather than in the atmospheres of the stars themselves, is beautifully demonstrated by observations of spectroscopic binaries. When the spectra of such stars are taken, the stellar lines undergo their periodic shifts, allowing the orbits of the two stars to be determined, but other absorption lines remain at a constant wavelength position. This indicates that the absorption occurs between the particular stellar system and the observer, i.e. in the interstellar medium.

The measured strength of the absorption varies from star to star. It depends upon the star's distance and also on its observed direction. Thus the absorbing material is not uniformly spread throughout space. There are localized concentrations, particularly in the plane of the Galaxy.

It is also noted that the absorption features may not lie exactly at the laboratory wavelength position, but show a Doppler shift. Thus we know that the absorbing material is in motion relative to ourselves. For some of the distant stars, multiplicity of the interstellar absorption lines is observed, indicating that the light from the star is passing through a series of absorbing clouds, each with its own relative velocity with respect to the observer. From such measurements of the interstellar absorption lines, information is provided about the composition of the interstellar clouds and their distribution and velocities. This information is then combined with our knowledge of the distribution of the stars to help build up the picture of our Galaxy.

13.3.3 Interstellar Reddening According to the absolute magnitude—deduced from the spectral type—and the measured distance, measurement of the apparent brightnesses of some stars reveals that they are not as bright as might be expected. Some of the energy which they emit is absorbed *en route* to the observer. For such stars the distance modulus (see Equation (7.2)) needs to be modified to read

$$m - M = 5 \log_{10} r - 5 + A$$

where A is the amount of absorption, expressed in magnitudes.

Magnitude measurements, using a range of spectral passbands, reveal that the absorption for any star is strongly wavelength dependent. At the blue end of the spectrum, the absorption is significantly greater than for the red. Consequently, as the energy contained in the red parts of the spectrum is able to penetrate the interstellar material more easily than the blue, the distant stars are likely to appear reddened. The amount of reddening that is observed depends on the wavelength distribution of the energy emitted by the star, i.e. on its spectral type. The variation of A is approximately proportional to the reciprocal of the wavelength, suggesting that the absorbing particles are approximately the same size as the wavelength of light.

If direct measurements of the colour indices of a cluster of stars are used for the plot of the HR diagram of the cluster, the distribution of the stars on the diagram is distorted if interstellar absorption effects are present. By comparing such an HR diagram with one which is not affected by absorption, the differences measured allow the effect of the extinction to be investigated—that is, the way in which it varies with wavelength. The curve relating extinction and wavelength is one of the starting points for the models which have been proposed as to the nature of the interstellar grains causing absorption.

13.3.4 Interstellar Polarization For the majority of the stars there is no reason for the light they emit to be significantly polarized. However, when measurements are made, it is found that the light from quite a number of stars exhibits polarization. The effect is usually found when the stellar spectrum reveals interstellar absorption lines. The amount of polarization recorded depends on the direction of the observed star and is likely to be greatest for the more distant stars and for stars which lie close to the plane of the Galaxy. The direction of vibration of the polarized light is usually almost parallel to this same plane.

Thus it appears that the particles of the interstellar material have an absorption which depends on the direction of vibration. When unpolarized light is passing through the material, resolved components are absorbed unequally. On emergence, one component is therefore stronger than the other and the light has become partially polarized.

To account for the differential absorption, it has been suggested that the particles are non-spherical. A needle-shaped particle, for example, would absorb more easily the resolved component which vibrated in the direction along the axis of the needle. If a beam of light passes through a region where the grains have their axes aligned, then the emergent beam will be partially polarized, with a direction of vibration at right angles to the axes. It has been suggested that the preferential alignment of the interstellar grains is brought about by the weak general interstellar magnetic field.

13.4 Planetary Nebulae

The name of this type of object is very misleading as the nebulosity has nothing to do with planets nor indeed with the Solar System. These objects may be briefly described as hazy shells of gas surrounding a central star. When viewed through a telescope or photographed, some of them present a disk-like appearance and it is for this reason only that they were given the name of "planetary". It is perhaps unfortunate that this name has remained attached to these objects.

About one thousand planetary nebulae are known. Some of them exhibit complicated structure in that the shell may not be exactly spherical or there may be a series of shells. The central star is very hot, usually presenting an O-type spectrum. The nebulous shell of gas exhibits emission lines of hydrogen and helium as well as the forbidden lines of doubly-ionized oxygen, already mentioned in context with the amorphous bright nebulae. An example of a planetary nebula is shown in Fig. 13.5.

Because of the high temperature of the central star, the peak of its emission occurs in the ultra violet. Consequently the star does not appear all that bright to the eye. The surrounding shell of material is seen as a ring. Part of the original star's light is scattered by the material forming the shell. In addition the shell absorbs ultra-violet energy from the star and re-emits it in the form of spectral lines in the visible region of the spectrum. There is little diminution of the star's brightness by the shell material that happens to be in the line of sight to the star.

Some of the nearby planetary nebulae allow trigonometrical parallax measurements to be made and then from their angular extent the diameter of the shell of material can be determined. Typical values are of the order of 20 000 AU.

For most planetary nebulae, however, the distances are difficult to measure. The absolute magnitudes of the central stars are usually not known with sufficient accuracy to enable a distance modulus to be calculated precisely from apparent magnitudes. Other methods of determining distances involve proper motions, angular dimensions, angular expansion and radial velocity. For example, the distance of a planetary nebula may be found from a comparison of its angular expansion (from photographs many years apart) and radial expansion (by spectrometry).

Measurements of Doppler shifts in the spectra taken from the light emitted by the shell indicate that there is irregular motion and turbulence in some of the planetaries. It is expected, however, that the radiation pressure of the light from the central star acting on the gaseous material would provide the shell with an expansion velocity of the order of a few kilometres per second.

13.5 Supernova Remnants

There are certain spectacular nebulae which are assumed to be the material ejected from supernova explosions. Perhaps the best-known is the Crab Nebula, almost certainly the result of a supernova explosion witnessed by the Chinese in 1054 A.D. The spreading nebulosity is luminous, the spectrum of the complex splashlike tangle of features being an emission one (see Fig. 13.6). Other radiation coming from the Crab is caused by relativistic electrons (electrons moving at speeds near the velocity of light), spiralling in the twisted

Blue

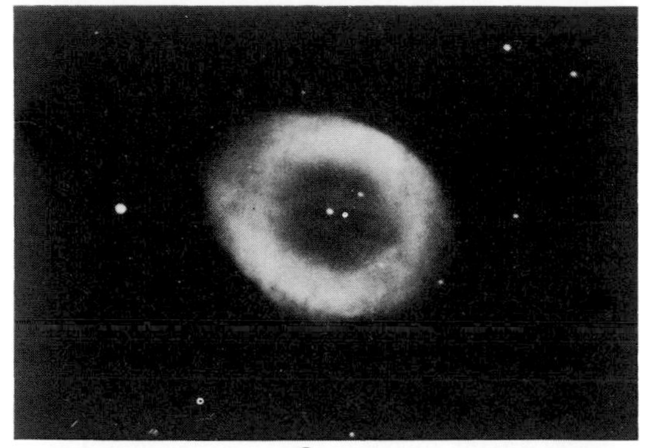

Green

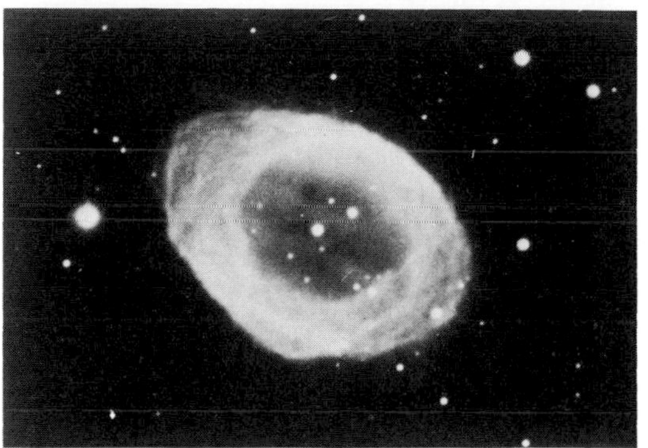

Yellow

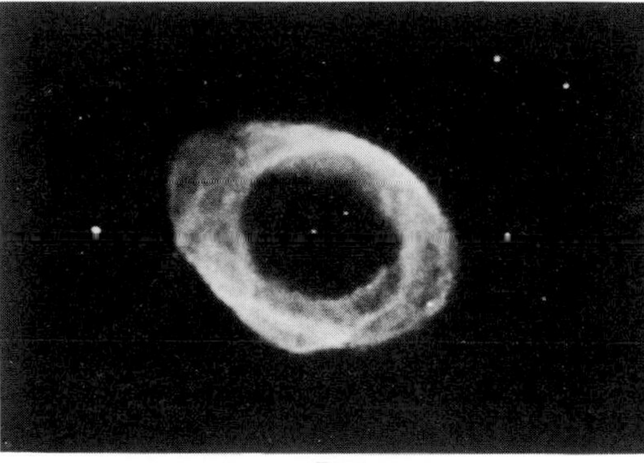

Red

Fig. 13.5 NGC 6720, the famous planetary nebula in Lyra, known as the Ring nebula, taken with four different colours. (By courtesy of the Mt. Wilson and Palomar Observatories)

skein of magnetic fields within the nebula. This type of radiation is called **synchrotron radiation**. The name arises from the fact that nuclear accelerators called synchrotrons cause electrons moving at high speeds in their powerful magnetic fields to emit this light.

Another nebula of this kind is the Veil Nebula, visible in the constellation of Cygnus (see Fig. 13.7, p. 221). Known also as the Cygnus Loop, it also is expanding, emits synchrotron radiation at radio wavelengths and is supposed to be the remnants of an ancient supernova outburst.

The supernovae observed by Tycho and Kepler in 1572 and 1604 have also recently been found to be associated with faint nebulosity emitting in the optical and radio wavelength regions.

By the end of 1974, the number of objects detected on photographic plates and identified as supernova remnants had risen to about 30.

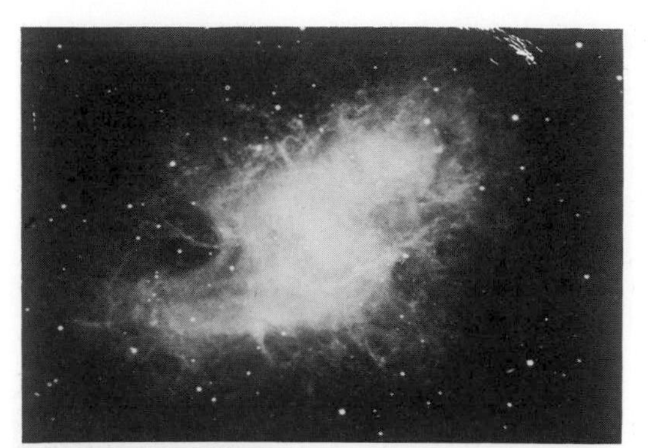

Blue λ3100–λ5000

Yellow λ5200–λ6600

Infra-red λ7200–λ8400

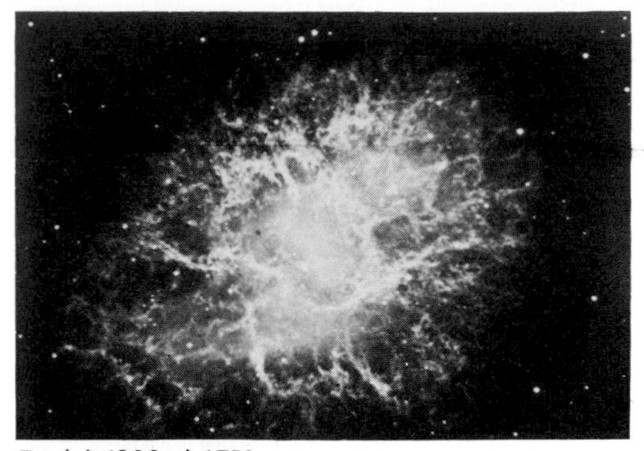

Red λ6300–λ6750

Fig. 13.6 Photographs taken of the Crab nebula through colour filters. Note the filamentary structure which is strongly evident in the photograph with the red filter which includes the Hα wavelength. (By courtesy of the Mt. Wilson and Palomar Observatories)

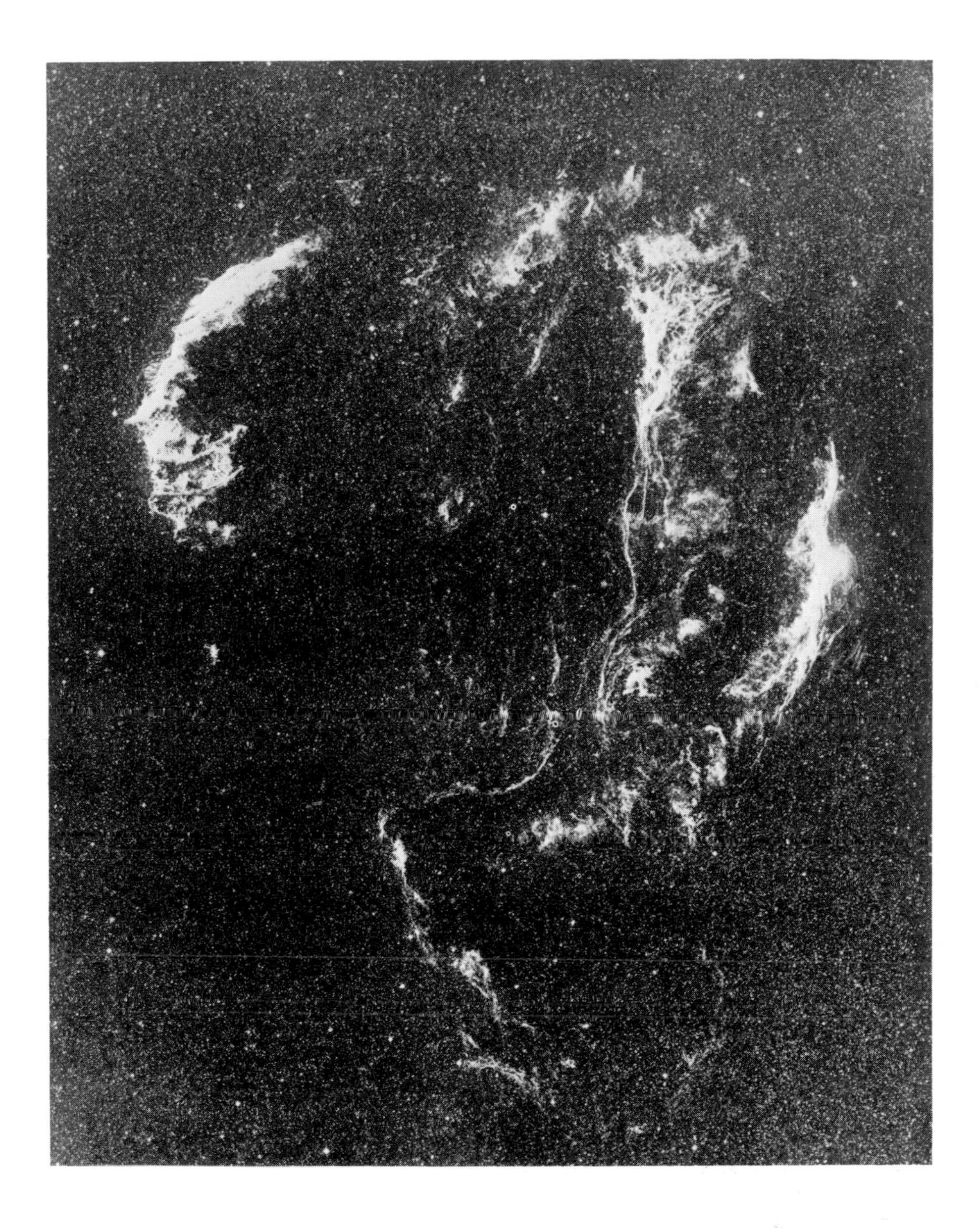

Fig. 13.7 NGC 6960-92, the Veil nebula, photographed in red light by the 48-in Mt. Palomar Schmidt telescope. (By courtesy of the Mt. Wilson and Palomar Observatories)

14 Celestial Objects Observed Outside the Optical Spectrum

14.1 Introduction

There are many familiar astronomical objects that emit radio waves with sufficient energy for their radiations to be detected by radio telescopes. For example, the radio noise from the Sun is relatively easy to detect. The Sun's radiation can also be measured at wavelengths corresponding to the infra-red, ultra-violet, X-rays and γ-rays.

In this chapter, however, we shall concentrate on those sources that have been brought to astronomers' attention whilst carrying out special sky surveys in different wavelength regions of the spectrum. As we shall see, many of these discovered objects have been identified with a visible object at a later date.

Some of the sources which radiate strongly at wavelengths outside the optical spectrum have natures that are far from being well understood. Such objects include the recently discovered quasars and pulsars. In the case of quasars, for example, very little is really known about their sizes or distances. At the moment we can do little more than just describe and present the measurements of their radiations. A clear and definitive interpretation of their natures may not be available for many years to come.

14.2 Galactic Radio Radiation

When Jansky, in 1932, investigated the noise levels that might be expected while using short wave radio receivers in conjunction with a directional antenna system, his aim was to study the sources of noise and to see how his equipment might be improved to reduce their effects. He found that there were two basic kinds of noise. One provided crackling sounds and was already known to be associated with thunderstorms; the other provided a continuous hissing noise.

Jansky found that the source of the latter noise was apparently in the sky and was not of terrestrial origin. He found that in one particular direction the noise had a maximum value and that this direction relative to the Sun changed with time. He was able to show that the direction was constant relative to the star background and that it was associated with the Milky Way. In fact the maximum noise corresponded to a direction in the constellation of Sagittarius, that is, the direction of the centre of our Galaxy.

This work was not immediately pursued by astronomers. However, a radio amateur, Reber, set up a home-made paraboloid dish and produced a contour map of the strength of the radio emission over the sky. He confirmed that the maximum radio emission came from the galactic centre and that the contours followed the belt of the Milky Way fairly closely. It is interesting to note that in addition he also detected a few emission peaks. In these directions there were no apparent optical sources with special features. These same peaks of emission were re-discovered together with new ones some twenty years later by the professional radio astronomers.

The radio emissions over the celestial sphere have now been mapped at several different frequencies. Each survey provides contour patterns which all show a general concentration of energy arriving from the Milky Way with the peak amount of energy coming from the centre of the Galaxy in the constellation of Sagittarius. It is not certain how all of the radio energy originates. Certainly it cannot all be ascribed to thermal or black-body radiation. It is thought that a major proportion of it is produced by the synchrotron process. The free electrons which are present in the hot gaseous regions in the Galaxy have motion through the weak galactic magnetic field. As a result, the electrons travel with helical motion and in doing so radiate by the synchrotron mechanism, producing radiation in the radio region of the electromagnetic spectrum. Thus it is thought that much of the general galactic radio emission comes from regions of hot, ionized hydrogen.

Just prior to the great surge of interest in radio astronomy, Van de Hulst was considering the problem that there might be radio spectral lines generated by transitions within atoms. As hydrogen is the simplest case to consider and also very abundant in the universe, he thought of this atom as being a possible source of radio emission. He was able to show that there is a small difference in energy for a neutral atom in its ground state depending on whether the magnetic fields (produced by spin) of the proton and electron are aligned or in the opposite direction to each other.

It can be seen intuitively that these two conditions have different energies when the atom is in its ground state. To excite the electron into an outer orbit will require a small extra amount of energy to combat the magnetic interaction when the proton and electron spins are in the opposite direction compared with the situation when they are parallel. Thus if the spins of the proton and electron are in the same direction, a reversal of the direction of the electron spin will release a low energy photon. The energy difference between these two states corresponds to a photon with a frequency of approximately 1427 MHz, or a wavelength close to 21 cm. Although the probability of the flip taking place is very small (the life-time of the slightly excited ground state being about $1{\cdot}1 \times 10^7$ years), it was thought that on account of the extremely large numbers of hydrogen atoms available within the Galaxy, the emission line might be observed. This indeed was the case in 1949, just five years after the theoretical prediction.

As the strength of the received energy depends on the number of atoms in the line of sight of the receiver, monitoring the 21 cm emission over the regions of the Milky Way allows the concentration of neutral hydrogen to be determined in particular directions. There is a more important application, however. Since the emission occurs with a particular wavelength, any motion of the neutral gas clouds would impose Doppler shifts, altering the apparent wavelength of the line. By using a receiver whose passband is scanned through the 21 cm line, line profiles can be obtained allowing differentiation between individual gas clouds which happen to lie in the same direction. Such measurements have allowed models to be built of the distribution and velocities of the clouds of neutral hydrogen within our galaxy. This method is described in more detail in Chapter 15.

The 21 cm line is sometimes seen as an absorption feature rather than in emission. Under this circumstance the detection of the Zeeman splitting of the line is easier. The splitting into left- and right-hand circularly polarized components produces a separation of 2·8 MHz per gauss. The radio astronomer's technique of being able to switch quickly between looking at the two polarization forms provides a sensitive means of detecting extremely weak magnetic fields. Measurements have shown that the galactic magnetic field strengths are typically between 10^{-5} and 10^{-6} gauss.

Other radio spectral lines have been predicted and indeed some of them have recently been detected. In particular the presences of the hydroxyl radical

(OH), ammonia, water, formaldehyde and formic acid have been observed within galactic material.

At the time of the general radio surveys of the sky, a number of positions were recorded from which the strength of the radio radiation had localized peaks. The objects emitting the radio energy were initially known as radio stars. When they were first discovered, the technique of measuring their positions was only moderately accurate. Their co-ordinates were not known with sufficient accuracy to allow them to be identified easily with optical objects.

In fact most of these radio sources are found not to have stellar natures, although it may be mentioned that some of the flares presented by UV Ceti stars have been monitored by radio telescopes.

There are now several different catalogues listing localized radio sources. Famous ones include the Cambridge 3C catalogue where objects are listed with a number in order of increasing right ascension (e.g. 3C 48) and the one obtained by the Parkes radio telescope (Austrália) where the first four numbers designate the right ascension, this being followed by the sign and the degrees of declination (e.g. PKS 1327 − 21).

In those early days, just twenty-five years ago, it was not known how distant were these radio sources of unknown nature. Except for the strong radio source in Taurus, which was very quickly identified with an optical object, it was a matter of conjecture for a number of years as to whether radio sources were within our Galaxy or were extragalactic bodies.

The radio source in Taurus appears in the same position as the Crab Nebula. It seems likely that the radio energy arriving from this source is generated by the synchrotron process. With this positive identification of one radio source, supernova remnants were searched for photographically in the directions of the other radio sources. In particular the brightest radio source, Cassiopeia A and another famous object, Cygnus A (see Fig. 14.1), attracted much attention. The outstanding problem at that time was the inability of radio measurements to resolve detail in the sources smaller than 5 minutes of arc and this in turn prevented accurate positions being available for the optical astronomers to use. Nevertheless a peculiar nebulosity was photographed in the Cassiopeia region.

Soon after the need for more accurate radio positional measurements was appreciated, interferometric methods were developed which provided the necessary precision to solve the problem. As well as determining accurate positions of the sources, the new techniques revealed that some of the sources had structure. Cassiopeia A was "seen" to be almost circular with an angular size of about 4 minutes of arc. It was later shown that the edges of the disk of this radio source were brighter than its centre, indicating that the body was perhaps a shell rather than condensed material. It also appeared that a jet of material extended beyond the disk. Photographs taken of the accurate radio position of Cassiopeia A recorded a pattern of short disconnected filaments. It turned out that the peculiar nebulosity previously photographed and mentioned earlier formed part of this complex structure. It now seems certain that this object is the exploded debris of a supernova. In the short time that this object has been known, it has already presented evidence that it is expanding. Measurements of the expansion suggest that the explosive event happened just less than 300 years ago. As far as is known, the event was not recorded and was apparently not seen by the astronomers of the day.

Thus we have evidence that two of the recorded radio sources are old supernovae and indeed the list has been extended, as both Tycho's and Kepler's supernovae have now been observed at radio wavelengths. Other expanded shells of nebulosity which can be recorded photographically have also been detected by radio telescopes.

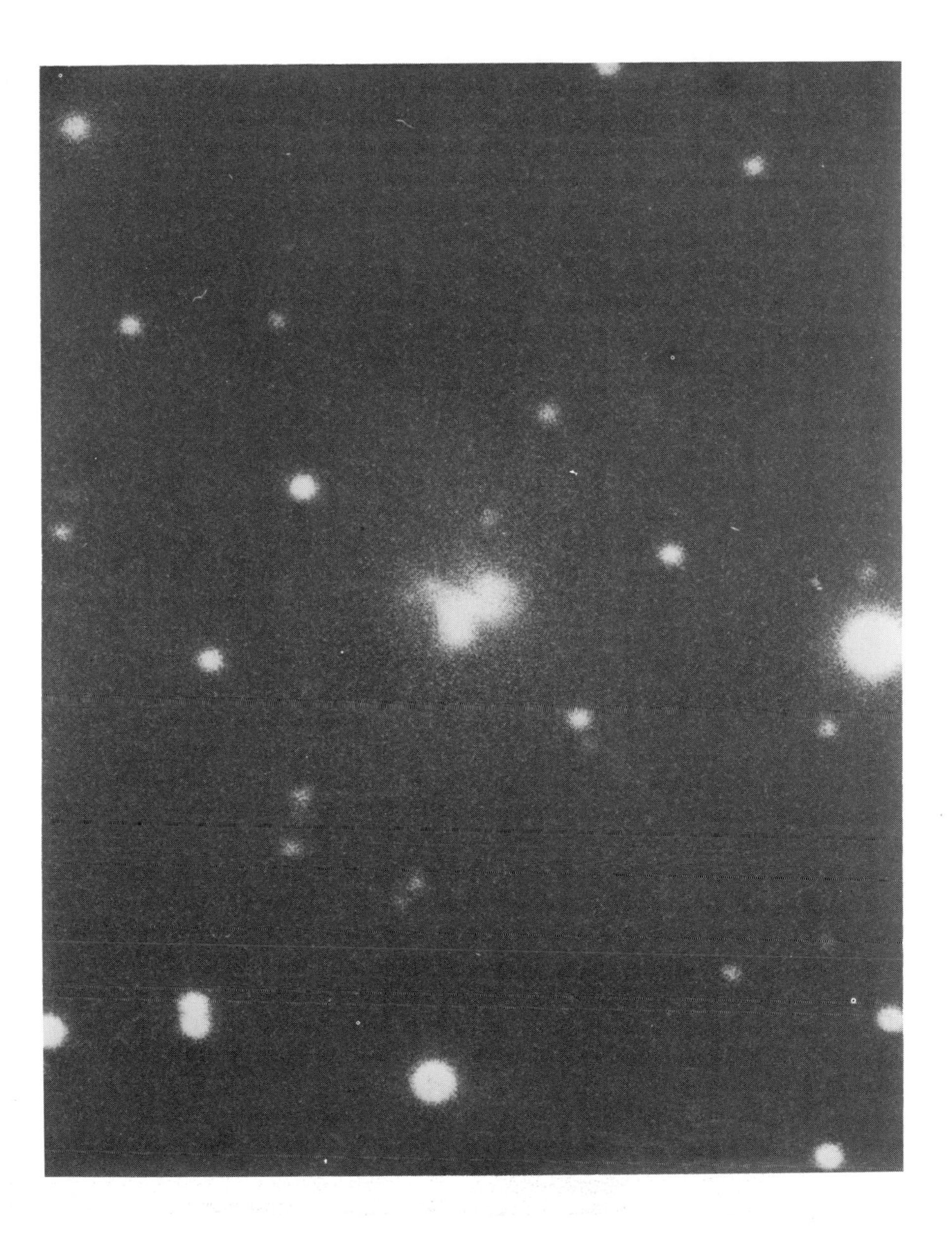

Fig. 14.1 The optical counterpart of the radio source Cygnus A taken by the 200-inch Mt. Palomar telescope. (By courtesy of the Mt. Wilson and Palomar Observatories)

However, not all the radio sources are identified with supernovae. Evidence that some of them are extragalactic began with the optical identification of the source Cygnus A.

14.3 Extragalactic Radio Sources

Interferometric measurements of Cygnus A revealed that it had a complex structure and that its radio appearance could be explained by considering it to be composed of two emission centres separated by about 1·5 minutes of arc. When its position was obtained with sufficient precision and then investigated by the Mt. Palomar 200-inch telescope, a photograph was obtained suggesting that the object comprised two colliding galaxies. A current hypothesis now takes the view that this remarkable object is a single galaxy and that there is a band of absorbing material across its middle, not seen clearly on the photograph, which gives the impression that the object is in two parts. Obscuring bands are a fairly frequent feature of spiral galaxies.

Many radio sources have now been identified with galaxies. In some cases the radio measurements present objects with two emission centres whereas the optical picture reveals only a single object. Cygnus A is probably a member of this class. Some of the radio sources that have been identified as extragalactic bodies have noticeable peculiarities when they are studied optically. For example, NGC 3034 (M82) is observed as a weak radio source. Photographs reveal that there are plumes of material emanating from the central condensation of this galaxy. By assuming the plumes or jets to be material in motion through the galactic magnetic field, the radio emission is again explained by the synchrotron mechanism.

Some of the nearer galaxies allow spatial detail to be resolved at radio wavelengths. The contours of equal radio emission may not match those obtained for optical brightness. In other words the radio shape can be different from the optical picture. For the Andromeda Galaxy (NGC 224, M31), radio observations show that the source of emission can be detected in the plane of the galaxy at distances from the centre where the galaxy cannot be seen optically.

This same galaxy also allows observations to be made of the 21 cm line, i.e. the distribution of neutral hydrogen within this galaxy can be investigated (see Section 16.7). Measurements of the Doppler shifts of this line have also allowed the rotational velocities of the spiral arms to be studied. Estimates of the mass of this galaxy have been obtained from these measurements. All the radio observations (coupled with the optical studies) suggest that NGC 224, though larger and more massive, is not significantly different in structure from our own galaxy.

Comparative measurements of the radio brightness and optical brightness allow us to apply what is known as the **radio index** to any radio source. The radio index is equivalent to a "colour" index over an extremely large difference in wavelength. The standard wavelength chosen for the radio magnitude measurement is 158 MHz.

It is found that the value of the radio index varies according to the type or nature of the galaxy that is observed. In some cases, such as Cygnus A, for example, the amount of radiated radio energy is greater than expected in comparison with the brightness of the optical part of the spectrum. For some radio sources no positive optical identification has been made. Thus the optical energy that we receive from these objects is insufficient to record them by the available techniques, yet we can observe them by using radio telescopes. These objects are included in the list of radio galaxies and it is believed that they are extremely distant galaxies, on the limit of our observable universe. This point will be taken up again in Chapter 16.

14.4 Quasars

A better appreciation of the discussion below may be gained after reading Chapter 16.

For some of the radio sources for which accurate co-ordinates are available, no galaxy can be seen. Nevertheless, in some cases, faint peculiar starlike objects have been photographed in those positions. The first to be discovered corresponded to the radio source 3C 48 and was photographed as a 16th magnitude object in 1960. The name given to these sources was "quasi-stellar radio source", this term now usually being abbreviated to **quasar**.

When the radio source 3C 273 (see Fig. 14.2) was investigated, the quasar turned out to be sufficiently bright (12·8 mag.) for reasonable spectra to be taken. Four broad emission lines were recorded but there was no immediate

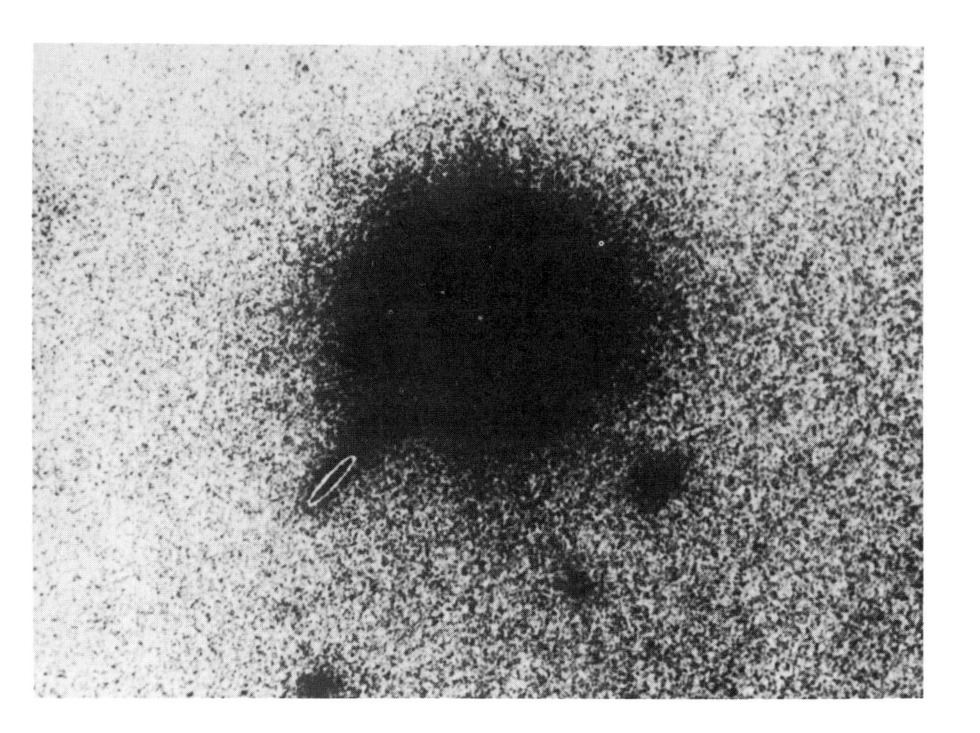

Fig. 14.2 The quasar 3C 273; a white oval encircles a jet of gas which is recorded on the photograph

explanation of their wavelength values until it was realised by M. Schmidt of Caltech that they corresponded to familiar Balmer line features which had been strongly colour-shifted. If the wavelength displacements are assumed to be Doppler, 3C273 exhibits a velocity of recession equal to 16% of the velocity of light. Such a velocity is very much in excess of even the fastest moving stars in our galaxy and immediately put in doubt the original assumption that quasars were members of our own galaxy (see Chapter 16 on the recessional velocities of galaxies). The radio source 3C48 was later scrutinized by optical spectrometry and wavelength shifts corresponding to a recessional velocity of 37% that of light were observed.

In addition to their remarkably broad emission lines, the quasars of the early investigations were found to have a large ultra-violet excess indicating that they are not radiating as black bodies. Colour index measurements therefore gave a preliminary means of identifying QSOs optically even without the initial prompting of a radio source position. Some of the extremely blue objects observed in specially commissioned surveys were shown to have the properties of quasars, except that at the time they were not detectable by the available radio astronomy techniques. As more and more quasars were discovered, the record for the recessional velocity was pushed to higher and higher values. It now turns out that the colour displacements can be so large as to shift the basic ultra-violet excess and the strong Lyman α line well into the visible part of the spectrum making the blue colour criterion for possible QSO identification no longer applicable for extremely high velocities. Calculation of the spectral shifts depends on the identification in the visible spectrum of standard ultra-violet lines such as that of triply ionized carbon ($\lambda_0 = 1550$ Å) and Lyman α ($\lambda_0 = 1216$ Å). During the 1980s the record speed for a quasar has been progressively advanced with a red shift of $z = 4$ (93% of the velocity of light) being broken for the first time in 1985. The present highest value was observed on the night of 12 September 1987 by the Anglo-Australian telescope when the quasar 0051-279 revealed a z value of 4·43. For such an object, Lyman α appears close to 6000 Å and the Lyman limit ($\lambda_0 = 912$ Å) appears well inside the optical part of the spectrum.

Quasars are observed to fluctuate in optical brightness, some of the changes occurring in relatively short periods of time. If the brightness changes are produced by the whole object undergoing a change, the time taken for the change can be no greater than the time it would take a light ray to cross from one extreme edge of the object to the other. Thus the brightness fluctuations indicate that quasars are relatively small, perhaps of the size of our Solar System.

Understanding of their nature and their relation to the universe is still at the stage of speculation. At present we still have no firm knowledge as to the distances of quasars and this is fundamental to interpreting other properties such as their sizes and luminosities. One school suggests that they are at extremely great distances and follow the Hubble law which describes galactic distances in terms of Doppler shifts. The spectral shifts of the high-velocity quasars are greatly in excess of those of the most distant galaxies and this reasoning (cosmological hypothesis) would place the high-velocity quasars beyond the galaxies. This in turn would require the objects to be extremely luminous and in consideration of their suspected size, the energy created per unit volume suggests states of matter and mechanisms for liberation of energy which are far beyond our present physical concepts.

Statistical arguments have been advanced to suggest that there are positional and associational relationships between quasars and galaxies with the proposal that galaxies might provide the source for quasars, imparting to them their high velocity on ejection. This non-cosmological theory is not widely accepted; various objections have been made, for example, that the behaviour and basis of the statistical tests are not wholly appropriate for the investigation.

Soon after their discovery, it was thought that quasars might provide another stepping stone in the advance of our knowledge of the nature of the Universe. This may indeed turn out to be so, but without more information about their distances and distribution, the step forward remains uncertain.

14.5 Pulsars

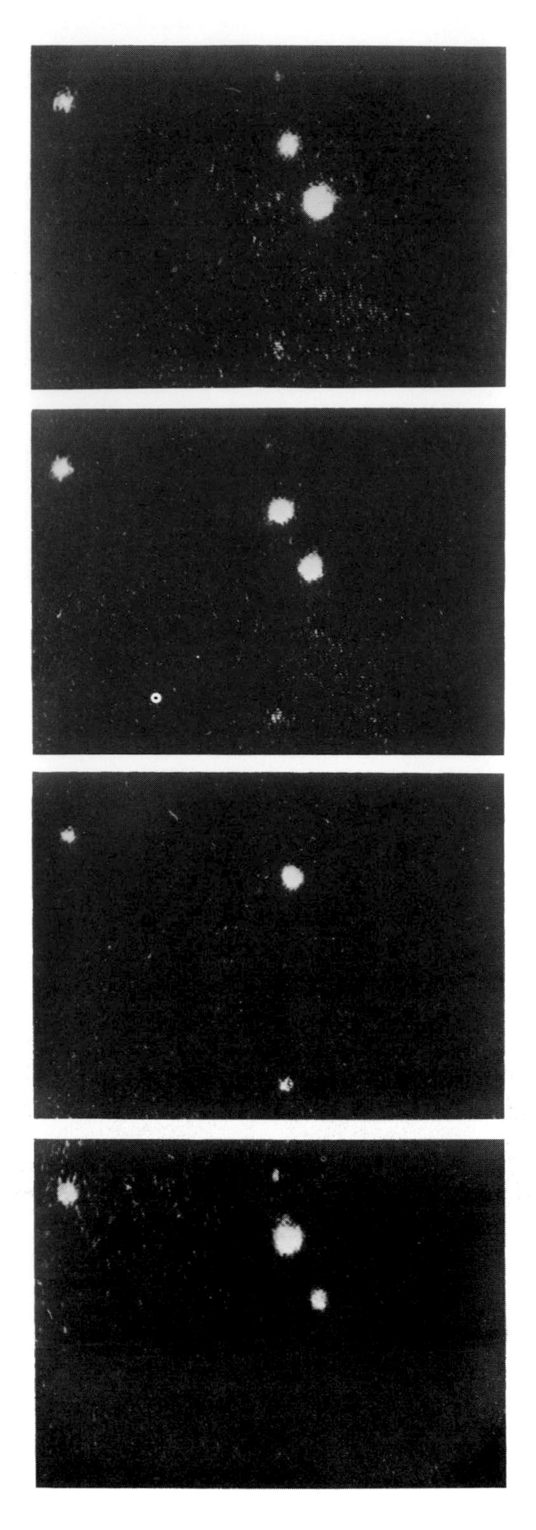

Fig. 14.3 Four different phases of Pulsar NP 0532 in the Crab nebula recorded on stroboscopic television photographs at the Lick observatory. (By courtesy of the Royal Astronomical Society)

From a series of observations commencing in 1967 and designed to investigate the scintillation imposed on radio sources of small angular extent by the interplanetary medium, a new kind of radio source was discovered. The object provides very little radio energy for detection. However, the emission is not constant but appears to have a pulse nature with a very regular time interval between the pulses. These objects have been called pulsating stars or **pulsars**. Once the time period has been determined, it is relatively easy to detect the source by integrating the signal into a series of channels, sweeping through the channels in time matching the periodicity of the pulses from the pulsar (see Fig. 14.3).

Since the discovery of the first pulsar was announced in 1968, a flood of papers discussing these objects has appeared in the research journals. In less than two years over 300 such papers have appeared. Already some of the ideas presented there are outdated and as the topic is in such a state of flux, it is only possible here to present the observational evidence and to exclude detailed discussion of the nature of pulsars.

The periods of pulsars range from a few hundredths of a second to a few seconds. Pulse widths are typically 5% of the period between pulses. By the integration process, an average pulse shape can be obtained for each pulsar. It is found that each pulsar has its own characteristic pulse shape. The radiation emitted in the pulse is strongly polarized.

From measurements of the pulse shape of the first discovered pulsar (CP 1919, i.e. Cambridge pulsed source at $RA = 19^h 19^m$), it has been

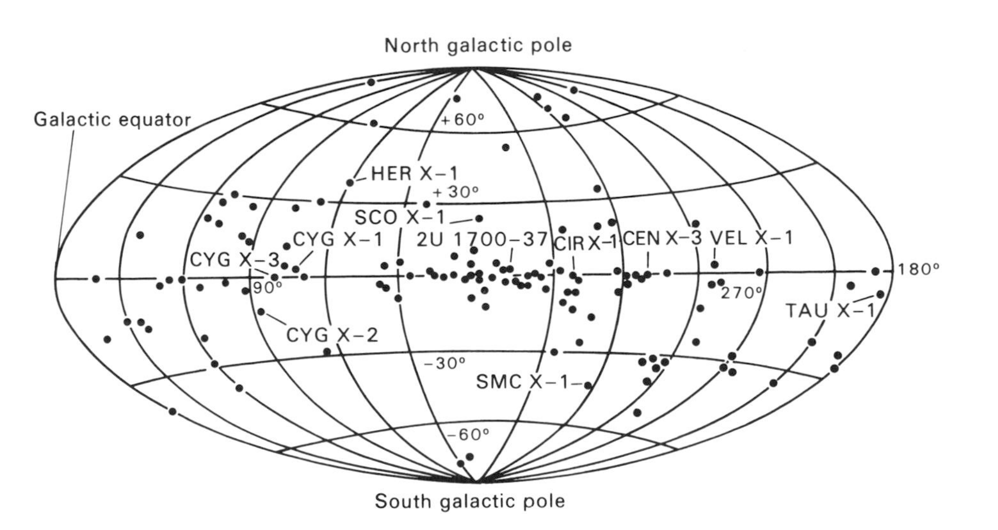

Fig. 14.4 The distribution of X-ray sources plotted on a galactic grid

estimated that the duration of the emission is only just greater than one hundredth of a second, so indicating that the source can be no larger than a few thousand kilometres.

There have been several attempts to obtain an optical identification of pulsars. The method used is analogous to the radio astronomer's technique. An optical telescope is directed to the pulsar's position and the signal from a photoelectric photometer attached to the telescope is scanned and channelled with the identical period determined for the radio pulsations. So far only two positive identifications have been made, these being the pulsar NP 0532 which corresponds to the central star seen within the Crab Nebula and PSR 0833 − 45 which is associated with the supernova remnant Vela X. Thus, we know the distances of two pulsars and know that they belong to our galaxy. Investigations of other known supernovae debris by both radio and optical techniques have failed to reveal pulsars.

Contrary to the initial suppositions, it has been found that all pulsars are gradually slowing down. Very occasionally a pulsar suddenly increases its pulse rate, to be followed by the normal decrease. It is as though the pulsar suddenly collapses down a little and, in conserving angular momentum, increases its rotational velocity. Such an event is referred to as a glitch.

The general regularity of the periodicity of pulsars is so precise that special relativity has to be invoked to describe its behaviour as observed on a moving Earth; the orbital velocity of the Earth is significant in relation to the velocity of light. As it turns out, one of the catalogued pulsars, PSR 1913 + 16, is a binary pulsar and its behaviour confirms the principles of general relativity in a number of ways.

The current consensus is that the pulsar is a collapsed remnant of an evolved or dying star. An accepted development for an old star is for it to collapse down to the white dwarf stage. In the theory of stellar evolution Chandrasekhar had shown that the masses of white dwarfs must all be below a critical mass equal to about $1{\cdot}2\ M_{\odot}$. During the collapsing stage, matter is blown off to form a planetary nebula or perhaps a supernova. In the process, it used to be thought that the mass loss by ejection would always be sufficient to reduce the core to below the critical mass. However current theory suggests that some stars may have sufficient remaining material to be above the critical mass, so enabling a remarkable metamorphosis to take place. As the collapse progresses the pressures become so enormous that the basic particles are forced to occupy identical spaces and coalesce. Electrons combine with protons and with the annihilation of charge, neutrons are produced. Thus the pulsar may well be synonymous with a *neutron* star of some 25 km diameter containing a mass of typically 2 $M_{\odot}$.

There are several theories related to the production of the observed pulses of radiation. A strong contender is the oblique rotator model. The rapidly rotating neutron star is likely to have a remnant magnetic field; electrons and protons would be accelerated from the poles of the field with polarized radiation being emitted. If the magnetic poles are not coincident with the rotational pole, the star's spin will cause a radiation beam to be swept across the line of sight to produce the regular pulse phenomena.

14.6 X-Ray Astronomy

All measurements of X-rays from celestial objects need to be made above the Earth's atmosphere. Early instrumentation was flown on rockets but great advances have come from the more stable platforms of Earth-orbiting artificial satellites such as Uhuru, Ariel V, SAS-3, OSO-8, and the HEAOs (high energy astronomy observatories).

From the more recent satellites, the positions of perhaps a few hundred sources are known to within ±20 arc seconds and this has given ample scope to allow firm identifications with optical sources.

Rather than describing the X-ray radiations in terms of wavelength, it is normal practice to refer to the energy (keV) associated with the photons. In the everyday terms of the medical world, X-ray machinery is normally operating with photons in the 50–100 keV range. However in astronomy the softer end of the X-ray range generally provides the main front for study, covering energies from 1–10 keV.

It is also common practice to describe the strength of a source in terms of its X-ray luminosity, this being determined from the received flux and knowledge of the source's distance. Hence one talks of low-luminosity galactic sources with luminosities of 10^{33}–10^{35} erg s^{-1} (10^{26}–10^{28} J s^{-1}) and extra-galactic objects emitting in the range 10^{42}–10^{45} erg s^{-1} (10^{35}–10^{38} J s^{-1}).

A wide range of objects act as galactic X-ray sources. Supernova remnants are seen to emit X-rays as such radiation has been detected coming from the Crab Nebula, Cassiopeia A and the radio object coincident with the position of Tycho's supernova of 1572. Many other galactic sources have also been identified with a range of optical counterparts. Nova Ophiuchus 1977 flared while HEAO-1 was operating and was found to emit X-rays. Bright stars such as β Per (perhaps the most famous of eclipsing binaries) and γ Cas (a well-known rapidly rotating Be star) are X-ray emitters as is the nearest star, α Cen, the radiation presumably associated with a solar-like corona.

Many of the galactic sources have been found to fluctuate in emission strength over relatively short time scales, these objects are now referred to as **bursters**. Typical behaviour is for the emission to increase suddenly by several times and then die down over about 10–20 seconds with a few such bursts being apparent during a day. The radiation is thought to be a result of irregular gas globules being transferred to a degenerate neutron star from a companion star; the temperatures that are achieved in the acceleration are sufficient to generate nuclear blasts when the material impinges on the neutron star's surface. The first bursters were found to be associated with globular clusters and at one time it was thought that the burster–globular cluster relationship was deemed to be all important to the understanding of the phenomena with talk of there being very massive ($M \geqslant 100\ M_{\odot}$) collapsed objects providing the necessary accelerations. Since the first discoveries, the general spatial distribution of bursters has been found not to match that of globular clusters and indeed several bursters have been associated with individual stars without globular cluster connections.

An important development in the field of astrophysics has ensued from the discovery that several of the X-ray sources comprised binary systems with the Uhuru satellite recording regular eclipses for four sources (Her H-1, Cen X-3, 3U 0900-40 and 3U 1700-37). Our general understanding for the production of the radiation (suggested on theory prior to the observational discovery) is based on mass transfer between two orbiting stars which have been subject to differing evolutionary time-scales with the material falling onto the collapsed star (white dwarf), being subject to very high acceleration and impacting with sufficiently high kinetic energy to produce the X-ray emission, much on the lines of the burster mechanism.

X-rays have also been found in association with the pulsar NP 0582 embedded in the Crab Nebula with a pulsation which matches the period obtained from both the optical and radio measurements. The energy contained in the X-ray emission from this source is very much greater than that contained in the remainder of the electromagnetic spectrum.

Extragalactic objects also make an exciting contribution to the list of X-ray sources. Over 40 identifications have been made with clusters of galaxies whose emission is probably generated in a hot ($\sim 10^8$ K) intergalactic gas. Other discrete sources are associated with active galaxies; over 30 Seyfert galaxies have been identified as X-ray emitters and about 30 other objects such as radio galaxies (M87 and Cen A), BL Lac objects, QSOs and emission-line galaxies have been studied. In order to model the X-ray emission strengths, non-thermal phenomena are required for the generation of the radiation in these intense, compact sources.

14.7 Black Holes

We know of two courses that a star can take during its later development. One possible evolutionary path leads to the production of a white dwarf star with material of extremely high density; a second more catastrophic possibility is that the outer layers may be stripped off completely in an explosive supernova event, leaving behind the concentrated mass of a neutron star or a pulsar.

From the theoretical work of Chandrasekhar, it has been well established that there is an upper limit to the mass of a white dwarf, this being at about 1·2 times the mass of the Sun. Emerging from the developing theories of neutron stars, an upper limit of mass is now being assigned; this limit is not yet definite because of our uncertainty of knowing how nuclear matter behaves at such high densities of the order of 10^{18} kg m^{-3}, but again it seems almost certain that the neutron star mass is below three solar masses.

For stellar cores or remnants containing more than a few solar masses, it has been postulated that when the nuclear sources of energy are exhausted, the internal pressures would be insufficient to maintain an equilibrium and that a "complete" gravitational collapse would occur. Such a body has been termed a **black hole**, in consequence of its surface gravity being so great that no radiation can escape from it.

One of the predictions made by the general theory of relativity is that electromagnetic radiation which emerges from a gravitational field is weakened and hence is red-shifted. In the case of a black hole, it is postulated that all the energy of the radiation is used up in trying to escape from the extremely strong gravitational field. In other words the red-shifts involved can be considered to be infinite.

Under the circumstances where radiation is considered to be composed of photons, a mass (rest mass) can be assigned to the photon. If the relativistic energy ($E = mc^2$) of the photon is smaller than the potential energy of a photon on the surface of a massive body, then the photon will not be able to escape. Thus, for a black hole to occur, we may write:

$$m_0c^2 - (GMm_0/r) \leq 0$$

where m_0 is the rest mass of the photon, c the velocity of light, G the universal constant of gravitation, M the mass of the body and r its radius.

By considering a black body of five solar masses, the above equation allows the deduction to be made that the maximum size of this body is approximately 7·4 km and that it has a mean density of about $5{\cdot}9 \times 10^{18}$ kg m^{-3}.

There is strong current speculation that our galaxy contains many black holes. One of the arguments for their existence is based on the evolution of stars in relation to the time scale of our galaxy. For example, in the galactic cluster known as the Hyades, white dwarfs with a theoretical upper mass limit of about 1·2 $M_\odot$ are found, even though there are main sequence stars in excess of the solar mass. Thus sufficient time has elapsed for the original more massive stars to have gone through the well-understood evolutionary phases, shedding sufficient mass to be able to collapse to the white dwarf state. If also there have been massive stars which have evolved in such a way as to be *unable* to shed sufficient mass, then they would now exist as black holes.

Methods of detecting the presence of black holes are being currently pursued. For example, some spectroscopic binaries in which only one component is observed directly and whose invisible companion exceeds the Chandrasekhar limit have been examined. In no case does the evidence point unambiguously to any such companion being a black hole; for any such suspect, there are also other plausible models explaining their natures.

Another approach to black hole identification comes from the prediction of the generation of X-rays liberated in the transfer of material from a star to its companion black hole in a binary system. The infalling matter is partially converted to radiation. Estimates of the amount of radiation produced and the relatively small space from which it radiates suggest extremely high temperatures and hence a concentration of radiated energy towards short wavelengths, i.e. X-rays. One X-ray object, Cygnus X–1, identified as part of the spectroscopic binary HDE 226868, seems to be an especially good candidate for a black hole. The X-ray emission undergoes rapid variation, indicating that the source is compact.

Finally, on this topic, it may be mentioned that black holes are being searched for, using "gravitation" detectors in the laboratory. During the collapse down to the black hole state, it is predicted that gravitational waves will be emitted, and efforts have been made to detect this radiation. No definite conclusions can yet be drawn.

There is uncertainty both on how many bursts of gravitational waves might be anticipated in our locality of the Galaxy and on the sensitivity of any apparatus to detect such bursts.

14.8 Infra-Red Objects

It is expected that many of the celestial objects beyond the Solar System should be detectable as infra-red sources by virtue of their thermal radiation. Indeed the spectra of some stars have been recorded photographically in the near infra-red.

Ground-based brightness measurements in the various atmospheric windows beyond the photographic infra-red region have been made of most of the types of object ranging from stars to extragalactic nebulae.

Measurements of some of the giant red stars have revealed that they have an infra-red brightness which is greater than expected. From measurements in the optical region of the spectrum, the temperatures of these stars can be determined. Using these temperatures, the amount of infra-red radiation can be predicted according to the appropriate black body curve. The infra-red energy received is higher than that predicted. It has been suggested that these stars are likely to have clouds of dust surrounding them. Some of the ultra-violet and visible radiation from the star is trapped by the circumstellar cloud and this raises the temperature of the cloud. It is because of the increased temperature of the outer layers of the star, not seen in the visible radiation, that we receive the extra quantities of infra-red energy.

One of the recent exciting discoveries made at a wavelength of 22 μm is that of an extended source within the Orion Nebula (NGC 1976, M42). It has been suggested that this object represents the early stages of a star during its contractional phase. Estimates of the time period of this phase suggest that over a period of years, the object might be seen to undergo change. Regular checks are to be made of this object. It will be remembered that one star (FU Orionis) within the same nebula, was seen to increase in brightness over a relatively short period of time.

Recently the application of two-dimensional infra-red detectors has allowed electronic pictures to be made of the Orion nebula. The radiation from the stars at the wavelengths corresponding to the detector's sensitivity is able to penetrate the gas and dust of the nebula and the stars can be seen vividly without the presence of the familiar (visual) obscuring material.

Mention may also be made that some of the galaxies radiate much more infra-red energy than is expected. The mechanism providing this radiation is not understood. As yet these infra-red observations, together with some others, have not been interpreted fully.

Infra-red observations have also been made by telescopes carried by balloons and artificial satellites. In particular, the Infrared Astronomical Satellite (IRAS) has provided a catalogue of measurements and discoveries. As well as the detection of many local bodies in the Solar System, as might be expected from the temperatures they achieve by insolation, the infra-red excesses (indicating the presence of dust) of many stars have been measured. Indeed, the spatial details of the dust envelopes and shaded clouds have been recorded. A remarkable discovery has been that of cool dust in the form of striated sheets and the term "infra-red cirrus" has been coined to describe the phenomenon.

Part 3
Galactic Structure and Cosmology

Chapters 15–18

PROGRAMME: The Galaxy is discussed in detail. Our picture of the universe is then completed by a description of the system of external galaxies, followed by a discussion on cosmology, cosmogony and the question of life in the universe.

15 The Galaxy

15.1 Introduction

Up till now we have not considered the overall distribution of stars throughout the space surrounding the Sun. The unaided eye sees about 6000 stars; with a pair of binoculars some 100 000 are distinguishable; photographic plates exposed in the focal plane of a large telescope reveal that stars are numbered in millions. Does the distribution of stars go on for ever throughout space or is there a boundary within which all the stars are contained, so forming a system of stars? Are there many such systems just as there are many globular clusters of stars? If so, do such assemblages of stars possess any recognizable structural features?

Even before the 20th century, when such questions received definite answers, astronomers had gone some way towards collecting evidence on which to base answers. We now look at the observations that have led astronomers to the conclusion that the Sun is a star lying in or close to the equatorial plane of a flattened assembly of 10^{11} stars called the Galaxy, the Sun's position being about two-thirds of the way outwards from the centre to the edge.

15.2 The Milky Way

Even primitive man was conscious of the fact that on a cold, clear night, the sky was divided into two by a band of light, variable in misty brightness along its length, that stretched across the heavens and kept its position among the familiar constellations of stars as the heavens revolved. With even a small telescope, this band of light, called the Milky Way, can be resolved into myriads of faint stars. Observations from a range of geographical latitudes sufficiently wide to take in all the sky show that the Milky Way follows a complete great circle. A first interpretation, that the stars form a slablike system in which the Sun is embedded, is confirmed by star-counts. The numbers of stars seen diminish sharply with increasing angular distance from the Milky Way. One of the first astronomers to carry out this method of measuring the distribution of the stars by star-counts was Sir William Herschel. He published an account of these researches in 1784. Earlier, in 1755, it may be mentioned, the philosopher Kant had already suggested that the Milky Way star system had a boundary and that certain faint nebulous patches of unresolved light might be other systems much further away. Herschel's work was confirmed and extended by subsequent astronomers. Not only does the number of stars on either side of the Milky Way fall off with increasing angular distance, but the number of faint and presumably distant stars varies significantly along the great circle band defined by the Milky Way.

This distribution in galactic longitude, as we would call it now, is such that there is a maximum density of stars in the direction of the constellation of Sagittarius. The numbers fall off with increasing angular distance in both directions along the Milky Way away from Sagittarius, until a minimum is reached in Auriga, 180° distant from Sagittarius.

The first interpretation, that the stars form a slablike system in which the Sun is placed, can therefore now be modified in that such star-counts suggest that the Sun is off-centre. Looking towards Sagittarius, with its densely-packed stars, is equivalent to looking towards the galactic centre; turning the telescope towards Auriga presents a view of the stars lying further out from the centre than the Sun. Very schematically, these ideas are embodied in Fig. 15.1 (see also Fig. 15.2).

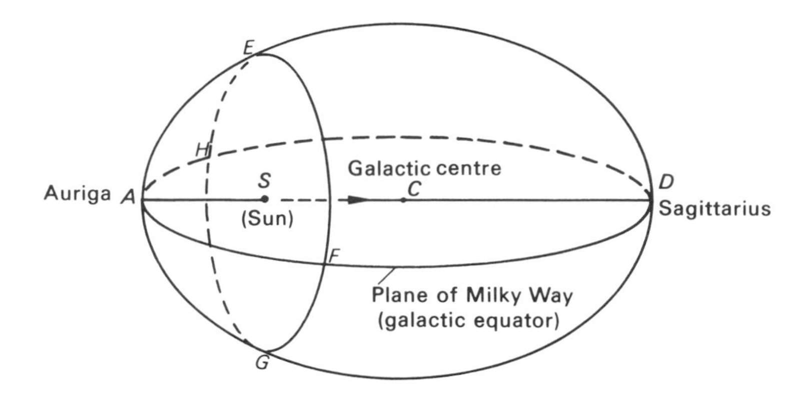

Fig. 15.1 A sketch of the main features of the shape of the Galaxy

Fig. 15.2 A photographic mosaic of the Milky Way from Sagittarius to Cassiopeia. (By courtesy of the Mt. Wilson and Palomar Observatories)

When different directions are viewed in the plane of the Milky Way, most stars are counted in the direction *SC*. The lowest number are seen in the opposite direction *SA*. Above and below that plane, the number falls off rapidly with increasing angular distance.

This off-centre position of the Sun is confirmed by the distribution of the globular clusters. The vast majority of them are found on the side of the sky defined by the galactic centre *C* and the plane *EFGH* at right angles to the Milky Way and passing through the Sun. The clusters are divided almost equally in number by the Milky Way plane.

As well as the great star fields of the Milky Way, there are regions within it where nebulous interstellar dust and gas clouds exist. As already mentioned, the densest of these contain matter sufficiently absorbing to hide completely any stars embedded in it or beyond it, and appear like dense smog-patches among the star fields. Others are bright, being illuminated by nearby stars. Like the stars, these nebulae are largely confined to the Milky Way and are particularly evident in the regions around and including Sagittarius. Their distribution is further evidence of the structure of the Galaxy sketched in Fig. 15.1.

Gross star-counts of the type first used by Herschel are inadequate to obtain more than a rough idea of the shape and structure of the Galaxy. Recourse must be had to the types of object found in the Galaxy, and in particular to the methods developed for measuring the distances of such objects. These methods are reviewed below in Section 15.3.

15.3 Review of Methods of Measuring Distances

At this point it is convenient to assemble the various methods described in earlier sections for measuring distances outside the Solar System. It is also useful to remind ourselves of their accuracies and ranges.

Distances of celestial objects may be found by

(*a*) trigonometrical parallax,
(*b*) secular parallax,
(*c*) dynamical parallax,
(*d*) spectroscopic parallax,
(*e*) the Cepheid period–luminosity relation,
(*f*) the cluster convergent method,
(*g*) the brightnesses of the main sequence stars of a cluster.

It goes without saying that in general not all methods can be applied to the same object.

(*a*) This method is limited in practice to stars within a range of 100 parsecs. Beyond that distance, corresponding to a parallax of 0$''$.01, the uncertainty, of the order of 0$''$.005, is attaining a substantial proportion of the parallax itself. Nevertheless, at least 6000 stars have had their distances measured by this method of trigonometrical parallax.

(*b*) In Section 10.11 we have seen that the method of secular parallax, although capable of being applied to stars more distant than those within 100 parsecs, is unreliable in the case of individual stars. It is impossible to distinguish between an individual star's own motion and that due to the Sun's motion. It is possible, however, to select a group of stars and obtain an average secular parallax for the group.

(*c*) This method is confined to binary systems. The large number of such systems has enabled it to be widely used. It will be remembered that the small spread of stellar masses enables an approximate value of two solar masses to be used to obtain a first approximation to the system parallax. Iteration using the apparent magnitudes of the components and the mass–luminosity relation improves the accuracy of the determined parallax. At present more than 2500 dynamical parallaxes have been determined. The range for this method is of the order of 1000 parsecs.

(*d*) The method of spectroscopic parallaxes depends upon the measurement of the distance modulus $(m-M)$, where M and m are the absolute and apparent magnitudes respectively. The apparent magnitude m is measured directly; M is found by establishing the giant or dwarf nature of the star from its spectrum, enabling it to be placed on the HR diagram.

The main advantage of this method is that the uncertainty in a spectroscopic parallax is about 10% of its value, *regardless of the star's distance.* Since the trigonometrical parallax uncertainty is always of the order of 0$''$.005, the methods become of comparable accuracy for stars with a parallax of 0$''$.05, corresponding to a distance of 20 parsecs. Beyond that distance, the spectroscopic parallax method is therefore much more reliable. The practical limit of the method is set when the stars are so faint that a reasonably detailed spectrum cannot be obtained.

(*e*) The Cepheid period–luminosity relation is one of the most widely-used and accurate methods of obtaining the distances of star clusters and galaxies. The cluster or galaxy, of course, must contain Cepheid variable stars. Variable stars which can be used for the purpose are found in globular clusters, in the galactic disk and in galaxies external to our own such as the Magellanic Clouds and the Andromeda Galaxy. Because Cepheids can be so luminous as to be visible for distances of up to 2×10^6 parsecs, they provide a method with a range far beyond any of the preceding methods. In addition, the presence of a number of Cepheids of different period within an object enables the distance to be determined accurately. (See the complete discussion of the method in Sections 12.2.3 and 12.2.4).

(*f*) In Section 10.6 and 10.7 it was shown how the parallaxes of stars in an open cluster could be found once the position of the cluster's convergent point had been determined. Unfortunately in practice only one or two clusters can have this method applied to them to obtain parallaxes of reasonable accuracy.

It may be noted that where an open cluster is the object whose distance is sought, more than one of the above methods can usually be employed, thus providing a check on their reliability.

(*g*) For the local main sequence stars, the variation of spectral type with absolute magnitude is well known, the HR diagram showing that these stars lie on a line. For a cluster of stars whose distance is not known, an HR diagram can still be obtained by plotting spectral type against apparent magnitude. By comparing the position of the main sequence stars with those on the diagram drawn with an absolute magnitude scale, the distance modulus is obtained for the cluster. Although some correction has to be applied for reddening and absorption of light, the method is a very powerful one.

In Table 15.1, the ranges in which the various methods may be used are given. It should be noted that the ranges are correct to order of magnitude only.

Table 15.1 Distance indicators

Method	Range in parsecs	Comment
(*a*) Trigonometrical parallax	$0–10^2$	Very inaccurate beyond 10^2 parsecs.
(*b*) Secular parallax	$0–10^3$	Very inaccurate for individual stars; applied to groups of stars.
(*c*) Dynamical parallax	$0–10^3$	Confined to binary systems.
(*d*) Spectroscopic parallax	$0–10^4$	More accurate than (*a*) beyond 20 pc. Reserved for stars for which a spectrum can be obtained. Of wide application.
(*e*) Cepheid period–luminosity law	$0–10^6$	Confined to Cepheid variable stars. Of very wide application (including nearby galaxies).
(*f*) Cluster convergent method	$0–10^2$	Limited to a few nearby open clusters such as Hyades.
(*g*) Cluster main sequence HR diagram	$0–10^4$	Applicable to many clusters.

15.4 Galactic Co-ordinates

In Roy & Clarke, *op. cit.*, Section 7.15, a set of co-ordinates frequently used in studies of galactic structure was defined and the student is referred to these definitions of **galactic longitude** and **galactic latitude**. It was seen that the origin of galactic longitude is the direction towards the galactic centre and that galactic longitude is measured along the galactic equator in the direction of increasing right ascension. Galactic latitude is measured north or south of the galactic equator, the north galactic pole being the galactic pole that lies in the northern hemisphere.

We can now make use of this system in considering more exactly the shape of the Galaxy.

15.5 Distribution of Objects in the Galaxy

Using the methods of Section 15.3, where appropriate, it is found that most of the open or galactic clusters are within a disk coinciding with the plane of the galactic equator and only 100 parsecs thick, though their distances from the Sun can be as great as 5000 parsecs.

Known planetary nebulae, in number over 500, have distances from the Sun ranging between 100 parsecs to greater than 10 000 parsecs. This number is almost certainly an underestimate, the total within the Milky Way being possibly as great as 10 000. The distribution of planetary nebulae is symmetrical about the galactic equator and shows a concentration towards the galactic centre, forming a flattened system.

Classical Cepheids of Population I lie in the galactic disk within the spiral arms; on the other hand Type II Cepheids, of Population II, are found in the central galactic bulge and in globular clusters.

Novae are found concentrated in the galactic disk, especially in the central bulge and also in globular clusters, as would be the case if they are Population II objects.

Finally, it may be added that in general, Wolf–Rayet and supergiant stars are also found for the most part in the galactic disk.

When we consider the distribution of these objects in galactic longitude, the vast majority are found symmetrically distributed about galactic longitude 0°. Like the star-count numbers, their frequency falls off from 0° to 180° and from 360° (0°) through decreasing longitudes to 180°.

The globular cluster distribution is different. No globular cluster is found near or in the disk inhabited by the open clusters, planetary nebulae, novae and supergiants. This does not mean that there are no globular clusters within the galactic plane. They are very distant objects. The nearest is 5000 parsecs away while the farthest is 100 000 parsecs from us, and it could well be that a large number of such clusters exist in the galactic plane, forever hidden from us by the great clouds of obscuring matter concentrated in that plane.

The distances of globular clusters are obtained from the Cepheids and RR Lyrae stars in them. When the cluster directions and distances are plotted, it is found that they form a roughly spherical distribution with a radius of 20 000 parsecs and centred in a direction corresponding closely to that of galactic longitude 0°. It is reasonable to assume that the centre of the globular cluster system coincides with that of the Galaxy.

The disk has a distinct bulge at the centre. This has been called the galactic nucleus and appears to be spheroidal in shape, its equatorial plane being coplanar with that of the disk. About 5000 parsecs thick, it has an equatorial diameter of about 10 000 parsecs. The rest of the disk is much thinner so that over most of its area it is not much more than 1000 to 1500 parsecs thick. Its radius is about 15 000 parsecs, with the Sun some 8500 parsecs from the galactic centre, according to the distribution of the distances of the objects in the Galaxy.

In Fig. 15.3 a more refined diagram of the Galaxy, seen edge-on, is given, in accordance with the above data.

A view of how our own Galaxy might look from the outside may be surmised by looking at Fig. 15.4, which depicts the Andromeda Galaxy.

15.6 Galactic Rotation

The shape of the Galaxy suggests that it is in rotation about the galactic nucleus just as Saturn's rings revolve about Saturn or the planets about the Sun. Whether in fact the rotation is such that each star pursues its orbit about the

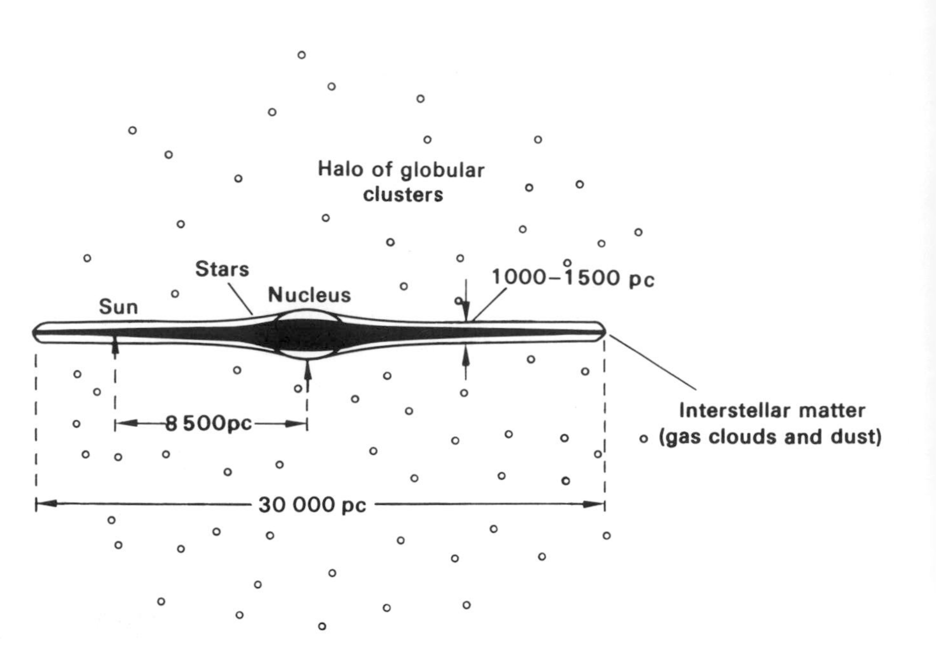

Fig. 15.3 The distribution of objects within the Galaxy

Fig. 15.4 The great spiral galaxy in Andromeda, NGC 224 (M31) with its two satellite systems NGC 205 and 221. (By courtesy of the Hale Observatories)

nucleus in accordance with Kepler's laws or the whole system rotates like a solid body, only observation can decide.

What sort of observational evidence for galactic rotation should be looked for? Let us take a simple example where we assume that under the gravitational attraction of the massive inner portions of the Galaxy, the stars in the vicinity of the Sun revolve in circular orbits about the galactic centre. Those nearer the centre than the Sun will have greater velocities; those farther out will revolve more slowly about the centre.

In Fig. 15.5(*a*) the situation is sketched. The region in which the stars are moving is taken to be small compared with the distance to the galactic centre so that we can assume that the small parts of their orbits shown are approximately straight lines. Then the stars X_1, X_2, X_3 and X_1', X_2', X_3' will revolve more slowly than the stars S_1, S, S_3 (S being the Sun) and they will in their turn be revolving more slowly than the stars Y_1, Y_2, Y_3 and Y_1', Y_2', Y_3' in the inner orbits. The lengths of the arrows in Fig. 15.5(a) represent the velocities.

Observing from the neighbourhood of the Sun, we would expect to see the Y-stars in the inner orbits overtake us while those in the outer orbits, the X-stars, fell behind. Those stars S_3 and S_1 ahead of and behind the Sun would neither overtake nor fall behind, having the same velocity as the Sun. This situation is sketched in Fig. 15.5(b). It can be seen that it would give rise to a systematic trend in the observed proper motions of such stars.

There would also be a resultant systematic trend in the radial velocities of the stars due to galactic rotation of this nature as seen in Fig. 15.5(c). By projecting the relative velocities of the stars on their heliocentric directions, the radial velocity vectors in Fig. 15.5(c) are obtained. As observations were made throughout galactic longitude from 0° to 360°, data confirming or denying the original hypothesis would be obtained. It would be expected that curves of the type sketched in Fig. 15.6 would describe the behaviour of radial velocity with galactic longitude. The negative velocities are of course velocities of approach. The amplitudes of the curves will also depend upon the distance of the objects from the Sun, as seen in Fig. 15.5(c).

It should be noted that measurement of radial velocities is more reliable in this investigation. The objects chosen are in general high-luminosity objects such as giants, supergiants, long-period variables, planetary nebulae and open clusters. The results show clear evidence that at least in the neighbourhood of the Sun the Galaxy rotates, the various parts pursuing their own orbits.

When the Sun's motion is measured in this way against (a) the system of globular clusters and (b) the nearer external galaxies, the orbital velocity of the Sun about the galactic centre is found. The Sun is travelling towards galactic longitude 90° with a speed of about 220 km s^{-1}. The period of revolution of the Sun about the galactic centre is therefore approximately 240 million years. Since the Earth is at least 5×10^9 years old, this suggests that the Sun has completed twenty, or maybe a few more, orbits round the Galaxy. The orbital period is sometimes referred to as a **cosmic year**.

These figures may seem to be in conflict with the values found for the solar motion in Chapter 10. It should be remembered, however, that the solar motion was measured with respect to the local group of stars. What we have been discussing in the present section is essentially the velocity of the local group with respect to the centre of the Galaxy. In other words, considering the

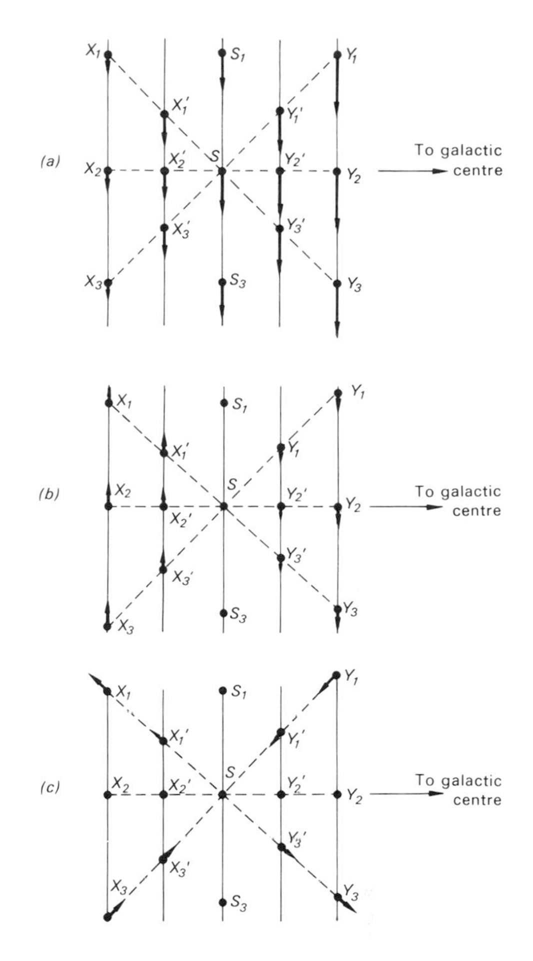

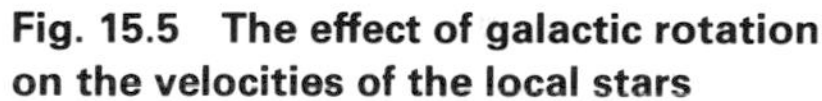
Fig. 15.5 The effect of galactic rotation on the velocities of the local stars

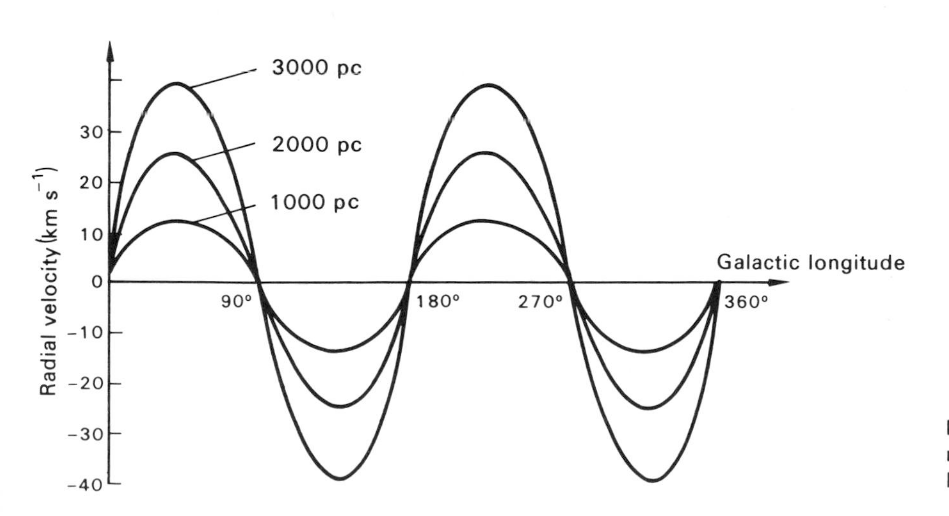

Fig. 15.6 The expected curves of stellar radial velocities plotted against galactic longitude

analogy made earlier of the stellar motions being like those of a flock of birds, the former solar motion was the velocity of one "bird", the Sun, within the flock; what we now have is the velocity of the flock itself.

15.7 The Mass of the Galaxy

Assuming that the stars pursue Keplerian orbits about the massive hub of the Galaxy, we can now obtain a rough estimate of the mass of the Galaxy. To a first approximation the mass outside the Sun's orbit, assumed circular, can be neglected and the mass, M, within can be taken to be acting as a point-mass situated at the galactic centre.

Let the mass of the Sun be $M_\odot$, its distance from the galactic centre be r and its velocity be V km s^{-1}. The Sun's sidereal period of revolution about the Galaxy is then T where

$$4\pi^2 r^3/T^2 = G(M+M_\odot). \tag{15.1}$$

But we also know that if $T_\oplus$ and a are the sidereal period of revolution of the Earth about the Sun and the semi-major axis of its orbit, we can write

$$4\pi^2 a^3/T_\oplus^2 = G(M_\odot+M_\oplus), \tag{15.2}$$

where $M_\oplus$ is the mass of the Earth.

Dividing Equation (15.1) by Equation (15.2) and remembering that $M \gg M_\odot$ and that $M_\odot \gg M_\oplus$, we obtain

$$\frac{M}{M_\odot} = \left(\frac{r}{a}\right)^3 \left(\frac{T_\oplus}{T}\right)^2.$$

Now

$$T = 2\pi r/V.$$

Hence

$$\frac{M}{M_\odot} = \frac{T_\oplus^2}{a^3}\frac{rV^2}{4\pi^2}. \tag{15.3}$$

If distances are measured in astronomical units and time in years, $T_\oplus = a = 1$, so that Equation (15.3) reduces to

$$M/M_\odot = rV^2/4\pi^2. \tag{15.4}$$

Now

$$r = 8500 \text{ parsecs} = 8500 \times 206265 \text{ AU}$$

and

$$V = 220 \text{ km s}^{-1} = \frac{220}{149{\cdot}6 \times 10^6} \times 31{\cdot}56 \times 10^6 \text{ AU year}^{-1}.$$

Hence, by substituting for r and V in Equation (15.4), it is found that

$$M/M_\odot \approx 10^{11}.$$

The mass of the Galaxy is therefore close to 10^{11} solar masses. The Sun is itself so average in stellar mass that the number of stars in the Galaxy must be of the order of 10^{11}, since it is estimated that not much more than 7% of the mass of the Galaxy is in the form of interstellar dust and gas.

15.8 The Spiral Arms of the Galaxy

When we view other galaxies, it is found that many of them exhibit a spiral structure, the spiral arms curling outwards from their galactic nuclei. It seemed unlikely to astronomers that our own Galaxy did not possess its own spiral structure and various investigations were conducted to obtain evidence one way or another.

Among the methods that have been used are (i) the photography of early-type hot stars and the clouds of ionized hydrogen associated with them, (ii) the use of radio telescopes to map the distribution of neutral hydrogen clouds in the Galaxy.

The first method is limited by the obscuration produced by dust—the interstellar smog. Nevertheless, the positions of the hot stars and the ionized hydrogen clouds in the neighbourhood of the Sun showed that they extended along two distinct arms with the possibility of a third. These arms have been called the Orion arm (the Sun lies on its inner edge), the Perseus arm and the Sagittarius arm, the last of which lies between 6 and 7 kpc from the galactic centre.

The second method is much more successful. The detection of the 21 cm (1420 MHz) radiation from the hydrogen clouds in the galactic plane has been used to reveal the distribution of hydrogen throughout the Galaxy to distances far greater than the limit of the optical method. Such regions containing neutral hydrogen are referred to as H I regions. In what follows we sketch the principles involved in this important research method.

In Fig. 15.7(a), let C be the galactic nucleus, S the Sun and H_1, H_2, H_3 three clouds of hydrogen. We assume for simplicity that the Sun and the hydrogen clouds are moving in circular, coplanar orbits about the galactic centre under the gravitational attraction of the massive galactic nucleus. There will of course be many other clouds in the Galaxy but we are concerned at the moment with the ones that lie in galactic longitude l. Then the arrows in length and direction at S, H_1, H_2 and H_3 denote the circular velocities of S, H_1, H_2 and H_3 respectively. In Fig. 15.7(b), a velocity equal and opposite to the Sun's velocity is applied to the Sun and the clouds so that, by forming the parallelogram of velocities at H_1, H_2 and H_3, the clouds' velocity vectors relative to the Sun can be formed. This is done explicitly for H_2 in Fig. 15.7(b).

In Fig. 15.7(c), these relative velocities are shown for H_1, H_2 and H_3. Also shown are the resultant radial velocities along the line of sight $SH_1H_2H_3$. It is seen that H_1 and H_2 are proceeding away from the Sun with different velocities while H_3 has a small velocity of approach.

If V and c are the radial velocity of a cloud and the velocity of light respectively, the Doppler formula, viz.

$$\Delta\lambda/\lambda = V/c,$$

will give the shift $\Delta\lambda$ in the wavelength λ, in this case of the 21 cm (1420 MHz) line. The amount of radiation received in the radio telescope from that cloud will depend upon its distance and also the amount of radiation it is emitting. This last quantity in turn will depend upon the size of the cloud. Assuming that all three clouds H_1, H_2 and H_3 are giving out equal amounts of radiation per second, we would expect that when a scan is made through the 21 cm line, a line-strength profile such as the one sketched as a dotted line in Fig. 15.7(d) will be obtained. The overall profile is of course a combination of the three outputs from the clouds H_1, H_2 and H_3.

Sampling in this way at intervals of a few degrees throughout galactic longitude from 0° to 360°, a set of such profiles is obtained. The problem then reduces to postulating a distribution of the interstellar hydrogen that would produce such a set of line profiles. When this is done it turns out that the Galaxy indeed has a spiral form, the hydrogen being arranged in long spiral arms coiling about the galactic nucleus. The Sun appears to be on the inside of one of these arms.

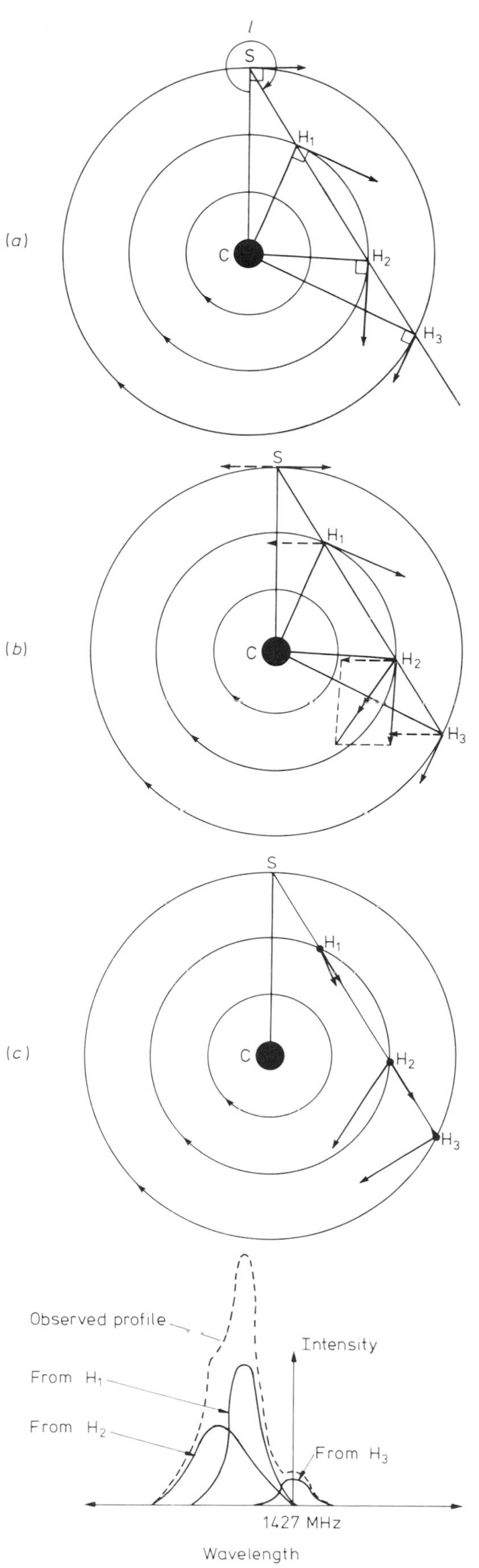

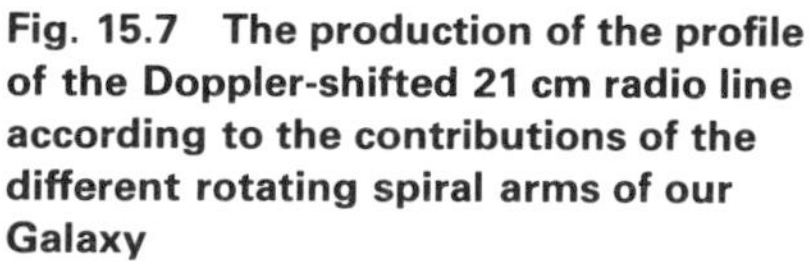

Fig. 15.7 The production of the profile of the Doppler-shifted 21 cm radio line according to the contributions of the different rotating spiral arms of our Galaxy

This investigation also showed that gas is streaming outwards from the galactic centre at speeds of up to 200 km s^{-1} and also revolving with the galactic rotation.

15.9 Stellar Populations

That stars can be divided into two broad population classes has already been discussed in Chapter 11. It will serve us here to recall the differences in the properties of the stars that allow them to be classified into populations.

The ideas of population classification were proposed by Baade in 1944 after studying the distribution of stars in the great galaxy in Andromeda. A summary describing the Populations is shown in Table 11.1. Besides the differences in their space velocities and abundances of elements (both briefly mentioned again below) the distributions of the populations in space are fundamentally different. This is very evident in the Andromeda Galaxy. The dust and gas-rich spiral arms of the Andromeda galaxy contained stars the brightest of which were blue main sequence stars; on the other hand, the central region of the Andromeda Galaxy (with little dust and gas) showed red supergiants as the brightest stars. Moreover, while the central region HR diagram resembled that of a globular cluster, the diagram for the stars in the spiral arms was similar to the one obtained for open clusters.

These differences also exist in our own galaxy. Baade subsequently gave the name of Population I to most of the stars in the vicinity of the Sun, the classical Cepheids, the T Tauri stars and the Wolf–Rayet stars. All inhabit the flat disk and spiral arms of the Galaxy where most of the dust and gas is also found. The name Population II was given to the halo objects—the globular clusters, the separate stars and the Type II Cepheids—and to the stars making up the galactic nucleus. Because of these differences the Galaxy observed from outside would appear bluish in hue in the spiral arms and reddish in the nucleus.

It was also found that in general the stars belonging to Population I show heavy element contents much higher than those in Population II. For example, the disk population stars have a heavy element mass abundance of 4% against a value of 0·3 to 1% for the halo population stars.

The fact that Population I contains objects much younger than those in Population II has already been noted. This age difference, it will be recalled, was deduced from a study of the distribution of turn-off points in their HR diagrams. Stars of Population II, forming the galactic nucleus, have space velocities which are, in general, much greater than those of Population I, these stars being found in the disk-like structure of the Galaxy. Thus it appears that the velocity a star has, depends to some extent on its age. The older the star, the more likely it is to have a high velocity.

All the above facts—the existence of the two populations, their different ages, their heavy element content, their occupation of regions relatively free of or rich in dust and gas and the differences in their space velocities—enable a tentative theory of the development of the Galaxy to be formed.

15.10 The Origin and Evolution of the Galaxy

The great mass of matter, predominantly hydrogen, that developed into our galaxy probably detached itself from the intergalactic medium some 10^{10} years ago. Like a protostar forming within a cloud of interstellar matter, the protogalaxy would shrink.

It is believed that Population II star formation began far from the plane that was to become the Milky Way and gave rise to the globular clusters in the halo about 6×10^9 years ago. Because of the scarcity of heavy elements in that era, the globular cluster stars have only a small proportion of heavier elements in their make-up.

The galactic nucleus may have been the next section of the Galaxy to form as a contracting, rotating system in which Population II stars not much younger than those in the globular clusters subsequently appeared. That these stars have the highest space velocities results from the greater kinetic energy that was available in the earliest times of the Galaxy. Star formation in both the halo and the nucleus presumably ceased when the available supply of dust and gas was used up.

The galactic disk resulted from the condensation and flattening of the remaining parts of the rotating protogalaxy. In this flat system, Population I stars were created and are still being created from the abundance of dust and gas found there. The lower end of the range of ages found for galactic clusters, namely 2×10^6 years, implies that such stars were formed very recently in the life of the Galaxy. The higher abundance of heavier elements in these stars is also in accordance with this idea. Previous generations of stars, at the explosive, unstable ends of their lives, have enriched the interstellar medium within the disk with the heavier elements created in their post hydrogen-burning phases. This cycling, sketched in Fig. 15.8, is still in progress.

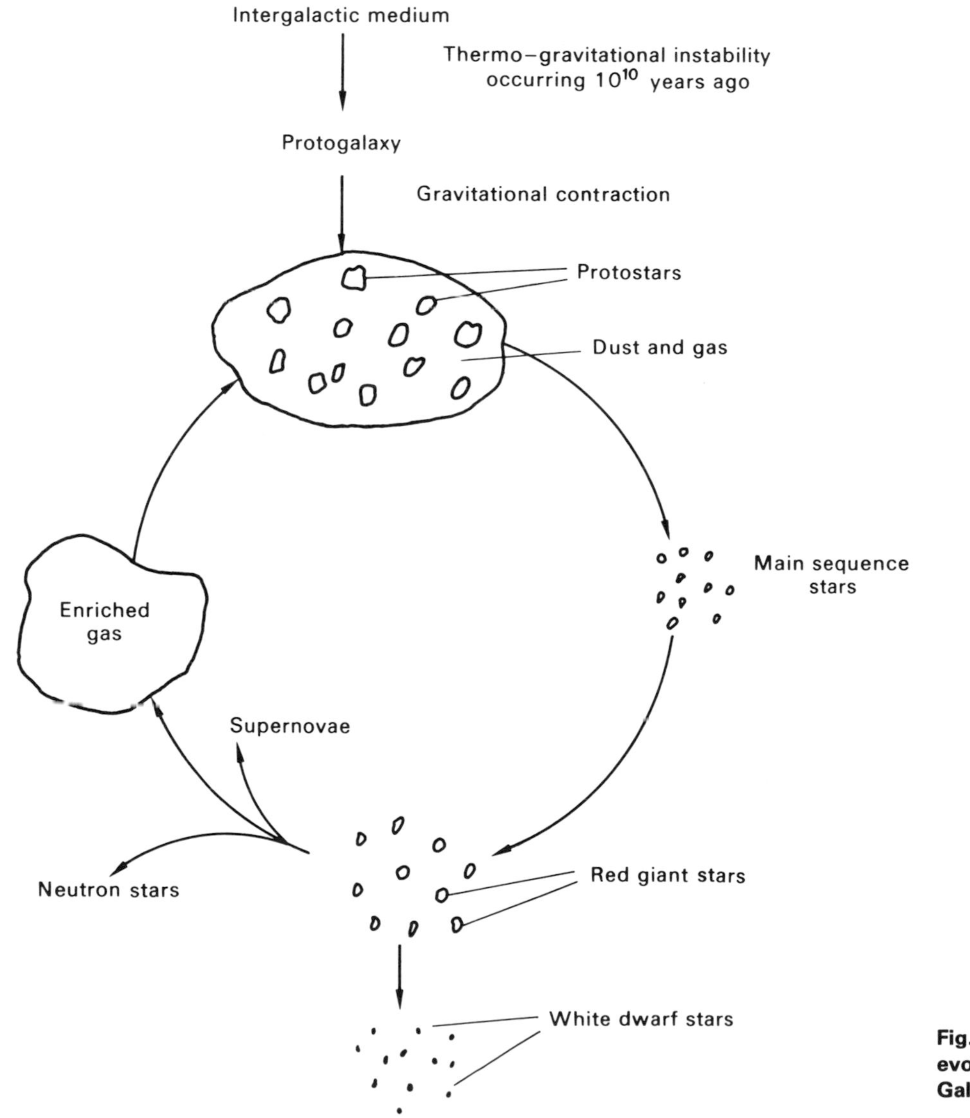

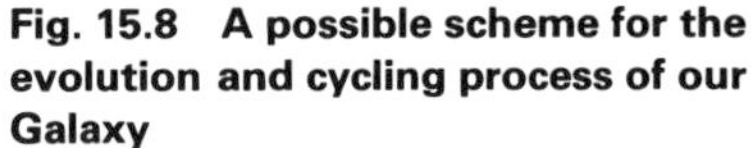
Fig. 15.8 A possible scheme for the evolution and cycling process of our Galaxy

The spiral arms that appear to spring out of the nucleus and coil round it, remain one of the unsolved problems of the Galaxy. There are dynamical reasons for believing that the present arms are young compared with the age of the Galaxy. It is possible that magnetic fields play a part in creating and maintaining them.

Very recently, astronomers have become aware that processes, as yet little understood, can occur at the centres of galaxies involving the liberation of large amounts of energy and it may be that the simple picture of the origin and evolution of our own galaxy outlined in this section will have to be modified drastically once a better understanding of these processes is reached.

15.11 Galactic Co-ordinates: Graphical Exercise

When investigating the structure of the Galaxy, it is useful to have a co-ordinate system which is related to its form and to the distribution of the various types of object. The chosen co-ordinate system uses longitude (l) and latitude (b). Galactic longitude increases in the same sense as right ascension. The positions of the galactic poles—north and south—correspond to the points on the celestial sphere where the great circles, which cut the galactic equator at right angles, all intersect.

The system currently in use is based primarily on radio observations of the distribution of neutral hydrogen in the inner parts of the Galaxy, the origin of longitude being taken close to the galactic centre. Prior to this system, galactic co-ordinates were based on the Ohlsson system which is more related to the apparent distribution of the local stars. The exercise below allows the galactic pole to be determined, giving a value close to that of the Ohlsson system.

It is a well-known fact that stars with spectral classifications W and O are usually found close to the plane of the Galaxy. By plotting their positions (right ascension and declination), a curve can be drawn through the distribution, this curve corresponding to the galactic equator.

The given list of stars in Table 15.2 has been taken from the *Henry Draper Catalogue* (*Epoch 1900*) and represents the list of O type stars from 7^h to 22^h RA. Plot the position of each star taking declination as the ordinate and right ascension as the abscissa (let 10 mm = 10°). Draw the curve corresponding to the galactic equator and determine the position in right ascension corresponding to the point $l = 0°$.

At the point where the galactic equator has a maximum or minimum value of declination, the great circle cutting the galactic equator at right angles is identical to the great circle describing points of equal right ascension. The galactic pole is, therefore, 90° away in declination from this point.

Estimate the value of right ascension where the galactic equator has a minimum value of declination—giving the RA of the galactic pole—and determine also the declination of the north galactic pole. (Check the answer against the value of RA $12^h\,40^m$, declination +28°.)

Galactic zero longitude on the Ohlsson system is chosen as the point of intersection of the celestial and galactic equators on the celestial sphere where declinations change from negative to positive values in the direction of increasing right ascension. In other words, the point corresponding to $l = 0°$ is the ascending node of the galactic equator.

Estimate the value of right ascension corresponding to the origin of galactic longitude. (Check the answer against RA $18^h\,49^m$.)

15.12 Galactic Structure: Exercise Using 21 cm Line Data

Our galaxy is known to have a spiral-arm structure. The details of the structure are built up from various kinds of observation, one of the important ones being

Table 15.2 Galactic O-type stars extracted from *Henry Draper Catalogue*

Star number	RA	Dec	Star number	RA	Dec
53667	$7^h\ 1^m$	$-8° 34'$	149038	$16^h 27^m$	$-43° 50'$
54662	7 5	−10 11	150135	16 34	−48 34
56925	7 14	−13 3	150958	16 39	−46 55
57060	7 15	−24 23	151804	16 45	−41 4
57061	7 15	−24 47	151932	16 45	−41 41
60848	7 31	+17 7	152147	16 47	−41 57
62150	7 37	−32 24	152233	16 47	−41 37
62910	7 41	−31 41	152270	16 47	−41 40
63099	7 42	−34 5	152386	16 48	−44 50
63150	7 42	−36 16	152408	16 48	−41 0
65865	7 56	−28 28	152424	16 48	−41 56
66811	8 0	−39 43	153919	16 57	−37 42
68273	8 7	−47 3	156327	17 12	−34 18
69106	8 10	−36 38	156385	17 12	−45 32
73882	8 36	−40 4	157451	17 18	−43 24
76536	8 52	−47 13	157504	17 19	−34 6
79573	9 10	−49 42	158860	17 27	−33 33
86161	9 52	−57 15	159176	17 28	−32 31
88500	10 7	−60 9	160529	17 35	−33 27
89358	10 14	−57 25	163181	17 50	−32 27
90657	10 23	−58 8	163454	17 51	−31 0
91824	10 31	−57 39	163758	17 53	−36 0
91969	10 32	−57 43	164270	17 55	−32 43
92554	10 36	−60 24	164492	17 56	−23 1
92740	10 37	−59 9	164794	17 58	−24 22
92809	10 38	−58 15	165052	17 59	−24 24
93128	10 40	−59 2	165688	18 2	−19 25
93131	10 40	−59 36	165763	18 3	−21 16
93162	10 40	−59 12	166813	18 7	−42 53
93250	10 41	−59 3	167264	18 9	−20 46
93843	10 45	−59 42	167633	18 11	−16 33
94305	10 48	−61 46	167771	18 12	−18 30
94546	10 50	−58 59	168206	18 14	−11 40
94663	10 51	−58 16	169010	18 18	−13 46
95435	10 56	−57 17	175876	18 52	−20 33
96548	11 2	−64 58	177230	18 59	−4 28
97152	11 6	−60 26	184738	19 31	+30 18
97253	11 7	−59 50	186943	19 42	+28 1
97434	11 8	−60 9	187282	19 44	+17 57
97950	11 11	−60 43	188001	19 48	+18 25
104994	12 0	−61 29	190002	19 58	+32 18
105056	12 1	−69 1	190429	20 0	+35 45
112244	12 50	−56 17	190864	20 2	+35 19
113904	13 2	−64 46	190918	20 2	+35 31
115473	13 12	−57 37	191765	20 7	+35 53
117297	13 24	−61 34	191899	20 7	+11 35
117688	13 27	−61 48	192103	20 8	+35 54
117797	13 28	−61 54	192163	20 8	+38 3
119078	13 36	−66 54	192639	20 11	+37 3
120521	13 45	−58 3	192641	20 11	+36 21
121194	13 49	−60 40	193077	20 13	+37 7
124314	14 8	−61 14	193576	20 16	+38 25
134877	15 7	−59 28	193793	20 17	+43 32
135240	15 9	−60 35	193928	20 18	+36 36
135591	15 11	−60 8	195177	20 25	+38 17
136488	15 16	−62 19	199579	20 53	+44 33
137603	15 22	−58 14	203064	21 15	+43 31
143414	15 55	−62 24	206267	21 36	+57 2
147419	16 17	−51 18	208220	21 50	+43 1

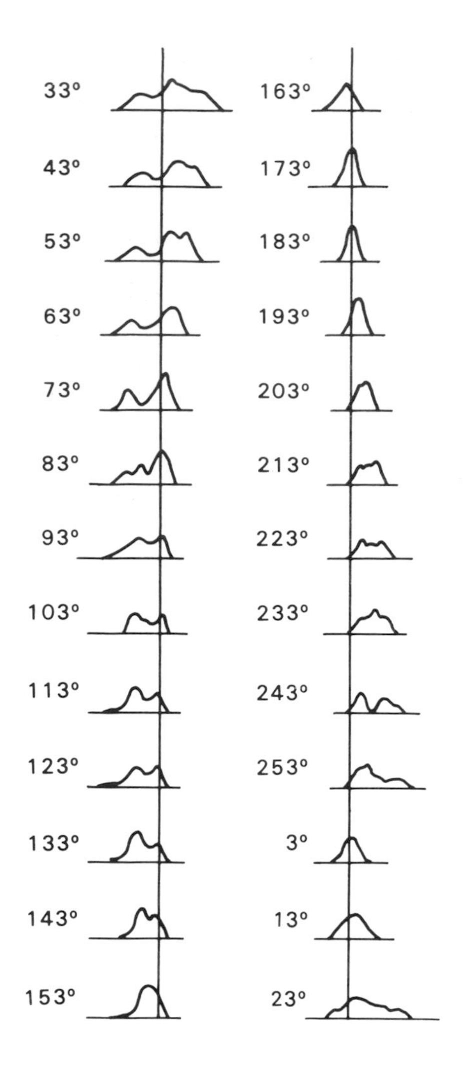

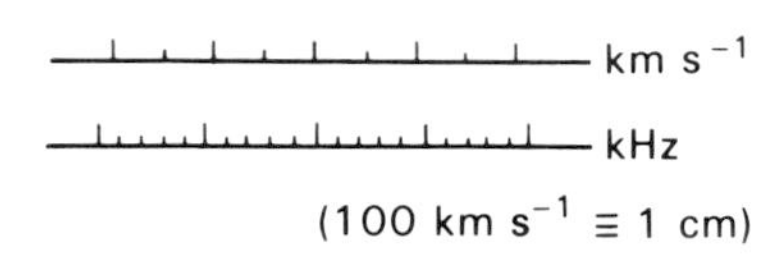

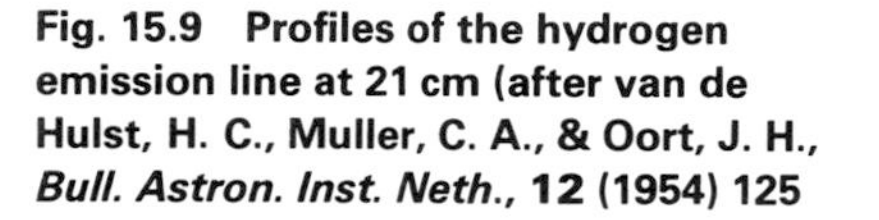

Fig. 15.9 Profiles of the hydrogen emission line at 21 cm (after van de Hulst, H. C., Muller, C. A., & Oort, J. H., *Bull. Astron. Inst. Neth.*, 12 (1954) 125

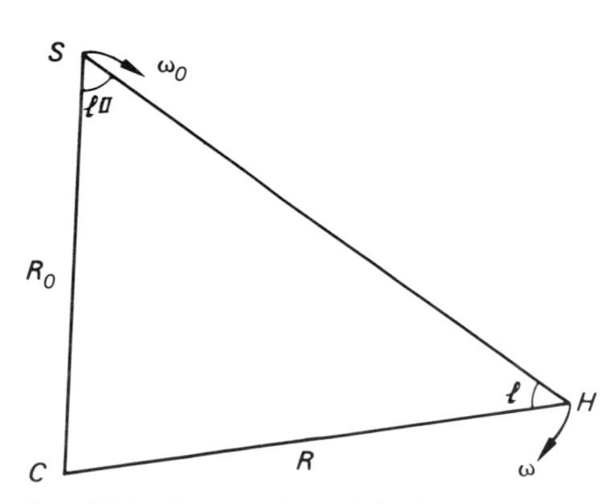

Fig. 15.10 The rotation of the Sun and a hydrogen cloud about the centre of gravity of the Galaxy

the radio astronomers' measurements of the 21 cm radiation which is emitted from the hydrogen clouds.

At any position along the galactic equator, the profile of this radio spectral line is observed to be complex indicating that we are observing in a particular direction several clouds at a range of distances, with different radial velocities and hence different Doppler displacements. (Some 21 cm line profiles are reproduced in Fig. 15.9.)

It is the aim of this exercise to make a simple analysis of some 21 cm line measurements and produce a model of the hydrogen cloud distribution in part of our galaxy.

To a good approximation, we can assume that objects in the Galaxy rotate about a common centre C in circular orbits (see Fig. 15.10). The way in which their angular velocities decrease according to their distance from the centre $\omega(R)$ is reasonably known, partly from observations and partly from theory (see curve in Fig. 15.11).

In Fig. 15.10, S represents the position of the Sun at a distance R_0 from the galactic centre and H is the position of a hydrogen cloud with galactic longitude l^{II} at distance R from the galactic centre. Suppose that the angular velocities of the Sun and the cloud are ω_0 and ω respectively.

Now the radial velocity of H with respect to S is given by the difference of the velocity components along the line SH, i.e.

$$V_{\text{rad}} = R\omega \sin l - R_0\omega_0 \sin l^{\text{II}}.$$

From the triangle SCH we have

$$R \sin l = R_0 \sin l^{\text{II}}.$$

Hence we can write the radial velocity of H with respect to the Sun as

$$V_{\text{rad}} = R_0[\omega(R) - \omega_0] \sin l^{\text{II}}.$$

By rearranging we have

$$\omega(R) = \frac{V_{\text{rad}}}{R_0 \sin l^{\text{II}}} + \omega_0.$$

According to an observed value of V_{rad} at a particular galactic longitude, and from knowledge of the constants R_0 and ω_0, a value of $\omega(R)$ may be determined. With the use of Fig. 15.11, the distance of the radiating hydrogen cloud from the galactic centre can then be evaluated.

For the purpose of this exercise, choose values of $R_0 = 10$ kpc and $\omega_0 = 20\ \text{km s}^{-1}\ \text{kpc}^{-1}$.

Figure 15.9 represents the profiles of the 21 cm hydrogen line at 26 different galactic longitudes separated by 10°. A radial velocity scale is also presented on the same figure. Taking the vertical line as the zero point, maxima to the left correspond to negative velocities and maxima to the right correspond to positive velocities. (The value of the sign of V_{rad} is obviously important when evaluating $\omega(R)$.)

To avoid possible ambiguities, measure only the *negative* maxima in the range $0° < l^{\text{II}} < 180°$ and only *positive* maxima in the range $180° < l^{\text{II}} < 360°$. Make measurements of the appropriate Doppler shifted emission peaks. (At some galactic longitudes, two or three peaks can be observed; each one should be measured.)

Using the formula above, determine values of $\omega(R)$ and then with the aid of Fig. 15.11, determine the corresponding values of R.

To plot the hydrogen clouds use a scale of 1 cm = 2 kpc.

Draw from S lines for the various galactic longitude directions. From C measure out the various determined values of R appropriate to the galactic longitudes, marking the points corresponding to the position of the hydrogen clouds. (Where there are two Doppler peaks, two clouds are obviously being seen at the same galactic longitude—they must be at different distances from the galactic centre.)

Sketch in a rough idea of the distribution of hydrogen, indicating the two spiral arms—the Orion arm containing the Sun and the Perseus arm some 2 kpc outside.

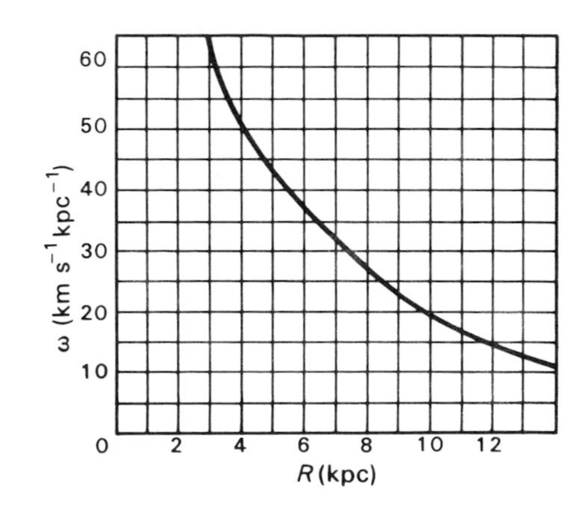

Fig. 15.11 The variation of angular velocity with distance from the centre of the Galaxy

Problem: Chapter 15

1. After correction for local solar motion, the maximum line-of-sight velocity in a direction making an angle of 30° with the direction to the centre of the Galaxy is $210\ \text{km s}^{-1}$. Calculate the approximate mass of the Galaxy in grams on the assumption that its mass is concentrated at its centre. (Distance of Sun from galactic centre = 8·5 kpc; constant of gravitation $G = 6{\cdot}67 \times 10^{-11}$ newton $\text{m}^2\ \text{kg}^{-2}$; 1 parsec = $3{\cdot}09 \times 10^{13}$ km.)

16
The Extragalactic Universe

16.1 Introduction

In the 18th and 19th centuries the term "nebula" was given to any celestial object of finite angular size that was not a member of the Solar System. As telescopes improved, these faintly glowing objects and dark obscuring regions were classified with names such as dark nebulae, planetary nebulae, diffuse nebulae or spiral nebulae. It was Lord Rosse, with his great 72-inch telescope, who first detected the spiral structure of the fourth class of nebulae. The largest members of this class were subsequently resolved sufficiently to reveal them to be stellar systems and it was reasonable to suppose them to be outside our own galaxy, indeed to be other galaxies. Until Miss Leavitt discovered the period–luminosity law for Cepheid stars, however, it was impossible to measure their distances.

The nearest galaxies are the Magellanic Clouds. It will be remembered that they are both resolvable into stars and that Cepheid variable stars are present in them. A knowledge of the distance of a number of Cepheids in our own Galaxy enabled Shapley to replace the apparent magnitude scale in the period–luminosity law by an absolute magnitude scale. He was then able to measure the distances of the globular clusters of our galaxy that contained Cepheids and to establish that the Magellanic Clouds, the Great Nebula in Andromeda and a few other spirals were well outside the Galaxy. This work, carried out in the early 1920s, has been confirmed, refined and extended since. The spiral nebulae and certain other nebulae are galaxies in their own right, of number 10^{11}, scattered through a volume of space compared with which the volume occupied by our own galaxy is infinitesimal. In the cathedral of the universe our own galaxy is but a speck of dust.

16.2 Catalogues of Galaxies

Because a number of the brighter galaxies were visible in the moderate-sized telescopes of the 18th century, they were included, with open clusters, globular clusters, planetary nebulae and other objects of finite angular size, in catalogues of that era, for example in the catalogue by Charles Messier of non-stellar objects visible from the northern hemisphere. About 30% of such objects are galaxies, the large Andromeda Galaxy appearing as M31. In 1888 J. L. E. Dreyer (see Section 13.1) published the *New General Catalogue of Nebulae and Clusters of Stars*, containing 7840 objects. Supplements called *Index Catalogues* appeared in 1895 and 1908. Objects in these catalogues are designated by NGC or IC followed by the catalogue number. Thus the Andromeda Galaxy is also object NGC 224 in the *New General Catalogue*. More recent catalogues are usually more specialized, such as Vorontsov–Velyaminov's catalogue of systems of galaxies (pairs, triples, and so on) showing interaction among the members.

16.3 Types of Galaxies

Galaxies are divided into three classes (i) elliptical, (ii) spiral, (iii) irregular. We will consider each in turn.

(i) *Elliptical galaxies.* In the telescope the nearer ones appear to have elliptical structure and are resolvable into stars. Some are lenticular in shape and highly flattened; others are almost circular. It is always possible that a circular one is either a disk seen face-on or a sphere. Statistical investigation shows that the observed distribution of shapes is consistent with the existence of elliptical galaxies of all shapes from spherical to highly-flattened.

These galaxies may be described by their degree of flattening. The letter E for elliptical is followed by a number from the series 0, 1, . . . , 6, 7, the largest number referring to the most flattened type.

Elliptical galaxies contain mainly Population II stars. They have little dust and gas present as if most of the original interstellar material has been used up in the creation of stars. Formerly thought to comprise about 25% of all galaxies, work on the relative intrinsic brightnesses of galaxies of different types has shown that the percentage of elliptical galaxies may be much nearer 50.

There is a wide range in size and intrinsic luminosity among galaxies. Among ellipticals, there exist giants and dwarfs. Giant ellipticals approach absolute photographic magnitude −21; dwarf ellipticals can be as faint as −14. In size, there is a correlation with total luminosity; diameters vary between 25 000 and 2000 parsecs.

(ii) *Spiral galaxies.* These objects form the largest class, although the percentage of spirals may be nearer 50 than 70, as was formerly believed. Our own galaxy belongs to this class.

Typically, a spiral galaxy has a lens-shaped nucleus, from opposite edges of which two arms grow. These arms coil round the nucleus in a spiral as in Fig. 16.1. The spiral arms lie in a disk coplanar with the equatorial plane of the nucleus (see Fig. 16.2).

The disk contains dust and gas, usually seen as a thin, dark band if the galaxy is presented edge-on to us. If the galaxy is resolvable into stars, those in the nucleus are found to belong to Population II while those in the arms are for the most part Population I.

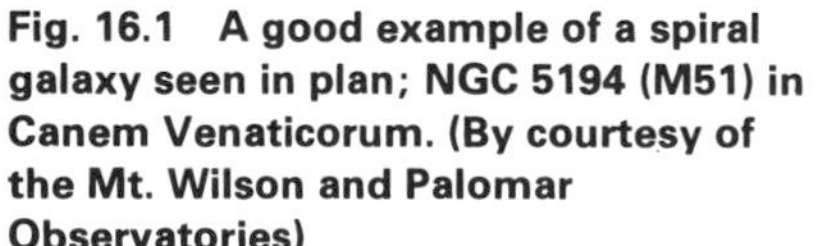

Fig. 16.1 A good example of a spiral galaxy seen in plan; NGC 5194 (M51) in Canem Venaticorum. (By courtesy of the Mt. Wilson and Palomar Observatories)

Fig. 16.2 An example of a spiral galaxy seen edge on; NGC 4565 in Coma Berenices. (By courtesy of the Mt. Wilson and Palomar Observatories)

Apart from our own galaxy, the best-known example of a spiral is the Great Galaxy in Andromeda (NGC 224, M31). It will be remembered that this galaxy is so near that, like the Magellanic Clouds, it is visible to the unaided eye as a faint hazy patch of light in the constellation of Andromeda. On photographs taken with large telescopes, its detailed structure can be studied.

It contains all the familiar features seen in the Galaxy—star fields, a system of globular clusters forming an almost spherical halo about it, regions of dust and gas in the disk containing the spiral arms, emission nebulae, giant stars, Cepheid variables, novae, bright blue stars, the same distributions of Population I and II objects found in our galaxy.

Recent and more accurate determinations of its distance and size reveal that it is over twice as large as the Galaxy. It is NGC 224 that has provided us with most of our understanding of all aspects of galactic structure.

There exists a sub-class of spiral galaxies containing some 30% of all spirals. These are known as **barred** in contrast to the remaining spirals which are referred to as **normal**.

In barred spirals, the nucleus has a bar-like extension of material on either side of it and it is from these extensions that the spiral arms spring (see Fig. 16.3).

(iii) *Irregular galaxies.* Forming no more than 5% of all galaxies, these objects have no particular form. Until recently both the Magellanic Clouds were placed in this class but it now seems more likely that they are more correctly placed in the barred spiral sub-class. The Clouds are probably satellites of our galaxy. About half of their matter is in the form of dust and gas and a bridge of material connects them.

Apart from the three major types of galaxies described above, other more specialized classifications are recognized. Certain galaxies, known as *Seyfert galaxies*, show nuclei that are starlike, have high-excitation emission line spectra, with very broad lines resembling those found in planetary nebulae. The gases in such nuclei are moving with large turbulent velocities measured in

Fig. 16.3 An example of a barred-spiral galaxy; NGC 1300 in Eridanus. (By courtesy of the Mt. Wilson and Palomar Observatories)

thousands of kilometres per second relative to the nuclei centres. Much of this material is being ejected from the nuclei. W. W. Morgan drew attention to another type of galaxy, the N galaxy, which is similar in some respects to the Seyfert galaxy, having a brilliant starlike nucleus containing most of the luminosity of the system. In addition most of the N galaxies studied are found to be strong radio sources.

16.4 Hubble's Classification of Galaxies

Hubble, who did much of the pioneer work in studying the extragalactic universe, arranged the elliptical and spiral galaxies in a particular order strongly suggestive of an evolutionary trend. The order is sketched in Fig. 16.4.

Proceeding from E0 to E7, we pass from elliptical galaxies that are almost spherical to highly-flattened systems. The normal spiral branch of the Hubble

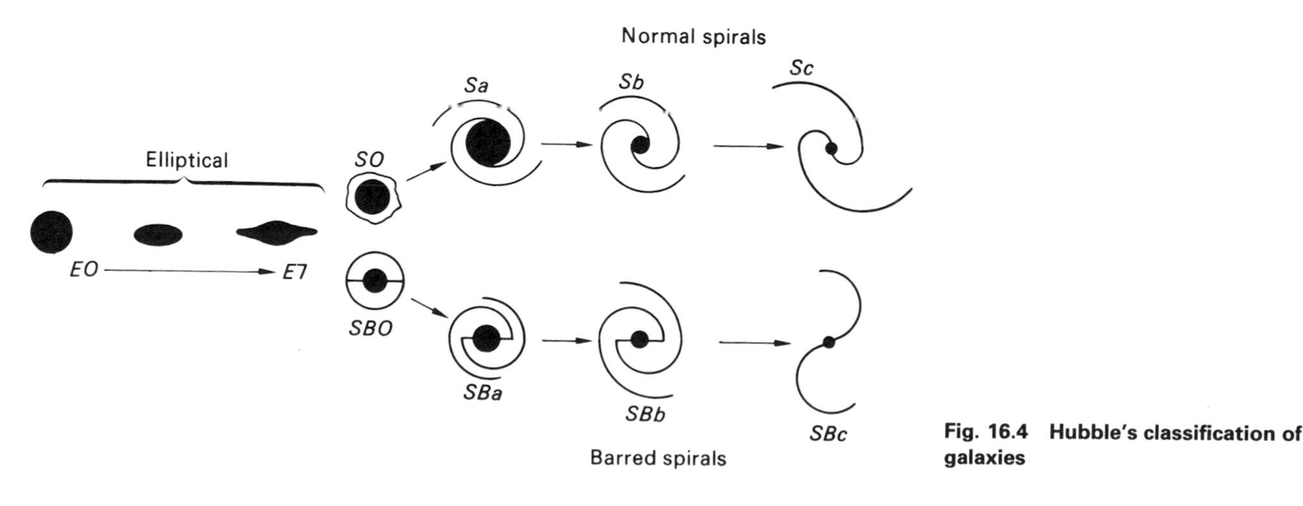

Fig. 16.4 Hubble's classification of galaxies

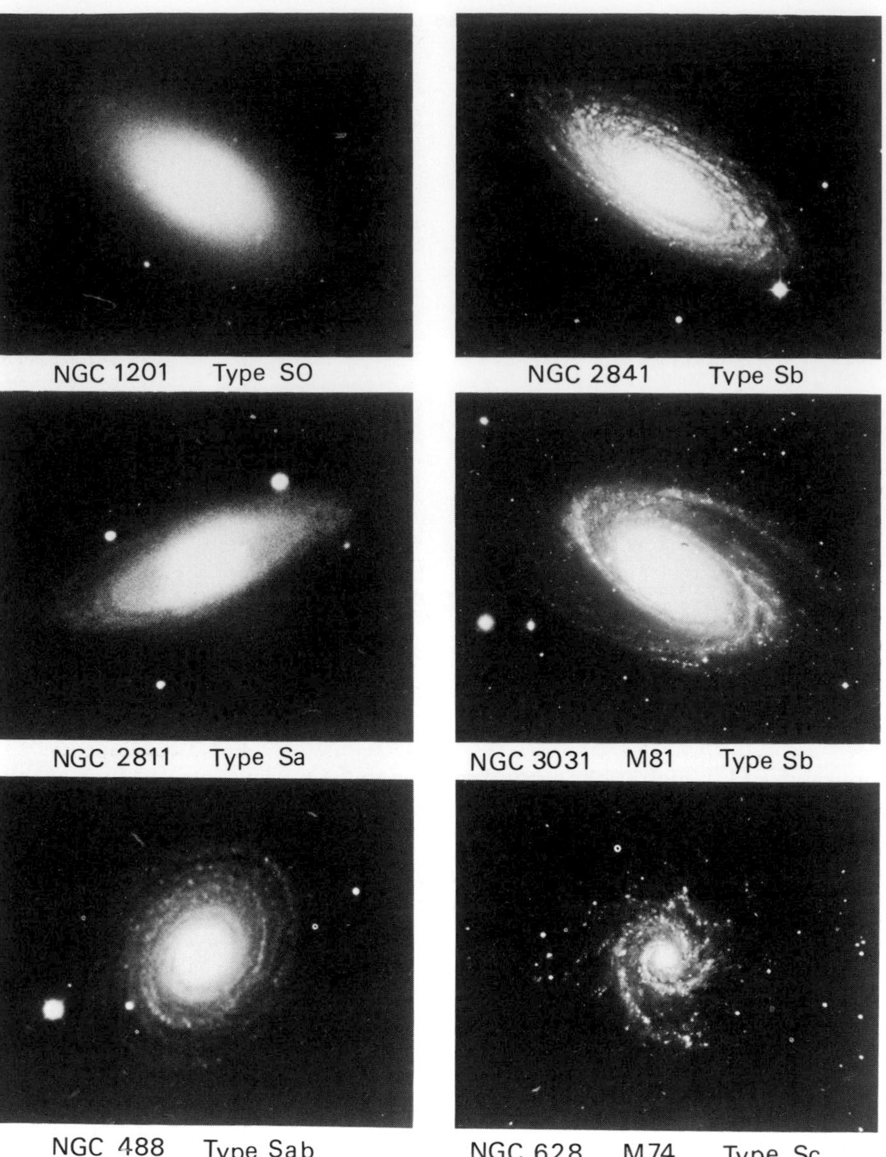

Fig. 16.5 Examples of normal spiral galaxies. (By courtesy of the Mt. Wilson and Palomar Observatories)

classification then takes us from S0 galaxies, with large nuclei and little or no evidence of spiral arms, through spirals with closely-coiled arms and smaller nuclei (types Sa and Sb) until we come to spirals of type Sc where the nuclei have almost disappeared and the arms are open (see Fig. 16.5).

A similar progression can be made along the barred spiral branch. Similar stages are reached in turn—SB0, SBa, SBb, SBc—from spirals with massive nuclei to those with tiny nuclei and arms trailing outwards from the bars (see Fig. 16.6).

The Hubble classification is still useful but the types of stars found in the various members of the sequence do not lend support to the idea that galaxies originate as E0 ellipticals and evolve with time to Sc normal or SBc barred spirals. We shall return to this question later.

16.5 The Distances of Galaxies

We are given a clue concerning the distances of galaxies when we investigate their distribution in the sky. It is found that there is an irregular band some 20°

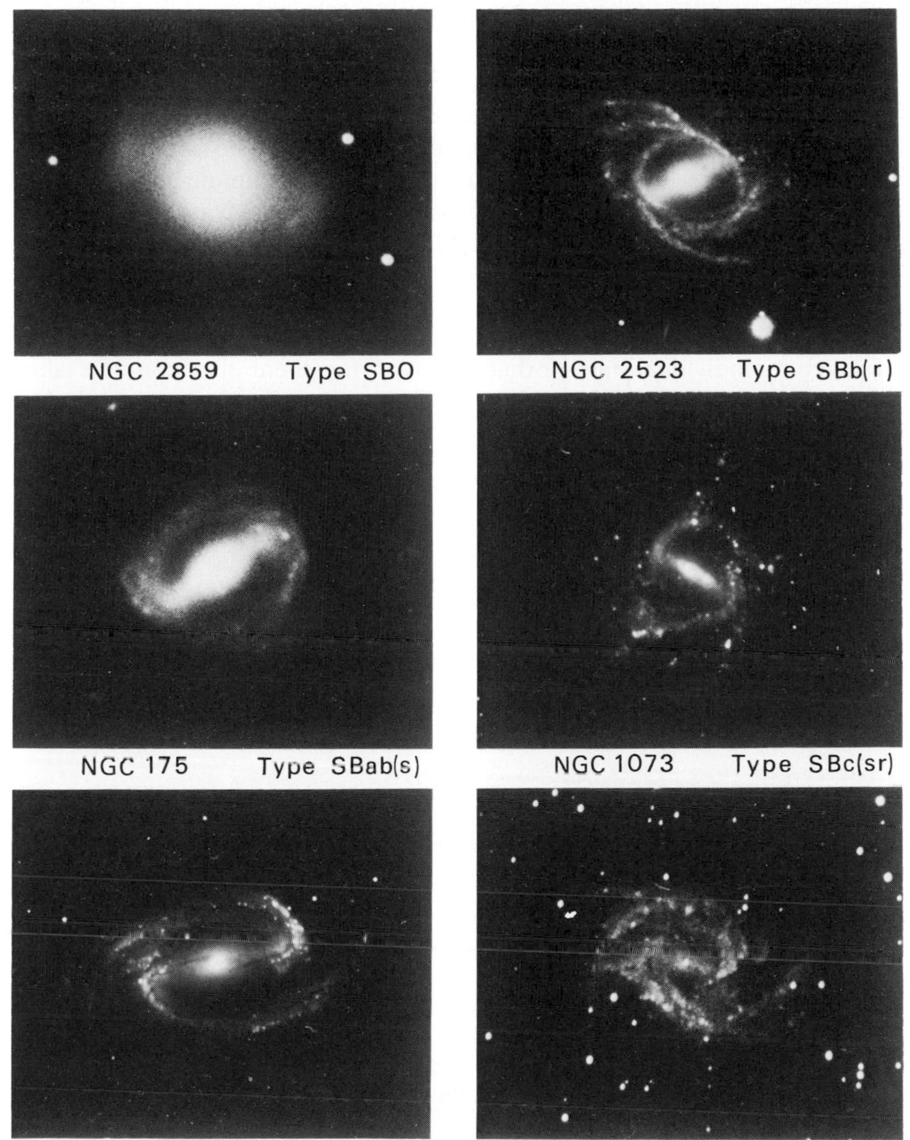

Fig. 16.6 Examples of barred-spiral galaxies. (By courtesy of the Mt. Wilson and Palomar Observatories)

broad running along the Milky Way, through the concentrations of the stellar distribution and dust clouds, in which galaxies are not found. The region is sometimes referred to as the **zone of avoidance**. If the galaxies are distributed randomly in space, then the zone of avoidance must be produced by absorption in the dust clouds of our own galaxy. The galaxies must be beyond this material.

About 150 galaxies are near enough to be resolved into stars using our largest telescopes. For these galaxies, stars of various kinds can be used to measure their distances.

One method already mentioned is the use of the Cepheid period–luminosity relation. Early estimates of the distance of NGC 224 (Andromeda Galaxy) established clearly its independence of our galaxy but put it at a distance such that its novae and globular clusters were much fainter than the corresponding objects in our own galaxy. The discovery, however, that Cepheids of Populations I and II have separate period–luminosity relations led to a re-assessment of the absolute magnitudes of the Cepheids in NGC 224 resulting in a doubling of this galaxy's distance. A corresponding re-assessment of the intrinsic brightnesses of the globular clusters in NGC 224 also followed, bringing them into agreement with those in the home galaxy. It also made NGC 224 larger than the Galaxy.

The Cepheid method of measuring distances is adequate for about a score of the nearest galaxies. For greater distances our knowledge of the absolute magnitudes of bright blue supergiants and novae is used. The former may be up to 100 000 times as luminous as the Sun; the latter at maximum brightness may shine with the luminosity of 30 000 to 200 000 Suns. Measurement of the apparent magnitudes of these bodies allows the distance of the galaxy to be determined by the distance modulus method.

Again early results were subject to errors of interpretation. Hubble pointed out himself that in distant galaxies, regions of ionized hydrogen could be mistaken for individual stars, or that groups of stars themselves could be taken as single objects. Both mistakes were made and subsequently discovered when the Hale 200-inch (5·08 m) telescope and better photographic equipment came into operation.

Occasionally a supernova (see Fig. 16.7) is discovered in a particular galaxy. A knowledge of the absolute magnitude to which the supernova rises enables its distance modulus to be computed. Its spectrum and subsequent decay in brightness, if available, allow some precision to be attached to the estimate of its absolute magnitude at maximum.

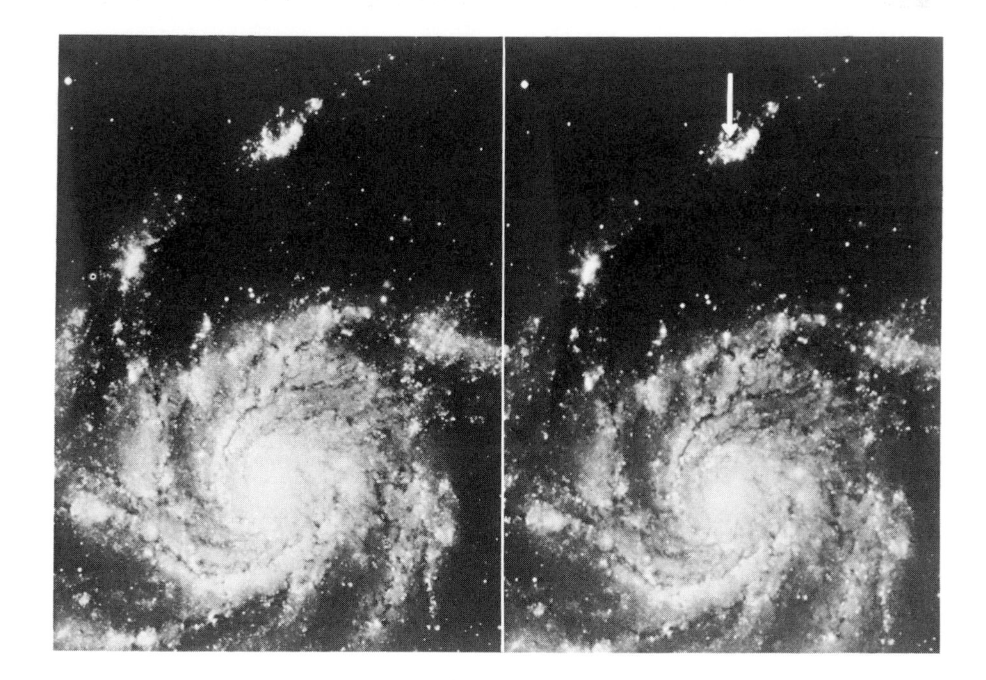

Fig. 16.7 The appearance of a supernova in a galaxy, NGC 5457 (M101); in the right-hand photograph (Feb 7th, 1951) the supernova is indicated by an arrow; the left-hand photograph (June 9th, 1950) shows the normal appearance of the galaxy. (By courtesy of the Mt. Wilson and Palomar Observatories)

Where the galaxies are so far away that no structural resolution is possible, statistical methods based on the galaxies' luminosities or apparent sizes are used. Confidence in such methods was gathered from the observation that galaxies are to be found in clusters (see Fig. 16.8) containing anything from a few dozen members to several thousand and that in the big clusters, galaxies of a particular class are in the majority. For example the giant clusters are almost entirely made up of elliptical galaxies.

Hubble enlarged his distance measurements by using these clusters. Examining the apparent brightnesses of galaxies in the Virgo cluster he found that their distribution in brightness was indeed narrow. The Pegasus cluster produced a similar result. But the mean brightness of the members of the Pegasus

Fig. 16.8 The cluster of galaxies in Hercules, photographed by the 200-in Mt. Palomar telescope. (By courtesy of the Mt. Wilson and Palomar Observatories).

cluster was only one-sixteenth the mean brightness of galaxies in the Virgo cluster. Hubble could therefore say that unless there was a considerable amount of intergalactic absorption of light, the Pegasus cluster was four times as far away as the Virgo cluster. It happened that some members of the Virgo cluster were near enough for estimates to be made of their distances from stars visible in them. In this way Hubble was able to tie in his new scale of distances obtained from the clusters of galaxies to the Cepheid distance scale.

We end this section with some examples of the distances of galaxies or clusters of galaxies.

The Local Group, of which our Galaxy is a member, forms a cluster of at least seventeen galaxies. They include the Magellanic Clouds, about 52 and 54 kpc from us, and the Andromeda Galaxy NGC 224, at a distance of 570 kpc. All the members of the Local Group are within a volume of space roughly 650 kpc in diameter.

Other galaxies, bright enough and near enough to register extended and spectacular images on photographic plates, lie between 1 and 11 Mpc. Among these are the Sombrero Hat, the Whirlpool and the Pinwheel at distances of 4·4, 2 and 3 Mpc respectively.

Clusters of galaxies such as Virgo (2500 members), Coma (1000), Gemini (200), Boötes (150) lie at distances between 11 and 400 Mpc. Their distances are fairly reliably known. For distances greater than 400 Mpc the uncertainties in measurement are such that little reliance can be placed on the figures obtained.

16.6 The Sizes of Galaxies

The sizes of galaxies vary from so-called dwarf galaxies to giant galaxies. Among the spirals, the Galaxy, about 30 000 pc in diameter, is a giant; NGC 224 (Andromeda), with a diameter of 60 000 pc, is among the largest galaxies known. Ellipticals are usually smaller, ranging downwards to dwarf ellipticals only 2000 pc across. The irregular galaxies are mostly small.

These sizes are illustrated in the galaxies forming the Local Group, comprising five spirals, ten ellipticals and two irregulars, the Magellanic Clouds being placed in the first class. It is possible that other members exist. For example, two faint, diffuse objects discovered in 1968 by Maffei in the Milky Way plane are now thought to be possible members of the Local Group. Maffei I is probably a giant elliptical galaxy as large as NGC 224, the nature of Maffei II as yet remaining uncertain. Detailed study of these objects is difficult to carry out because of their close proximity to the Milky Way, the estimated absorption being of the order of 5 magnitudes.

We have seen that the Local Group is contained in a volume of space roughly 650 000 pc in diameter. The ranges in size are given in Table 16.1. The distribution is in one way atypical in that the elliptical class contains more than its fair share of dwarf galaxies.

Table 16.1

Type of galaxy	Range in diameter size (kiloparsec)	Number
Spirals	8 to 60	5
Ellipticals	1 to 3	10
Irregulars	3 to 4	2

16.7 The Masses of Galaxies

The masses of galaxies are determined in a number of ways. In Section 15.7 we have seen that the mass of our own galaxy, of the order of 10^{11} solar masses, was found by treating the Sun as a satellite of the Galaxy. A similar method can be used to measure the masses of nearby galaxies by observing their speeds of rotation. The spectral lines of various parts of a nearby galaxy on either side of its nucleus will show Doppler shifts with respect to the spectral lines of its nucleus. A careful study of NGC 224 by optical and radio measurements reveals that although its central regions rotate like a solid body, the outer parts revolve at different angular velocities about the centre, the angular velocity decreasing the farther out the part is. NGC 224 is therefore behaving like the Galaxy, rotating about an axis perpendicular to its disk and passing through the nucleus, the spiral arms trailing. Equating centrifugal force to gravitational force then gives an estimate of its mass. For NGC 224 a mass of the order of 3 to 4×10^{11} solar masses is obtained.

A number of galaxies occur in pairs gravitationally connected, or in trios, or in greater numbers. For example the elliptical galaxies NGC 221 and NGC 205 are satellites of NGC 224; it has already been mentioned that the Magellanic Clouds can be treated as satellites of our own galaxy. Assuming the galaxies to be point-masses with respect to their gravitational effects upon each other, estimates of their masses can be made by studying their velocities with respect to each other. Results by this method usually give larger masses than those derived from the study of the internal rotation of the individual galaxies. This is perhaps not surprising. The faintest outer portions of these galaxies may not have been detected and these, or the possible existence of gas and dust in a spherical halo, would not have much effect on the rotational speeds studied. The mass derived by this rotational method is therefore likely to be an underestimate. In addition, the simple equating of centrifugal force to gravitational force may be too simple—other forces such as magnetic fields may play their part.

The smallest spirals seem to have masses about 5×10^9 solar masses. The Small and Large Magellanic Clouds have masses of the order of 2×10^9 and 7×10^9 solar masses respectively. According to recent work on double galaxies, the average mass of spirals in 14 such systems is about 10^{10} solar masses, while that of ellipticals in 27 doubles is some 3×10^{11} solar masses. It should be noted that the uncertainties in such figures are high.

From brightness studies, it seems that ellipticals are less luminous per unit mass than spirals.

16.8 The Distribution of Galaxies

Just as stars of our own galaxy were counted to elucidate their distribution, so galaxies can be counted in various sample areas of the sky.

The first finding is that there is a marked zone of avoidance coincident with the Milky Way in which relatively few galaxies are to be found. The small numbers and low apparent brightnesses near the plane of the Milky Way are readily explained by the obscuring effects of the dust clouds in and around this plane.

When the local effect is allowed for, it is found that (*a*) most galaxies belong to clusters, (*b*) there are obscuring dust clouds *between* galaxies, (*c*) there is evidence of the existence of supergalaxies—clusters of clusters of galaxies. De Vaucouleurs has suggested that the supergalaxy to which the Local Group of galaxies belongs is an ellipsoidal system 15×10^6 pc across and of order 10^6 pc thick. From a study of radial velocities and the distribution of the 50 000 galaxy members of this supergalaxy, de Vaucouleurs finds it to be rotating and expanding, its centre being the large Virgo cluster of the order of 10^3 galaxies. Other supergalaxies have been found, also rich in members.

On this large scale, out to the limit of our present instruments, there are no boundaries to the distribution of galaxies. They appear in all directions, faint smudges on the photographic plates that may have been exposed for many hours in the focal planes of carefully-driven telescopes. It is a salutary experience to look at these smudges and realize that the light that produced such images has been travelling for anything up to 30 000 000 000 years.

16.9 The Spectra of Galaxies

Most galaxies are very faint objects and long time exposures are required to register their tiny spectra. Nevertheless the lines in such spectra are capable of yielding a great deal of information.

For most galaxies the spectra are produced from the integrated light of the millions of stars within them. They are similar to those of main sequence dwarf G stars, perhaps the most common stars in existence. The galaxies are therefore formed of the elements familiar to us on Earth and in much the same abundance as is found in the solar atmosphere.

If the galaxy is rotating, the shape of the spectral lines will reveal that rotation. In addition, the lines will be displaced towards the red or the blue end of the spectrum if the galaxy is receding from or approaching the Earth. Making due allowance for the Earth's movement round the Sun and the Sun's orbital velocity about the home galaxy's centre, we obtain the other galaxy's velocity with respect to our own along the line of sight joining them.

In the 1920s, a far-reaching discovery was made concerning the spectra of galaxies, a phenomenon subsequently called the **Red Shift**. It is important to distinguish carefully between what is observed and what is deduced.

When the photographs of many hundreds of galaxies are arranged in order of decreasing apparent brightness and size, it is found that with few exceptions their spectra have been arranged such that the spectral lines are displaced more and more towards the red end of the spectrum. (See Fig. 16.9.)

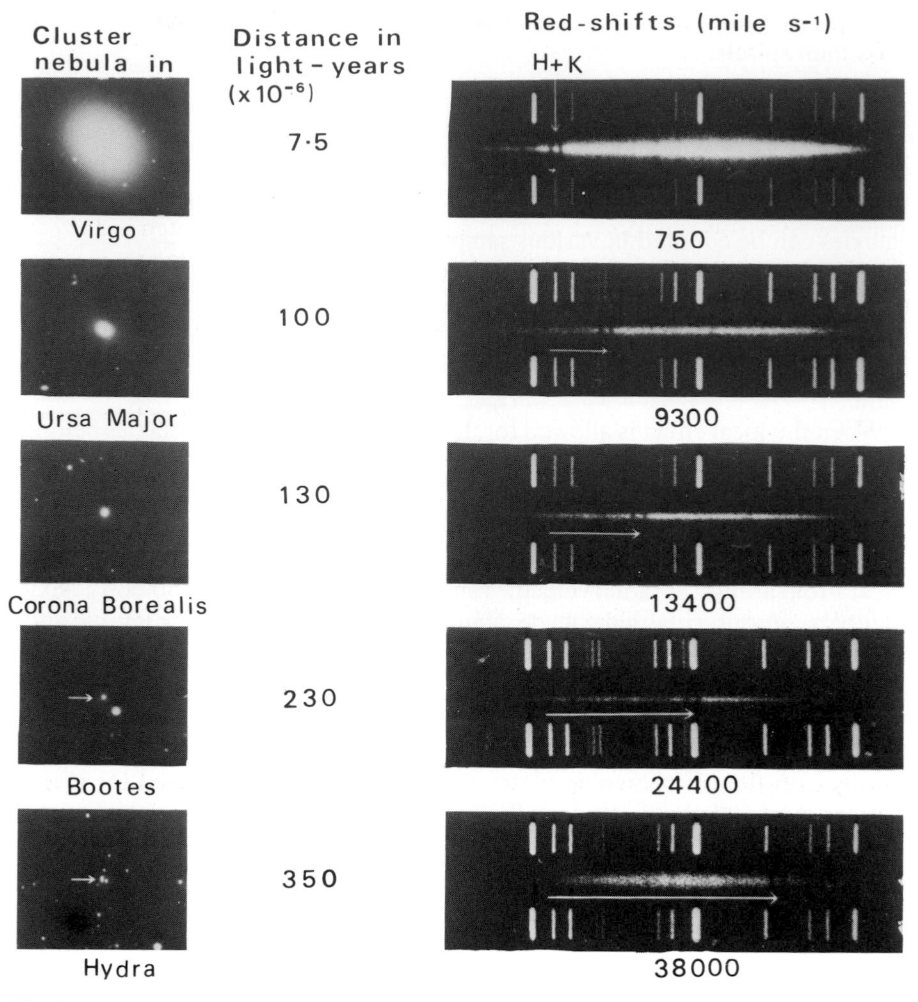

Red-shifts are expressed as velocities, $c\,d\lambda/\lambda$.
Arrows indicate shift for calcium lines H and K.
One light-year equals about 6 trillion miles, or 6×10^{12} miles.

Fig. 16.9 Examples of the observationa evidence for Hubble's law giving the relation between the red-shift and distance for a range of galaxies. (By courtesy of the Hale Observatories)

If we make two assumptions, namely that the fainter a galaxy is the farther away it is, and that the shifts in the spectral lines are Doppler in origin, then we are saying that in general, the farther away a galaxy is, the faster it is receding from us. Let us look at each assumption in turn.

Methods of measuring the distances of galaxies have already been discussed in this chapter. A great deal of confidence can be placed in these methods, especially where more than one method can be employed as in the case of the nearby galaxies. The first assumption is therefore supported by all the available evidence.

With respect to the second assumption that the Red Shift in the spectral lines of galaxies is Doppler in origin, it should be said that there is no plausible alternative theory. In any case, interpreting such spectral shifts as Doppler in cases such as orbital motion of planets or the rotation of Saturn or Jupiter or in stellar motion studies has always been supported by other independent evidence.

The Red Shift is therefore accepted as indicating that the velocity of recession V of a galaxy is strictly proportional to its distance d. Or,

$$V = Hd,$$

where H is a constant of proportionality called **Hubble's constant**.

Recent evidence suggests that

$$H \approx 55 \text{ km s}^{-1} \text{ Mpc}^{-1},$$

in other words, for every additional million parsecs of distance from our galaxy, the recession velocity is increased by about 55 km s^{-1}. At the time of writing, velocities of over 120 000 km s^{-1}, or over two-fifths that of light, have been measured from optical spectra of very faint galaxies. It may be noted in passing that even in the radio range of wavelengths, the distance–velocity relation holds for distant objects. For example, the radio source Cygnus A emits the 21 cm line of neutral hydrogen. The source can also be observed optically and it is found that both optical and radio spectral lines are shifted by the same amounts proportional to wavelength. Thus if $\Delta\lambda$ is the shift at wavelength λ, $\Delta\lambda/\lambda$ is a constant throughout a wide range of wavelength as it should be if the shift is Doppler in origin. Indeed the most recent radio Doppler shifts refer to objects that lie outside the volume of space reached by the most modern optical methods.

It is also noteworthy that the Red Shift is not dependent on direction over the celestial sphere once allowance has been made for distortions imposed upon recession velocities by the rotation and expansion of the local supergalaxy.

16.10 The Ages of Galaxies

The obvious method of obtaining the age of a galaxy is to look for the oldest objects within it. It is then reasonable to assume that the galaxy is at least as old as these objects. In the case of our own galaxy, the globular clusters were found to be about 7×10^9 years old, with little variation on that figure. A figure of the order of 10^{10} years for the age of the Galaxy follows. Similarly NGC 224 with its globular clusters is probably at least 10^{10} years old.

Although the Magellanic Clouds are classified as spirals, they have a much smaller proportion of Population II stars. In this respect they resemble irregular galaxies. Radio observations of the 21 cm line emission of neutral hydrogen shows that the Large Magellanic Cloud has over 5% of its material in the form of gas. The figure for the Small Magellanic Cloud is at least 27%. It can be estimated that it would take about 5×10^9 years for 73% of the original gas to condense into stars.

For the ellipticals, the estimated ages also average out at 5×10^9 years. In them, dust and gas are absent or present in very small amounts. Their stars are all old Population II, for in the absence of interstellar material, no new stars can be born.

With the large uncertainties in our present methods of estimating the ages of galaxies, it is impossible to arrange with any real confidence the various types of galaxies in order of ascending age. Hubble's classification remains a useful device for describing the main classes of galaxies but is not by any means accepted as an indication of the way in which galaxies evolve.

If we use Hubble's constant, giving it the value assigned to it in the previous section of 55 km s^{-1} Mpc^{-1}, we can calculate the time in the past at which all the galaxies were gathered into a small volume. It must be assumed of course, that the rate of expansion given by Hubble's constant has not changed.

Then the "age" of the universe, t, will be given by

$$t=\frac{10^6\times 206265\times 1\cdot 5\times 10^8}{55\times 31\cdot 56\times 10^6}\text{ years,}$$

putting 1 parsec $=206265$ AU, 1 AU $=1\cdot 5\times 10^8$ km and 1 year $=31\cdot 56\times 10^6$ s.

The value obtained is $1\cdot 8\times 10^{10}$ years.

Although this figure is in fair agreement with the oldest ages for galaxies, it must be noted that the value for Hubble's constant is still not known to high accuracy.

Indeed its measured value has changed markedly in the past 40 years. Again measuring in km s^{-1} Mpc^{-1}, Hubble's original value, about 530, was revised when the distances to the Andromeda Galaxy and further galaxies were almost doubled on correction of the Cepheid period–luminosity law. This brought the numerical value down to 290. Better absolute magnitude criteria for the brightest stars and for field and cluster galaxies decreased H still further to 180. Sandage in 1958 derived a value of 75 from his analysis of the brightest stars in resolved galaxies. More recent work has led astronomers to propose a value nearer 50. With each new revision of Hubble's constant, therefore, the deduced size and age of the Universe is obliged to alter in sympathy!

Finally, it may be remarked that observational data are being accumulated so rapidly with respect to the study of galaxies that it seems likely that within ten years, many of the uncertainties abounding in this field will have been removed.

17 Cosmology

17.1 Introduction

We are now at a stage in our studies where we can consider the universe itself as an entity composed of parts such as the galaxies and can ask and try to answer questions regarding its structure, its age, its evolution, its future and so forth. The branch of astronomy that deals with such matters is called **cosmology**.

All through the history of astronomy, cosmologists have tried to account for the universe, making use of their carefully-collected observational data and applying their ideas of the "laws" and "forces" at work in the world about them. In the earliest days these "laws" and "forces" were strongly anthropomorphic, involving the actions of gods and demons; the observational data were quite scanty and inexact and so the cosmologies or theories of the universe these early cosmologists created were inadequate at best. With the discovery of the scientific method and the brilliant achievements of Newton and his successors in accounting for an enormous range of natural phenomena in terms of the law of gravitation and Newton's laws of motion, the implicit belief arose among scientists that the universe was a physical structure, subject to natural law, capable of being measured and understood by the methods and rational thought of man.

Such a belief, the mainspring of human scientific endeavour, has of course been enormously fruitful in the past three and half centuries, but thoughtful scientists in all ages have realized that this belief may be naïve and false, if not downright arrogant.

For example, it could well be that the brain of man is not a powerful-enough thought-engine to understand the universe. Many of its attendant concepts may be beyond the ability of man to formulate, just as we believe the brain of a chimpanzee to be structurally inadequate to design an electronic computer, or produce the equivalent of Newton's *Principia*. Nevertheless man's curiosity about the universe into which he is born urges him to make the effort.

A jumping-off place is to make a list of questions along the following lines.

1. How big is the universe?
2. Is it finite or infinite in volume?
3. How old is it?
4. Is it evolving?
5. If it had, or had not, a beginning in time, will it have, or have not, an end in time?

It may be as well, before adding to the list, to stop and consider the above questions more closely. They are, in fact, the questions most often asked by the man-in-the-street about the universe.

Before applying them to the universe, it is illuminating to apply them to the Earth's surface by replacing the word "universe" in Question 1 by the words "Earth's surface". People living in the Dark Ages asked such questions. With respect to the shape and size of the Earth's surface we could have a variety of possible answers.

For example, the Earth could be flat and finite and so would be of limited area, having an edge over which people could fall. Many of Columbus's sailors believed in this "model Earth" and urged their captain to turn back in case their ship reached the edge. This type of Earth is sketched schematically in Fig. 17.1(a).

Or the Earth's surface could be flat and infinite, continuing outwards in two dimensions so that a man could journey for ever and never reach familiar territory (Fig. 17.1(b)).

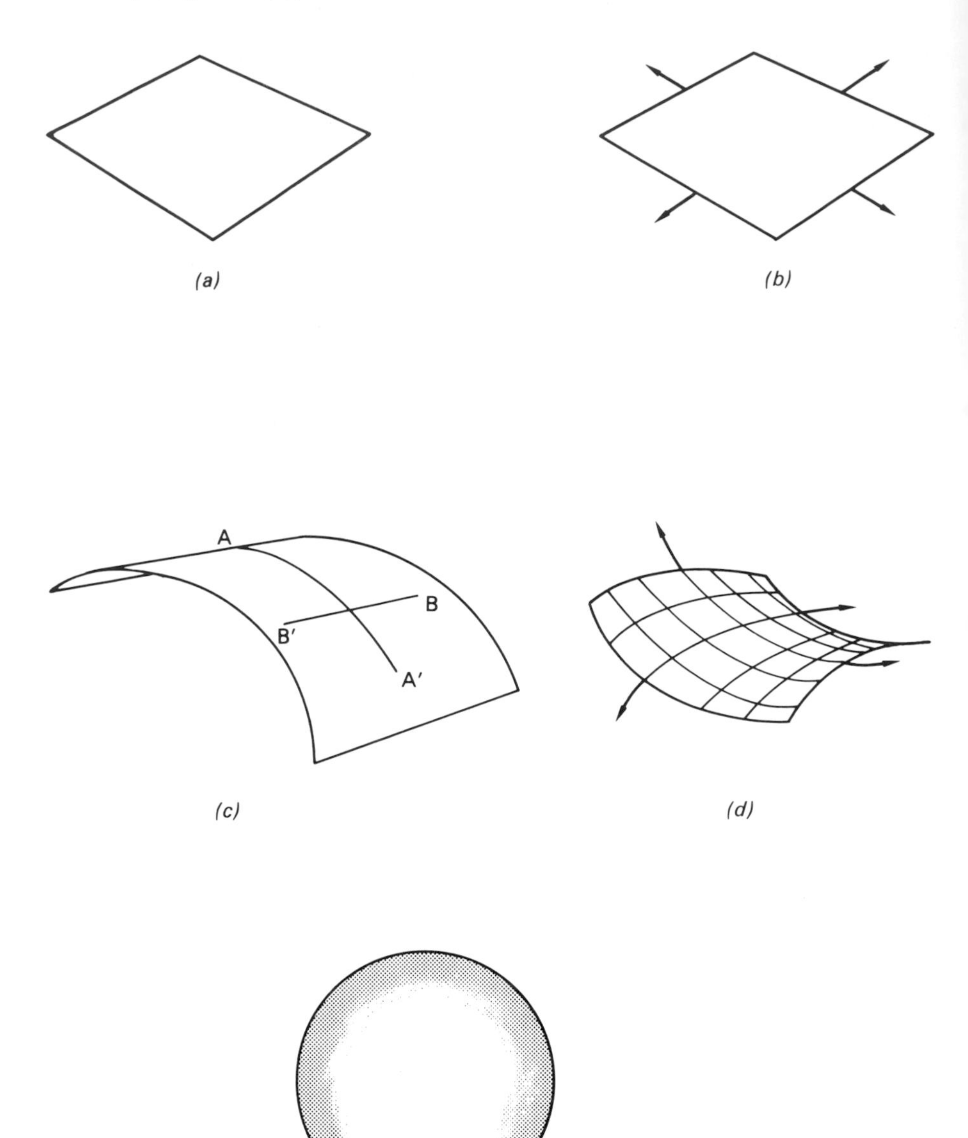

Fig. 17.1 Model Earths. (*a*) A flat, finite Earth. (*b*) A flat, infinite Earth. The arrows indicate that the edges are taken to infinity. (*c*) A curved, finite Earth. (*d*) An infinite, saddle-shaped Earth. (*e*) A curved, finite Earth

Or the two-dimensional Earth's surface could be curved into a third dimension. This introduction of a third dimension gives a new set of model Earths depending on the type of curvature. In Fig. 17.1(c), the flat, finite Earth of Fig. 17.1(a) is simply bent. In Fig. 17.1(d), the bending is not so simple, resulting in a saddle-shaped Earth that may, or may not, be finite in extent. In Fig. 17.1(e) we have a spherical Earth, one example of a family of finite, closed Earths.

The first problem facing the Earth cosmologists would be to find ways of discovering which model Earth the real Earth resembled. If it could be assumed that light travelled in a straight line through *three*-dimensional space and was not confined to the *two*-dimensional surface of the Earth, they would be able to say that significant differences should exist between the observational behaviour of objects such as sailing-ships in, for example, the worlds depicted in Figs. 17.1(a) and 17.1(c). A ship sailing away from the observer in the flat world (a) would simply get smaller and smaller until it vanished in a point. A ship in world (c), which is curved, would behave likewise if it followed a course such as BB' but would behave otherwise if its course lay along a line AA'. The observer at A would not only see the ship diminish in size but would see it vanish, hull-down, as it dropped below the observer's horizon-plane, the masts disappearing last of all. By making such observations on a fleet of ships following a variety of courses from the observer, the curvature, and type of curvature, could be detected and measured.

Other experiments could be performed. The geometry of the sphere—a two-dimensional surface curved into a third—is different from the geometry of a plane. Thus in a spherical triangle, the sum of the angles is always greater than 180° while in a plane triangle the sum is always equal to 180°. Again, the areas of circles drawn on a sphere behave differently from those of circles drawn on a plane.

If, in Fig. 17.2(a), the circles, centre O, have radii such that $OA = r$, $OB = 2r$, $OC = 3r$, then the areas of circles 1, 2 and 3, taken to be A_1, A_2, and A_3 respectively, are

$$A_1 = \pi r^2,$$

$$A_2 = \pi(2r)^2 = 4\pi r^2 = 4A_1,$$

$$A_3 = \pi(3r)^2 = 9\pi r^2 = 9A_1.$$

In Fig. 17.2(b), consider three circles drawn on a sphere. Let their radii be such that $O'A' = r$, $O'B' = 2r$, $O'C' = 3r$, where O' is the pole of the circles.

Now consider Fig. 17.3. The area of a spherical cap of a sphere, centre C and radius R, the cap being bounded by the small circle $KALK$, is well known to be given by

$$A = 2\pi R^2[1 - \cos(r/R)],$$

where r is the distance of the pole O from the small circle.

In Fig. 17.2(b), then, we can write down the expressions for the areas of the circles 1, 2 and 3, viz A_1, A_2, and A_3, where

$$A_1 = 2\pi R^2[1 - \cos(r/R)],$$

$$A_2 = 2\pi R^2[1 - \cos(2r/R)],$$

$$A_3 = 2\pi R^2[1 - \cos(3r/R)].$$

Then A_2/A_1 is not equal to 4, nor is A_3/A_1 equal to 9, as in the case of the circles drawn on a flat plane.

Suppose we in some way draw such circles on the Earth's surface, in the Sahara desert, say. Then—in principle at any rate—we could measure the relative areas of the circles by counting the number of grains of sand they contained, to a depth of one centimetre. It would in this way be possible to measure the curvature of the Earth's surface from the results obtained, but only if our circles were large enough for them to be affected appreciably by the curvature.

Thus, if r/R is small enough, we may write with sufficient accuracy

$$A_1 = 2\pi R^2\{1 - [1 - \tfrac{1}{2}(r/R)^2]\} = \pi r^2 = A_1,$$

$$A_2 = 2\pi R^2\{1 - [1 - 2(r/R)^2]\} = 4\pi r^2 = 4A_1,$$

$$A_3 = 2\pi R^2\{1 - [1 - \tfrac{9}{2}(r/R)^2]\} = 9\pi r^2 = 9A_1,$$

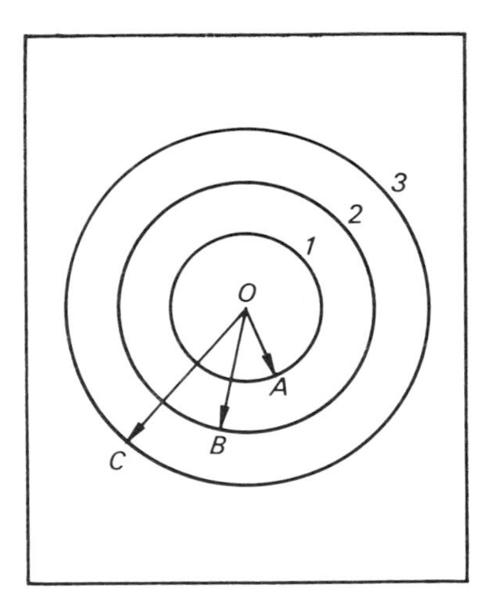

(a)

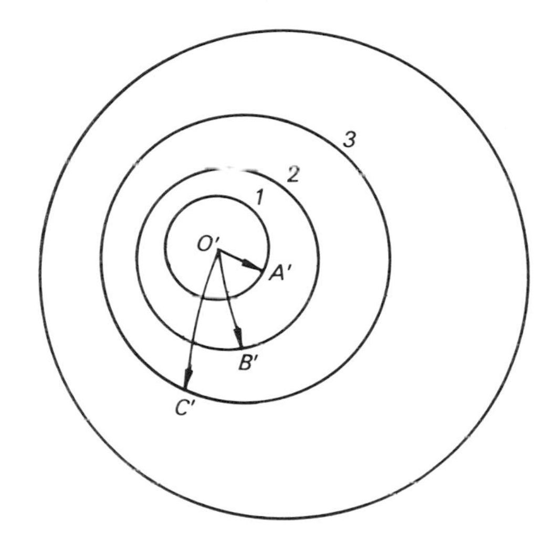

(b)

Fig. 17.2 Circles of various radii drawn on (a) a plane, (b) a sphere

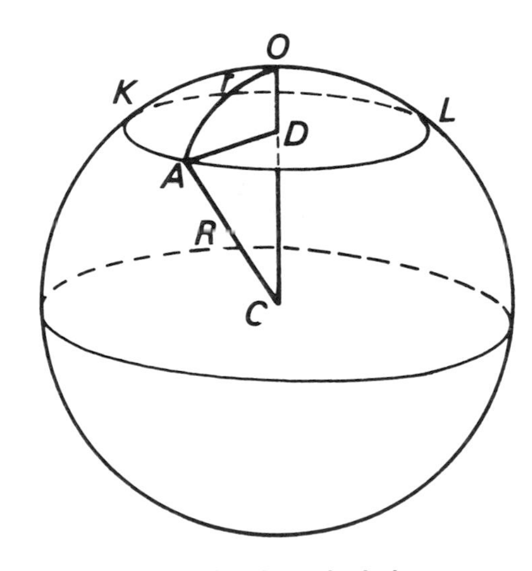

Fig. 17.3 Sphere showing spherical cap of radius r

making it seem as if the circles are drawn on a flat plane. The important lesson to be learned from this is that our sample of the territory to be investigated must be large enough for differences to exhibit themselves.

A second important lesson to be appreciated is that we must have at our disposal some means of measuring distance so that we may know the radii of the circles we draw. In the terrestrial case there is no problem. A graduated rod, or a steel tape, or some other measuring device is easily obtained and is sufficiently inflexible to provide reliable answers.

Having established the shape of the Earth, other investigations, physical, chemical, geological, would, one might hope, provide answers to the questions regarding the Earth's past, present and future.

17.2 Observational Data in Cosmology

Before we consider the possible application of analogous methods to the problem of the structure of the universe itself, we will look at the sort of observational data available and with which any satisfactory cosmological theory must agree.

Unlike the example of the Earth, we are presented with data not only belonging to the present epoch but also belonging to past eras. That this is so is due to the finite velocity of light. When we observe the galaxy NGC 224 in Andromeda, we believe that we are viewing it as it was some 1 600 000 years ago. Our views of more distant galaxies are chronologically earlier still if, and we must be conscious of that "if", our belief that we know their distances is correct. Assuming for the moment that our belief is justified, we are able to see earlier epochs of the universe's history back to a point in time some ten to thirty thousand million years ago. Naturally, as we observe further and further back into time and outwards into space, detail is progressively lost but it is at least a possibility that a search for differences in the objects seen at various distances could provide clues as to the evolution of the universe.

Perhaps, for example, there were more galaxies per cubic megaparsec at earlier epochs than there are now. Or less. Or the number may have remained constant. Again, perhaps the proportion of irregular galaxies to regulars was greater in the past than it is now. Or there were more ellipticals. Careful studies carried out to successively greater distances could in principle decide the matter because each further sample of galaxies belongs to an earlier epoch in the life of the universe.

Thus counting and comparison of the galaxies linked to distance and time comprise important observational data.

Just as important is the so-called Red Shift. It will be remembered that roughly speaking, the fainter a galaxy is, the larger is the displacement of its spectral lines to the red end of the spectrum. Thousands of galactic spectra have contributed to the establishment of this law. It should also be remembered that the most generally accepted interpretation of the Red Shift is that it is Doppler in origin and that the fainter a galaxy is the farther away it is. On this interpretation, the Red Shift becomes the velocity–distance relation, the galaxies receding from our galaxy with velocities proportional to their distances from us.

A further major piece of observational data is that the Red Shift and the distribution of galaxies (allowing for clustering effects) is isotropic, that is, the direction in which you observe over the celestial sphere is unimportant.

The final piece of evidence is the universe's microwave background radiation. The existence of this radiation was predicted by Alpher, Bethe, Gamow and Herman in 1948. From their cosmological research, which included the problem of the creation of the elements, they showed that in the early moments of an expanding universe, hot black-body radiation would have an important part to play; they further appreciated that a remnant of this radiation might still exist at the present day. From various considerations the radiation temperature was calculated to be within a few degrees of absolute zero. The first observational evidence that supported this view came from workers using the 20-foot horn-reflector system at Holmdel, New Jersey, in 1965, when they announced the discovery of an isotropic radiation at a wavelength of 7·35 cm and with a temperature of about 3·5 K. Measurements at various other wavelengths have established the essentially black-body nature of the radiation spectrum and revised the temperature to 2·8 K. The isotropic nature of the microwave radiation is now confirmed to about 1 part in 1000.

17.3 The Cosmological Principle

Over the past few centuries a certain assumption has been accepted by most astronomers. This assumption, sometimes called a **cosmological principle**, is that the human race's home or observational post is nothing special in the universe. The earliest form of this assumption involved the displacement of the centre of the universe from the Earth to the Sun by Galileo; a more recent version displaced the centre from the Sun, recognized to be but a run-of-the-mill star, to the centre of the Galaxy. More recently still, a more sophisticated version not only realized the Galaxy to be one out of thousands of millions of such objects, but jettisoned the idea of a universal centre, at least in the sense that there is a point in space that can be said to be the centre of the universe.

So the cosmological principle in one form states that to an observer free to range in space and make astronomical observations of distant objects as we do from the Earth, the observational data he acquired would remain the same on average, leaving aside the inevitable statistical fluctuations in the distribution and type of the galaxies he encountered. From our view-point in the home galaxy we see the red shift–apparent brightness relation, or in the usual interpretation the velocity–distance relation, implying that all other galaxies are receding from ours and with velocities proportional to their distance from us. The cosmological principle states that to an observer in *any* other galaxy, the universe would exhibit the same behaviour on the large scale, with all galaxies external to the observer's own showing the red shift–apparent brightness relation.

Various cosmological theories have been constructed that incorporate such a cosmological property and they will be considered later.

A different version of the cosmological principle—sometimes known as the "perfect cosmological principle"—extends the idea of sameness to time, as well as space. The **perfect cosmological principle** states that to an observer free to range in space *and time* and to make astronomical observations of distant objects from any chosen viewpoint, the observational data he collected would remain the same on average apart, that is, from unimportant statistical fluctuations. Thus the number of galaxies per cubic megaparsec throughout space would not change with time and there would be no evolution of the universe as a whole.

Again a number of cosmological theories have been based on this perfect cosmological principle, producing what are called "steady-state universes".

17.4 The Velocity of Light

It will be recalled that the first reliable measurement of the velocity of light was made by Roemer in 1676 when he deduced that the discrepancies between the

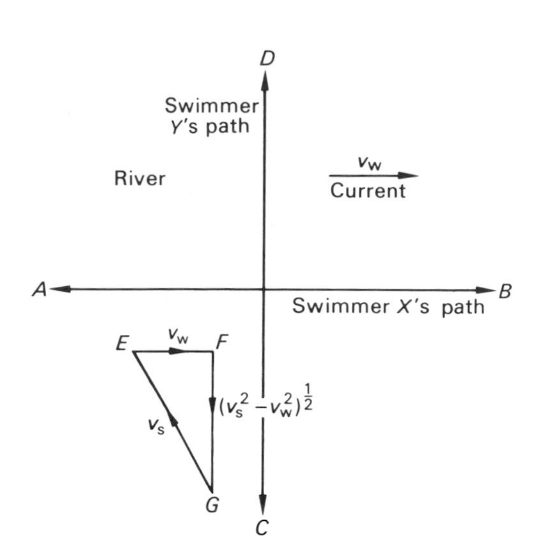

Fig. 17.4 Two swimmers, one moving parallel to the current, the other moving at right angles to it, cover the same distance there and back (*AB* = *CD*) in different times

predicted times of eclipse of Jupiter's Galilean satellites and the observed times were due to the different lengths of time it took light to cross the varying distance between Jupiter and the Earth. By the middle of the 19th century other methods of measuring the velocity of light over parts of the Earth's surface were being produced. Fizeau and Michelson, among others, refined these methods and the value is now known to very high accuracy. The measurements have indeed been made over a wide range in wavelength and it is accepted that the velocity of light in a vacuum is quite independent of wavelength.

In 1881 Michelson and Morley designed an experiment to try to detect and measure the velocity of the Earth through the luminiferous ether, the medium postulated by scientists as pervading all of space and as the substance that carried light waves. If the velocity of the Earth was v km s^{-1} with respect to the ether, it should have an effect on the velocity of light. Part of a ray of light was made to travel along a path in the direction of the ether drift past the Earth, being then reflected by a mirror back along its original path. Another part of the ray was sent along a path of equal length but at right angles to the direction of ether drift and again was reflected back. Both beams were then allowed to interfere producing a fringe pattern. If the instrument, which, with its platform, floated in mercury, was now rotated, the fringe system should have shifted because the orientation of the rays with respect to the direction of travel of the Earth through the ether had been changed. No such shift was ever detected.

The situation was somewhat analogous to the case of two swimmers X and Y who swim the same measured distance there and back in water which is moving. Swimmer X swims along a path ABA in Fig. 17.4 while swimmer Y swims along a path CDC where $AB = CD$. If they both swim equally well and therefore have the same speed in static water, one would expect swimmer Y to win the race.

That this is so may be seen from the following argument. Let v_w and v_s be the velocities of the water and the swimmers in static water respectively. Then the resultant velocity of swimmer Y is $(v_s^2 - v_w^2)^{1/2}$, given by the triangle of velocities EFG, in *both* directions during his swim. From A to B, swimmer X's velocity is $(v_s + v_w)$; it is $(v_s - v_w)$ while he is returning. If t_x and t_y are the times taken by X and Y respectively during their swims, it is easily seen that

$$t_x = AB\left(\frac{1}{v_s - v_w} + \frac{1}{v_s + v_w}\right)$$

and that

$$t_y = 2CD/(v_s^2 - v_w^2)^{1/2}.$$

Since $AB = CD$, it transpires that $t_x > t_y$, and that

$$t_x = \frac{t_y}{[1 - (v_w/v_s)^2]^{1/2}}.$$

If the current now changes and flows in a direction parallel to CD, the result of a new race should be different with X, still swimming a path ABA, now winning. The negative outcome of the Michelson–Morley experiment, however, implied that there was neither winner nor loser. The result was a draw.

To emphasize the importance of this conclusion, let us consider a light source S with two observers O_1 and O_2 observing a ray of light from the source. Let the velocity of light be c and let the observer O_2 be stationary with respect to the source. The other observer O_1, however, is moving towards the source with velocity v. Then the experiment shows that to *both* observers the velocity of light received from S is of value c.

This result was quite contrary to all previously-held ideas about time, space and measurement. Consider both observers to be using rulers along which the ray of light travelled and clocks to measure (in principle at any rate) the time it took the photons of light to move from one end of the rulers to the other. Before the experiment, the observers had got together and carefully checked that the rulers were of equal length and that the clocks had the same rate. During the experiment the observers were equally careful to give themselves a known relative velocity v, the relative velocity being parallel to the direction of the light ray as in Fig. 17.5. Yet, although the observer moving relative to the source would have expected to measure a velocity of light of magnitude $(c+v)$, his result had the same value c as measured by the stationary observer. The only conceivable explanation was that something happened to the scales of the rulers and the clocks due to there being a relative velocity between the observers. If indeed the moving observer changed his radial velocity v with respect to the source to a new value v', he would find that the measured value of the velocity of light was c as before.

Fig. 17.5 The velocity of light emitted by a source *S* is measured to be the same value *c* by two observers O_1 and O_2, even although O_1 has a velocity relative to *S*

It appeared in fact that the observers' respective results in measuring time and space depended upon their velocities in that their scales of time and space automatically adjusted themselves to ensure the constancy of the velocity of light. This startling conclusion that the observed length of a ruler and the rate of a clock could depend upon their velocities, though dismaying at first and contradictory to "common-sense", became the jumping-off place for a number of physicists who produced revolutionary and immensely fruitful theories about the universe we inhabit. Among them were H. A. Lorentz and A. Einstein.

The so-called Lorentz transformation was introduced by the former. It consists of relations that must exist between the co-ordinate systems and time-scales used by two observers with a relative velocity v. When the x_1, y_1, z_1 system of observer O_1 in Fig. 17.6 is moving with velocity v in a direction parallel to the x-axis of the x, y, z system of observer O, the Lorentz transformation states that

$$x_1=\frac{x-vt}{[1-(v/c)^2]^{1/2}}, \qquad y_1=y, \qquad z_1=z, \qquad t_1=\frac{t-(xv/c^2)}{[1-(v/c)^2]^{1/2}},$$

where t_1 and t are the times measured by O_1 and O respectively and c is the velocity of light.

The Lorentz transformation satisfactorily accounted for the Lorentz–Fitzgerald contraction, the name given to the explanation simultaneously put forward by Lorentz and Fitzgerald that the observed dimensions of a material object depended upon its velocity through space. It should be noted that if v is much less than c, so that (v/c) can be neglected, which is the case for all ordinary speeds we encounter,

$$x_1=x-vt, \qquad y_1=y, \qquad z_1=z, \qquad t_1=t,$$

which is the classical set of relations. If the velocity of light had been much smaller in our universe, our common experience of objects and observers moving at ordinary speeds would have made us familiar with such sights as an arrow shortening in flight or a running man appearing thinner than he was when standing still, or the seconds hand of a rapidly-moving clock turning more slowly than it did when the clock stood on a mantelpiece. Time and space would have shown themselves to be related in the above ways at a much earlier stage in the development of scientific thought. As it turned out, the necessity for a re-assessment of the nature of space–time was not revealed until the Michelson–Morley experiment was conducted towards the end of the 19th

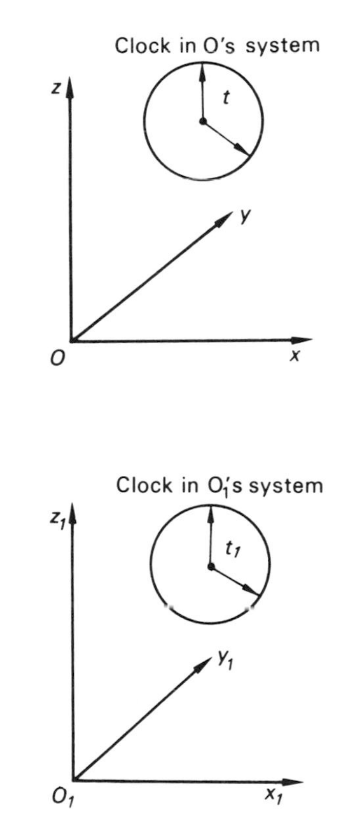

Fig. 17.6 The Lorentz transformation. The relations between the time- and distance-scales of observers O and O_1 depend upon the observers' relative velocity

century. The consequences of this new approach were described by Einstein in his theories of relativity.

17.5 Einstein's Theories of Relativity

17.5.1 The Special Theory of Relativity This theory, published by Einstein in 1905, was based on two postulates.

1. *The Relativity of Uniform Motion.* The laws of nature are the same for all systems in uniform rectilinear motion with respect to each other.
2. *The Constancy of the Velocity of Light.* The velocity of light in any reference system is not a function of the velocity of the source but is a true constant of nature.

Both principles imply that the velocity of light does not depend on the relative velocity of source and observer.

Einstein's special theory of relativity thus removes the paradoxes concerning the velocity of light. In developing the theory, Einstein found it necessary to modify the laws of classical mechanics and in so doing, he formulated his famous equation

$$E = mc^2,$$

where E is the energy equivalent to a mass m, c being the velocity of light. This relation, as we have seen, is an integral part of the atomic processes producing the stars' radiant energy. It is also, of course, the exchange rate underlying man's ability to extract power from nuclear reactions.

In addition to this law, the special theory predicted that the observed mass of an amount of matter would depend upon its velocity relative to the observer. Thus, if m and m_0 are the masses of a particle when it has velocity v and is at rest, it is found that

$$m = \frac{m_0}{[1-(v/c)^2]^{1/2}}.$$

This relation has also been verified in the laboratory for atomic particles moving at high speed.

17.5.2 The General Theory of Relativity This extension of Einstein's original work was published in 1915. It reformulated the concept of a gravitational field, producing a dynamics that has effects indistinguishable from Newton's for most practical cases but significantly different in three instances. In all these instances the differences between Einsteinian and Newtonian dynamics, though very small, were capable of experimental test.

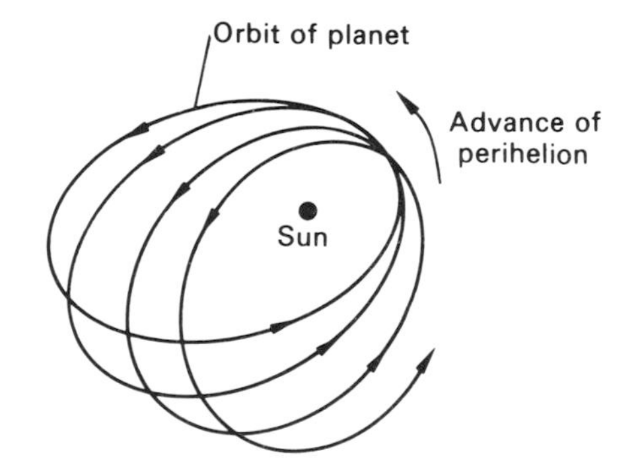

Fig. 17.7 The orbit of a planet according to Einstein is not a fixed ellipse but a path very similar to that swept out by a planet moving in a slowly-rotating ellipse

1. *The advance of the perihelion of Mercury.* It had been known that the perihelion of Mercury, even after due allowance had been made for the perturbations of all the other planets, was advancing faster than predicted, by about 40 arc seconds per century. Newtonian dynamics was quite unable to explain this discrepancy; Einstein's generalized theory, applied to a planetary orbit, predicted that it would not be a closed ellipse but would rotate as in Fig. 17.7. The amount could be calculated and, in the case of Mercury, was found to be very nearly 40 arc seconds per century.
2. *The deflection of light rays in the vicinity of the Sun.* According to the generalized theory, a ray of light passing by a large mass should be bent towards the mass. For a planetary mass the bending, even for a grazing ray, is imperceptible but for a mass such as the Sun's, the effect on a ray just grazing the surface was predicted to be of order 1·75 arc sec. This prediction was first

tested at the total solar eclipse of 1919. The positions of stars near the Sun and rendered visible during totality were photographed. At a different epoch, when the Sun's longitude was sufficiently altered for the same stars to be visible at night, their positions were again photographed. In Fig. 17.8, it is seen that during totality, the bending effect of the Sun's gravitational field would cause

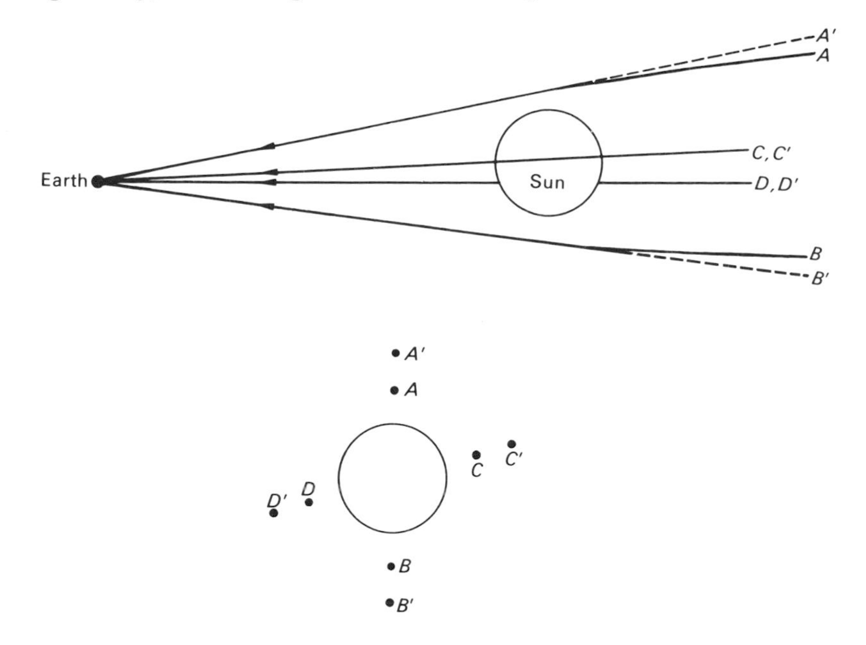

Fig. 17.8 Rays of light from stars in directions *A, B, C, D* are bent slightly in the vicinity of the Sun causing their apparent positions to be displaced outwards to *A′, B′, C′, D′*

the stars' observed positions to be displaced outwards from the positions they normally were seen in at night. The measurements made at this and subsequent solar eclipses were in good agreement with prediction.

3. *Reddening of light emitted from a source in a strong gravitational field.* Attempts have been made to observe this very slight lengthening of the wavelengths of light when emitted or absorbed by atoms in a strong gravitational field by careful study of the spectra of the Sun and white dwarf stars. In the former case detection of the shift in the spectral lines is doubtful, other processes tending to mask it. It is, however, accepted that the effect is apparent in the latter where the enormous densities encountered enhance it. Indeed the effect has now been detected in the laboratory using the Mössbauer effect.

Thus all three astronomical tests—the advance of Mercury's perihelion, the deflection of a light ray near the Sun, the reddening of light from a source in a strong gravitational field—support the general theory of relativity.

17.6 The Einstein Universe

Einstein applied his general theory of relativity to the problem of the structure of the universe. The mathematical techniques required are not easy and are beyond the scope of this text, but the type of approach is not entirely dissimilar to the methods we have already encountered in dealing with the internal structure of a star.

There, we saw how certain laws—the perfect gas laws and the law of radiative transfer—could be written in the form of differential equations. By various mathematical procedures, these equations could be solved. By attaching values to the parameters or constants appearing in the equations, different kinds of stellar models could be obtained. The way in which a given stellar mass was related to the star's radius and luminosity could therefore be found, as well as a detailed knowledge of the distribution of mass, temperature, pressure and density throughout the star. The stellar models could then be compared with real stars and their properties.

Taking into account the postulates of relativity and the particular way in which space and time are related, the Einsteinian dynamics enabled a set of equations—the so-called **field equations**—to be written down. The solution of these equations provided a great variety of model universes with different types of spaces, depending upon the values attached to the parameters involved.

It is quite impossible to visualize these spaces and their properties in common terms. The words in our everyday languages have acquired meanings related to our everyday experiences; we have no words to describe adequately the properties of the mathematical model universes that are the possible solutions of the field equations. This inability of a language such as English to deal with certain concepts is easier to understand if we remember that the languages of some primitive tribes on Earth are totally inadequate to cope with ideas that English can take in its stride. Nevertheless, mathematicians are able to operate quite easily with spaces of any number of dimensions and obtain mathematical descriptions of their properties that can be described loosely in everyday terms as *three-dimensional flat space*, such as that of Euclidean geometry, or *three-dimensional curved space*, and so on. Curved space can be open or closed, depending upon the type of curvature.

Some understanding of these models may be obtained by re-considering Fig. 17.1. There, we saw how a *two*-dimensional surface could be described as flat (Fig. 17.1(*a*), (*b*)) or curved (Fig. 17.1(*c*), (*d*), (*e*)) depending upon it being indeed flat or being curved into a *third* dimension. Two types of curvature, open or closed, were sketched in Fig. 17.1(*d*) and Fig. 17.1(*e*) respectively and in both cases we could always assign to any point of the surface a radius of curvature. In the closed, spherical case, only one radius of curvature was required.

Our model universes are analogous to these examples. For example the closed "spherical" universe would be finite in *volume* (just as the Earth's surface Fig. 17.1(*e*) is finite in *area*) and "curved" into a *fourth* dimension with a definite radius of curvature. We can in fact talk about the radius of the universe in that sense. If we could measure in some way the degree of curvature, then a value could be found for the length of the radius. On the other hand, although the open curved universe analogous to the Earth model depicted in Fig. 17.1(*d*) would have at each point within it a definite radius of curvature, its volume would be infinite.

So far we have talked as if a model universe is not changing with respect to time in its geometrical space properties. Most of the families of universes obtained from the solution of the field equations were found to be non-static, however. The values assigned to a so-called **cosmological constant** appearing in the field equations dictate whether, for example, the model universe we have termed a closed spherical universe begins life with a small radius and expands until it is infinitely large, or begins large and contracts, or oscillates in size with a certain period.

Einstein's original model universe was a static one. He found that even this was in unstable equilibrium and, once disturbed, it would begin to expand or contract. The objects in it would recede or approach each other because of the expansion or contraction of the space in which they were embedded.

If we again consider the closed, finite Earth-model drawn in Fig. 17.1(*e*), we can imagine cities and towns scattered over its surface. An observer in a particular city measures the distance along the Earth's surface between his city and those others visible to him. If the Earth were to expand, the observer would find that the other cities were receding from his with speeds proportional to their distances from his city. He would at first consideration think that his city was in a privileged situation. Comparing notes with other observers in other cities, however, he would discover that they observed the same type of recession for cities other than their own.

Einstein's expanding, closed, finite universe model therefore would produce effects very similar to those we observe when we study the spectra of galaxies of different brightnesses and, presumably, distances. For an observer in our Galaxy, the other galaxies would recede with speeds proportional to their distances, producing the observed Red Shift. This model universe, and similar expanding universe models, are sometimes lumped under the colloquial term of "Big Bang universes". It is important to note, however, that in Einstein's cosmology the cosmological principle would not be broken. The Galaxy is not the centre of this expansion; hypothetical observers in other galaxies would also measure velocities of recession for galaxies not their own that increased linearly with respect to the galaxies' distances.

Different cosmologies were subsequently proposed by other workers in this field, notably de Sitter and Milne. More recently the steady-state universe, developed by Bondi, Gold and Hoyle, has arisen as a cosmological theory of a fundamentally different nature from the others and will be briefly discussed in the next section.

17.7 The Steady-State Universe

The steady-state theory is based upon the perfect cosmological principle which, as we have seen, states that an observer free to range in space and time would find that no overall change in the universe was taking place, although of course, various parts of the universe would differ from each other in detail, and would move according to the laws of dynamics. On the overall scale, however, the universe would present an *unchanging* aspect. The steady-state universe is therefore in stark contrast to those cosmological models that include as a property the evolution of the universe. In the steady-state universe, no evolution takes place and there is neither a beginning nor an end to time or the universe.

In seeming paradox to the above statements, the theory predicts the expansion of the universe, that is, the galaxies should appear to be receding from any observer with velocities proportional to their distances. The argument is bound up with the extreme thermodynamic disequilibrium of the universe and Olbers' paradox.

It is an observational fact that the matter in the universe, for example, stars, is emitting greater quantities of radiation than it is receiving and that there is an enormous range in temperature from the hundreds of millions of degrees at the hearts of novae to the very low effective temperature of interstellar space. In a static universe this disequilibrium should have disappeared long ago.

In 1826, Olbers discussed a question that at first glance seems absurdly trivial. *Why is the sky dark at night*? Olbers' discussion, however, showed the question to be a highly significant one.

His argument was based on very reasonable assumptions, namely that the average number of stars per unit volume and their average luminosity were constant throughout time and space.

In Fig. 17.9, a shell of radius r and thickness dr is centred at the observer O. Then if ρ is the number of stars per unit volume throughout space, the number of stars in the shell is dn, given by the product of ρ and the volume of the shell. Hence

$$dn = 4\pi r^2 \rho \, dr.$$

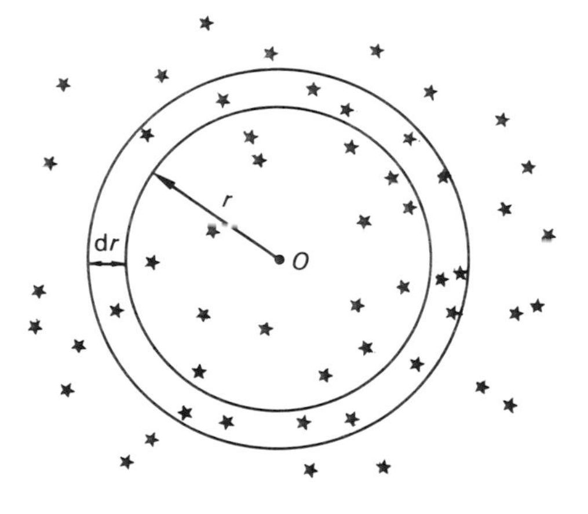

Fig. 17.9 The number of stars (number density ρ stars per unit volume) in a shell of radius r and thickness dr is equal to $4\pi r^2 \rho \, dr$

The intensity of starlight at O, due to the stars in the shell is then $\mathrm{d}I$, where

$$\mathrm{d}I = L\ \mathrm{d}n/4\pi r^2 = L\rho\ \mathrm{d}r,$$

and L is the average luminosity of a star.

Additional shells, each of radius $\mathrm{d}r$ can then be added indefinitely and it is seen that the intensity of the starlight arriving at the observer must be infinite—an absurd result, quite contrary to our experience. Even if allowance is made for the interception by other stars that happen to be in the way of a proportion of the rays of starlight, the argument shows that at any point O, the intensity must be equal to the average brightness at the surface of a star. In other words, the night sky brightness should vaporize the Earth.

We now know, unlike Olbers, that the stars we see are assembled into the Galaxy, but we are going to apply Olbers' argument to galaxies and not to stars.

If we replace stars by galaxies, the result turns out to be the same. If the perfect cosmological principle is to be retained, it may be shown that one way out of the impasse is to assume that the universe is expanding, as indeed the Red Shift obervations indicate. But this, however, is not in accordance with the perfect cosmological principle of a universe unchanging in time and space. And so a further necessary property of the steady-state cosmology is that matter is continually being created to make up for the decrease in density of matter in space.

This *continual creation of matter* was formerly assumed to take place at random positions throughout space, a figure of order 10^{-43} kg m^{-3} s^{-1} being the order of magnitude of the rate of creation required to maintain the *status quo*. This is a very small rate, quite impossible to observe, being the equivalent of the creation of one gram of matter in each volume equal to that of the Earth every 10^{11} years. More recently, the possibility has been considered that the continual creation of matter, if it does occur, takes place at the centre of galaxies rather than throughout space. In this context a remark made by Jeans in 1928 is of interest. Speaking of the galaxies, then more often called "spiral nebulae", he wrote:

> "The type of conjecture which presents itself, somewhat insistently, is that the centres of the nebulae are . . . points where matter is poured into our universe from some other, and entirely extraneous, spatial dimension so that, to a denizen of our universe, they appear as points at which matter is being continually created."

Recent infra-red and radio studies showing that the centres of some galaxies are indeed the seat of powerful and unexplained forces may be of relevance here.

Thus in the steady-state universe, individual galaxies condense out of intergalactic matter, evolve and recede while new matter enters the universe to keep the overall density constant. There is no beginning or end to this universe: the overall picture remains the same throughout eternity.

17.8 Observational Tests

To decide which one of the various cosmological theories provides a model universe that describes the real universe we live in, recourse must be had to observation and measurement. We saw in Section 17.1 how by elucidating the type of geometry allowed by the Earth's surface, an observer could deduce the type of curvature followed by that surface and indeed measure the radius of curvature. But of course, in drawing his circles and measuring their areas he had to cover a large enough sample of the territory or his results were inconclusive. Even if the Earth had been evolving, for example expanding, he could still understand what model Earth best described it, from his measurement of the recession of points at various distances from him and his belief in the cosmological principle.

Where the universe is concerned, one approach by the observer would be to attempt to discover the type of geometry followed by the space about him. Instead of circles, he would draw spheres of various radii about him and measure their volume. Now the volume of a sphere of radius r is $\frac{4}{3}\pi r^3$ in "flat" Euclidean space. A sphere of double that radius would therefore be 8 times the volume of the first; if three times that radius, its volume would be 27 times, and so on, since the volume is proportional to r^3. But in "curved" space of the closed finite variety, the volume increases less rapidly than r^3. On the other hand, for negative space curvature of the Fig. 17.1(*d*) type the increase is more rapid than r^3.

The grains of sand counted by the experimenter in the Earth-surface problem of Section 17.1 are now galaxies. The problem is complicated by the finite velocity of light so that it is one not only of space but of time. The Red Shift produces additional complications in that it affects the luminosities of the galaxies. The distance measurements may also be systematically affected, as well as inherently inaccurate.

Yet it is by the study of radio and optical objects at various distances that the observational cosmologist will finally be able to judge with fair certainty the model universe that best describes our own. The main types of measurement possible are (i) apparent magnitudes, (ii) red shifts, (iii) numbers of objects, (iv) angular sizes. In any given cosmological theory, the relations among these parameters may be stated. Perhaps the fourth requires elaboration. In Euclidean flat space, the angular size of an object is proportional to its distance. But in curved space, objects would appear to diminish in size with distance then increase in size. Our closed finite Earth-surface model illustrates by analogy this peculiar property. In Fig. 17.10, three objects B_1, B_2, B_3 of the same linear size are at increasing distances from an observer O. Light from these objects travels along the great circles from them to O and it is seen from the diagram that whereas B_2 is of smaller angular size than B_1, object B_3 is the largest in angular size, though the most distant.

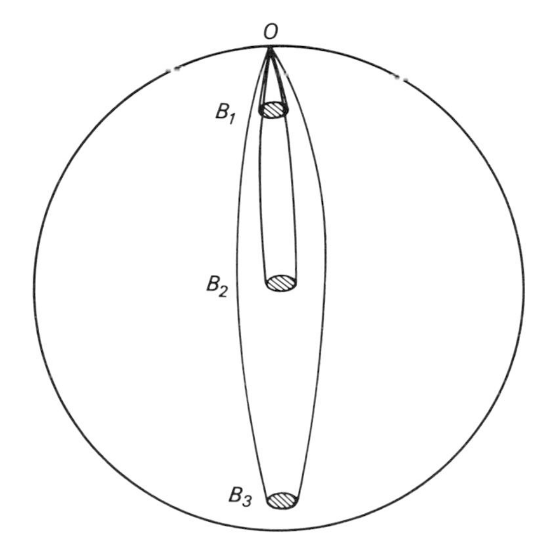

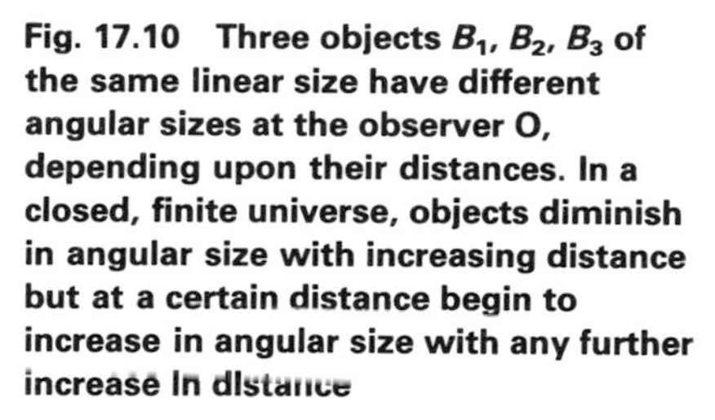
Fig. 17.10 Three objects B_1, B_2, B_3 of the same linear size have different angular sizes at the observer O, depending upon their distances. In a closed, finite universe, objects diminish in angular size with increasing distance but at a certain distance begin to increase in angular size with any further increase in distance

It now appears that measurements of radio sources extend our sample of the universe far beyond that accessible to optical telescopes. If it is confirmed that the newly-discovered objects known as quasars with their extreme red shifts are indeed among the most distant bodies accessible, they may enable us to decide finally between steady-state and evolving theories of the universe. As it is, evidence seems to be accumulating that favours an evolving universe rather than a steady-state one. Certainly the discovery of the uniformity of the microwave background black-body radiation is one of the strongest indications yet that our universe is a "Big Bang" one. If future observations support this, it will then be the task of observational cosmologists to decide which of the many evolving universe models our universe resembles.

18
Cosmogony

18.1 Introduction

In this chapter problems regarding the origin of the universe, the galaxies, the stars, the elements and life itself are discussed briefly. **Cosmogony** is the name given, strictly speaking, to the theory of the creation of the universe but in this chapter we shall also include these other topics. Indeed the names **galactic cosmogony** and **stellar cosmogony** are sometimes given to the problems of the origin of the galaxies and the stars respectively. In passing it may be mentioned that *gonia* is of Greek origin meaning *begetting*.

18.2 The Origin of the Universe

Very little can be said in the present text about this problem beyond making a few general remarks. In the evolving universe case where there was a definite beginning—for example, the model that begins in an indefinitely small volume—it has been postulated that the matter of the universe was originally hydrogen. How that hydrogen was created is impossible to imagine scientifically. The creation at one moment in time of all the matter in the universe is no easier to comprehend than the continual creation of matter postulated by the steady-state theorists. Indeed in one way it is more difficult. With the steady-state universe there is neither a beginning nor an end to time whereas with a universe that has a distinct beginning or creation, the idea of time before that event is meaningless. In this respect it may be remarked that for all our scientific progress, we are still as ignorant of the true nature of time as our remote ancestors.

18.3 The Origin of the Galaxies

The problem of the formation of our own galaxy has already been discussed in Section 15.10. Various theoretical studies have been made to clarify the problem of galaxy formation. Most of them give reason to believe that in a universe composed of dust and gas of low density, gravitational instabilities would exist leading to the setting-up of regions in which the random velocities of the particles would in general be less than the overall escape velocity. These regions in the course of time would develop into **protogalaxies** from which the irregulars, spirals and ellipticals would form. The detailed dynamical processes are not as yet well understood and until we can attach accurate ages to galaxies of different kinds, theories of galactic origin and evolution are rather speculative. Indeed the most reliable of our observational data lie in the now well-established properties of Populations I and II stars, in particular the different proportions of heavy elements the stars in these classes have. Problems of this nature, in fact, are bound up not only with the birth of stars and their evolution but also with the problem of the origin of the elements.

18.4 The Origin of the Elements

If the elements had been created an infinite time ago, the radioactive members of the periodic table of the elements would by now have decayed into their stable final products. This argument, from the presence of radioactive elements like uranium, shows that the elements we observe in the universe were created or synthesized in the finite past.

When "big bang" theories of the origin of the universe were put forward, it was at first thought that in the early days of the expanding universe, pressures and temperatures would be high enough to form the heavier elements out of the original hydrogen. Various processes and initial conditions were postulated that would produce the elements in the relative abundances observed today. Such theories were ingenious but complicated.

Later, when stellar interiors were investigated and it was found that the conditions at the centres of stars could maintain nuclear reactions, attention was turned to the possibility that all the elements were produced within stars from hydrogen. It was seen, too, that the older stars in the Galaxy contained heavy elements in abundances far smaller than those found in the younger second- or third-generation stars. This also supported the idea that the stars produced the heavier elements and then, in the catastrophic stages in their lives, sprayed them into interstellar space to become part of the make-up of the succeeding stellar generation when it condensed from the enriched interstellar medium.

In general, as a star ages, its central regions become hotter and denser. The first reactions destroy very quickly the low abundances of the light elements deuterium, lithium, beryllium and boron. Their presence constitutes a problem but it is possible that they are created with the heaviest elements during supernova explosions. An alternative explanation suggests that they are the result of **spallation reactions**, these being cosmic ray bombardment of small boulders only a few metres in diameter adrift in the protostar's environment.

The main reaction succeeding the destruction of the light elements is the fusion of hydrogen into helium. In the more massive stars, helium is then, as we saw in Chapter 9, used to form carbon which may combine with more helium to form oxygen, neon and magnesium. By now the central temperature of the star is around 6×10^8 K. By the time it has risen to some 5×10^9 K, elements as massive as iron will have been formed after a series of different nuclear reactions.

Elements heavier than iron can be created in stars from the iron group—whose members include titanium, copper, nickel, iron and zinc—in reactions known as **slow** or **rapid neutron capture processes**.

All these elements are in one way or another in due course of time returned to the interstellar medium through the violence of novae and supernovae or continuously from Wolf–Rayet stars, P Cygni stars, and so on.

The close agreement between the abundances predicted from nuclear studies dealing with the formation of the elements in stellar interiors and the observed abundances creates confidence in the basic proposition that nucleosynthesis does take place in stars. Indeed, if the steady state model universe describes our universe, the interiors of stars are the only places we know of where conditions of pressure and temperature exist sufficiently high to synthesise the elements from hydrogen.

18.5 Life and the Universe

Having asked how old the universe is, how it has evolved, what was the mode of origin of the elements and similar questions, it is inevitable that we ask the question: "Is there life elsewhere in the universe?" It is a question worth discussing. The answer must be "yes" or "no". Whatever the answer is, it must have the profoundest philosophical implications for mankind. In this last quarter of the 20th century the question is particularly relevant and there are grounds for believing that before the century has passed, the answer may well be forthcoming.

In a sense the astronomical wheel has turned full circle. In the early days of astronomy, man believed that by studying the movements of stars and planets, information about the lives of human beings could be gleaned. As astronomical knowledge accumulated, however, it was realized that the universe was fashioned on a remote, more aloof scale than formerly believed, with the Earth but a tiny component of it. So belief in astrology among enlightened men died. And yet, we will now see that by turning his telescopes and other pieces of equipment to the heavens, the astronomer is enabled to make certain statements that have direct bearing on the human race, its life-span and position within the universe. These statements are three in number. They sum up the astronomer's experience of the universe, indeed what we have learned in this course.

18.6 The Astronomical Facts of Life

18.6.1 The Size of the Universe We have learned that the universe is vast. How vast it is impossible to appreciate if we consider the actual distances. But if we use and extend the analogy we introduced before, we can obtain some idea of its extent.

If a scale is taken such that the known universe is as large as the United Nations building, it may contain of order 10^{11} specks of dust, average distance between neighbouring specks being about one centimetre. These dust-specks are the galaxies. If we choose a particular one, our galaxy, and expand it until it is the size of the continent of Asia, we find that it too is made of specks of dust, the average separation of these dust-motes being 100 metres. There are about 10^{11} of them. These are the stars in the Galaxy. Choose the one that is the Sun. Then a penny dropped on it covers the Solar System on that scale. Expand the penny-sized Solar System to the size of the same continent and the Earth is a ball about the height of a two-storeyed house.

18.6.2 The Age of the Universe The universe has been in existence for at least 10^{10} years, possibly as long as 3×10^{10} years. Again we cannot grasp such durations of time. If, however, we fix upon a figure of 3×10^{10} years as a characteristic "age" for the universe, and adopt a scale on which 10^3 years is represented by 1 second, then the universe is about one year old. By January 1st of that year, the processes leading to our present-day universe were well under way. By June our galaxy was in existence; on or around November 1st the Earth was created. Life appeared on the Earth about December 14th. Man evolved into much his present form about a quarter of an hour before midnight on December 31st. The earliest civilizations of which we have any record came into being some 20 seconds before that midnight and the last three and a half centuries of our scientific and technological age have occupied the last one-third of a second of that year.

18.6.3 The Sameness of the Universe One additional major fact we have found in our study of the universe is its universal sameness. Over most of its volume, the same elements appear in much the same abundance–hydrogen dominates, followed by a lesser proportion of helium, with all the other elements represented in very small quantities. (We have seen that the apparent anomaly of the Earth and other similarly-sized planets having less than their fair share of light elements may simply be a consequence of their mode of formation.) The same chemical and physical processes operate throughout the universe, the same nuclear reactions act, magnetic and gravitational fields manipulate material in their own special ways. These processes have been going on in all parts of the universe so that the conditions we find in our tiny corner of the cosmos are repeated endlessly—as they must have been long before human beings arrived on the scene.

18.7 Life As We Know It

Before considering how life arose originally, let us attempt to estimate the number of places in the known universe where life as we know it could survive. By "life as we know it" we mean the kind of life that is found on the Earth. Study of that life in its many forms reveals that it consists of complex organisms composed of carbon, nitrogen and hydrogen with only a 1% addition of other elements. To survive, most of these organisms must be supplied with water, radiation, the correct gravity, atmospheric composition and pressure, temperature in a particular range, food, and so on.

An important factor in all this is the extraordinarily wide adaptive powers living organisms possess. For example certain blue-green algae are found in water at a temperature of 85° C. Some plant seeds can survive temperatures as low as −190° C. Germs of *streptococcus mitis*, inadvertently transported in Surveyor 3's television camera to the Moon in 1967, survived in the harsh lunar environment until they were brought back to Earth 950 days later. Complicated life-forms have been found at the bottom of the deepest ocean trenches on Earth where sunlight never penetrates and the pressure is many tons to the square centimetre.

We have seen that it is a cardinal astronomical principle to assume that the home of mankind is in no way special in the universe or that the Sun is an unusual star. Indeed all the available evidence is that the Sun is one of the commonest types of star in the universe. Yet we cannot assume that because the Sun has a planet Earth possessing life, other Sun-type stars will in all probability have life-bearing bodies in their planetary systems or indeed have planetary systems at all. We must proceed step by step and try to estimate the probabilities that certain necessary conditions for "life as we know it" exist.

A suitable procedure would seem to be to look for Earth-type planets in the universe orbiting Sun-type stars at distances such that the molecule H_2O can exist as water and is not always steam or ice. In doing this, we are probably selecting the most suitable places in the universe where life as we know it could survive.

We therefore begin by estimating the number of suitable planets in the known universe.

18.8 Suitable Planets for Life

In Chapter 6 the origin of the Solar System was discussed. It was mentioned that views currently held suggested that a reasonably high proportion of stars should have planetary systems. All modern theories of the Solar System's origin are variations of the one theme that if a star possesses a cloud of interstellar material, a planetary system will probably develop. In recent years evidence has accumulated that several of the nearest stars do have planets.

The presence of Jupiter and Saturn in the Sun's system of planets could most readily be detected from a planet orbiting any of the nearest stars if a long and accurate series of observations of the Sun's proper motion was made. Over scores of years, the Sun, observed as a star, would pursue an arc on the celestial sphere that would, if long enough, be seen to be wavy. The centre of mass of the

Sun–Jupiter–Saturn system (neglecting the other planets in the Solar System) would follow a great circle arc, but the Sun would in general oscillate about that line. By analysis of the oscillations' amplitudes and periods, masses and orbits for Jupiter and Saturn could be obtained without ever seeing these planets.

This method has been applied to a small number of the nearest stars whose proper motions are larger than average.

The most convincing case is Barnard's star. With its extra-large proper motion, its path against the stellar background can be studied very precisely by photography. It is found that deviations in its path from a straight line can best be explained by postulating the existence of two planets in almost circular orbits about it and with masses somewhat less than Jupiter's.

A variation of this method is to examine the orbits of certain binary stars. Again, if the orbit of one component of a visual binary about the other is well-determined, it is sometimes possible to detect an additional waviness suggesting the presence of a third invisible body in orbit about one of the components. Thus, the binary system 61 Cygni has a body about 16 times the mass of Jupiter orbiting one of its components. Again, the binary 70 Ophiuchi is accompanied by an unseen companion with a mass about 8 Jupiter masses.

With the increase in sensitivity of radial velocity measurements, the effects of planetary motions about stars can be detected from cyclic variations in stellar spectra. Evidence of planets not more than ten times the mass of Jupiter is available from such studies of several stars including ε Eri and γ Cep. Speculation on the existence of planetary systems has recently been enhanced by measurements by IRAS (Infrared Astronomical Satellite) which has revealed the presence of dust around stars with spectral type which made the discovery unexpected. As a follow-up to the IRAS survey, several stars have been scrutinized using special techniques to highlight their local environment and one star in particular (β Pic) has been shown to be surrounded by a dust ring, possibly primordial material for planetary formation.

If therefore we say that about 10% of stars have planetary systems, we are being safely conservative in our estimate.

Since the number of stars in the known universe is of order 10^{22}, the number of planetary systems is of order 10^{21}.

Now the fraction of main sequence stars in the spectral classes F, G and K, a range of stars possessing properties not too far removed from those of the Sun, is about 0·25. Again let us be pessimistic and reduce the fraction to 0·10.

Then the number of planetary systems orbiting suitable stars is 10^{20}.

In our own solar system there are nine major planets of which five—Mercury, Venus, Earth, Mars and Pluto—could be said to be terrestrial in size. Of these only two, Earth and Mars, orbit the Sun at distances such that a reasonable temperature range exists, that is, a range of temperature within which H_2O can be liquid for part of the time in some regions of these planets. A figure of one Earth-sized planet per ten planetary systems at a distance from its star where water can exist is therefore an underestimate. Nevertheless it still leaves 10^{19} Earth-sized planets orbiting main sequence stars of the right spectral classes at the right distances. We can be confident that these planets will not be made from unknown elements but will be comprised of the familiar elements found on Earth, on the Moon and in meteorites, present indeed in much the same proportions.

18.9 The Origin of Life

Given the conditions described above, which we have seen must exist in many places throughout the universe, would life originate under these conditions?

Various lines of research are being conducted today to answer that question. There are grounds for believing that the Earth's original atmosphere was quite unlike the one it possesses today. It may have consisted of a mixture of carbon dioxide, water vapour, ammonia and methane. With no free oxygen, there would have been no ozone layer shielding the Earth's surface from solar ultra-violet radiation. Lightning storms would be as frequent as they are now, probably more frequent.

Scientists have passed ultra-violet radiation and electric discharges through sterilized flasks containing mixtures of carbon dioxide, water vapour, ammonia and methane. In this way they have synthesized the amino acids and purine and pyrimidine compounds that comprise the nucleic acids, all complicated molecules forming the building-blocks of living cells. Although they have not yet created life, many scientists engaged in this work believe that in the conditions present on the primitive Earth, such organic substances would grow in numbers. Further interactions would produce even more complex molecules until the first self-reproducing organism appeared. Thereafter, evolution would act, mutations would arise and after perhaps 10^9 years, the Earth would be populated by its multitude of species of life. In the process, the Earth's atmosphere was transformed by photosynthesis, the method by which plant-life uses sunlight to create food from carbon dioxide and water. The oxygen liberated in this process entered the atmosphere.

It could well be, in fact, that the original amino acids and other complex molecules existed long before the Earth was formed. In 1970, analysis of the Murchison meteorite revealed the presence of 17 amino acids together with a number of hydrocarbons similar to those produced in experiments with primitive atmospheres. It is extremely unlikely that these acids and hydrocarbons were contaminants of the meteorite since NASA scientists also found that among the amino acids there was an almost equal division of spirally-shaped left- and right-handed molecules. Except for those amino acids made artificially, most terrestrial amino acids are left-handed, that is, they rotate in a left-handed direction a beam of polarized light passed through them. It could be, therefore, that when the Solar System was being born, amino acids already existed in space.

Until we create life in the laboratory or find life elsewhere in the universe, we cannot be certain that the above theory concerning the development of life on Earth is correct. It may never be possible to synthesize life under laboratory conditions. But if we find any form of life on a body other than Earth and can reasonably be certain it originated there, then we would have to accept that given the right conditions, life will develop. And we have seen that a conservative estimate of the number of planets where such conditions exist is 10^{19}.

18.10 The Search for Extra-Terrestrial Life

This is a major goal of space research. Before 2000 we shall probably know if life exists on the planet Mars, the most suitable candidate after Earth as a home for life as we know it. Even if life does not exist there now, any geological study of the Martian rocks that revealed the fossils of extinct life would have far-reaching consequences. A range of versatile life-detection equipment was developed in connection with the Viking missions to the surface of Mars. Such equipment should have been capable in principle of detecting not only present life but also signs of past life and conditions leading to the appearance of life.

Among the equipment was an automated amino acid analyser capable also of nucleic acid identification. Again, by means of pyrolysis and chromatography equipment, it should have been possible to identify all of the other molecules of biological significance. In pyrolysis, substances are heated until they are gaseous. Their affinity for various absorbent materials, measured by chromatography, gives clues as to their nature.

Experimental results from Vikings I and II were ambiguous. They may be interpreted as being due to highly unusual chemical reactions or to the presence of life. Future experiments will undoubtedly clarify the matter.

By 1990 all the planets in the Solar System except Pluto will probably have been visited by instrumented probes. If it were found that no life existed elsewhere in the Solar System, the search for extra-terrestrial life would have to be left to the radio astronomers. Present-day radio telescopes are sensitive enough to detect transmissions from similar telescopes out to a distance that includes the hundred nearest stars. Even more powerful telescopes are being developed. It has indeed been suggested that an ideal place to build a radio-telescope is the other side of the Moon which is always turned away from our radio-noisy Earth.

The search for life would then be a search for intelligent life, carried out by seeking to discover artificiality in any radio signal received. Could we recognize the imprint of intelligence in such a signal?

At various international conferences on the problem of extra-terrestrial life, the question has been discussed and it has become clear that not only is there a chance of recognizing that a signal is of intelligent origin but it would also be possible to enter into communication with the intelligent species and exchange information.

For example, the following signal is obviously artificial. In Fig. 18.1 the message received by a radio-telescope pointing to a particular radio source obviously consists of the numbers 1, 2, 3, 4, , 6. The sender wishes the receiver to transmit in return the sequence 1, 2, 3, 4, 5, 6; the 5 inserted in the reply would be "message received and understood".

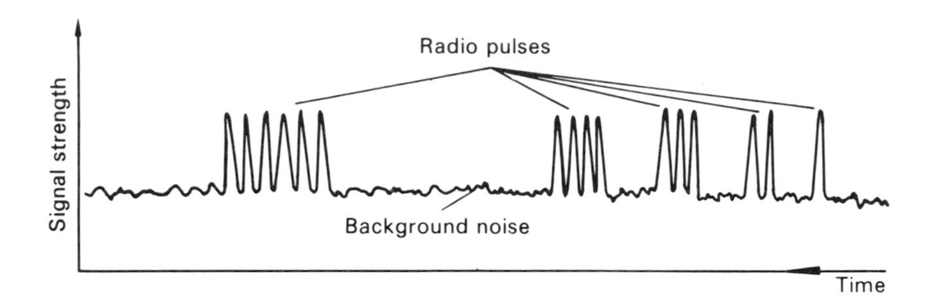

Fig. 18.1 Signal received by a radio-telescope and obviously of artificial origin

Once contact had been made, a common vocabulary would have to be built up. First steps would be the transmission of simple sums. Thus in Fig. 18.2 we have the sums

$$2\times3=6,$$
$$4\times2=8,$$
$$1\times5=5.$$

A series of such sums would ensure that the patterns decided upon for the multiplication sign and the equality sign were understood. In this way a set of mathematical symbols could be learned and used even to the extent of making logical statements of considerable complexity. It might be thought that little progress would follow from that stage, but this is not the case.

In the translation and understanding of ancient Egyptian hieroglyphics, a major breakthrough was achieved when the Rosetta Stone was discovered. This was a slab on which was engraved the same message in Egyptian hieroglyphics and ancient Greek. Fortunately there is a universal Rosetta Stone. When a civilization has developed to a certain scientific level, it realizes that hydrogen, helium and the other elements can be arranged in the *Periodic Table of the Elements*. It does not matter in what corner of the Universe or in what era of time the civilization resides. There is only *one* Periodic Table, a vast body of nuclear, physical, chemical, numerical information that enables a common vocabulary to be established between the two corresponding races. From then on progress would be rapid, subject only to the finite speed of propagation of electromagnetic radiation.

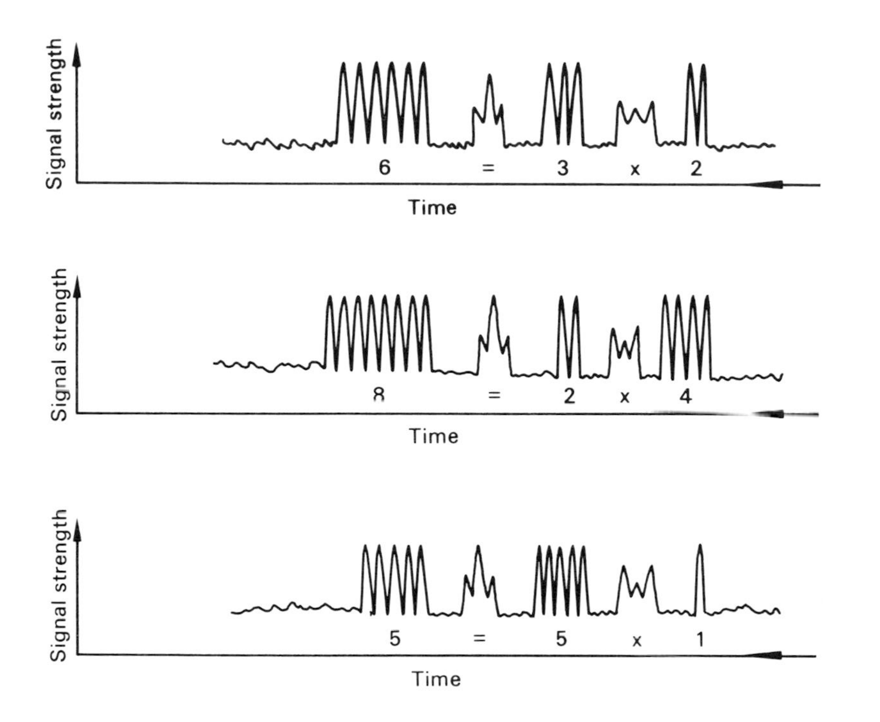

Fig. 18.2 A set of multiplication sums transmitted in order to establish recognition that two different radio patterns represent the multiplication sign and the "equal to" sign

This limit, in fact, makes it necessary for the interstellar conversation to be not between individual human and individual alien but between human race and alien race. So one generation would ask its questions and hope that its descendents would think the answers worth having. That this is so for many questions at any rate is beyond doubt. Other races, with other senses perhaps and other histories, would surely have perspectives on this mysterious universe worth sharing. The only trouble is that the second major fact of the universe, its age, and our late arrival on the scene, makes it probable that we are among the most primitive natives of the lot, with little or nothing of value to barter in exchange.

The argument behind this assessment of the human race's status is not merely the fact that the race has existed only during the past million years, or one thirty-thousandth of the universe's life. Nor does it depend mainly upon the additional fact that it is only during the last one three-thousandth or so of the human race's life that it has shown any marked degree of scientific skill. It is the additional factor of *acceleration* in the rate of scientific and technological progress, so evident in the past forty years, that clinches the matter. In all fields of human skill we have barely arrived, compared with what, hopefully, we may achieve. Though there may be other races at our level, therefore, all earlier arrivals must have advanced far beyond us or have succumbed to savagery through inability to solve their problems.

If this classification of the human race in a prekindergarten stage seems a daunting or depressing prospect, consider the alternative, which is the possibility that no life exists elsewhere.

If our search for extra-terrestrial life were to fail, then the universe would increasingly become a waste-land for the human race, a darkness in which there flickers momentarily in one small corner the tiny candle-light that is terrestrial life.

The astonishing thing is that we are alive in an era that could have this tremendously important question answered during the last part of the 20th century. So in a sense the wheel has come full circle. The ancients thought that Sun, Moon, planets and stars influenced the lives of individual men; we smile at such beliefs but know now that by understanding the universe, the thought of Man cannot help but be profoundly influenced by it, its size, its age, its sameness. No matter if the human race is alone in the universe or is but one of innumerable intelligent species, it is the cosmic equivalent of a baby taking its first steps.

Already that baby is reaching out to the universe about it with telescope and spacecraft. In 1960 Frank Drake used the 85-foot radio telescope at Green Bank, West Virginia, in a preliminary attempt (Project Ozma) to detect intelligent radio signals from the directions of the nearby stars *Epsilon Eridani* and *Tau Ceti*. Although no signals were detected, more systematic searches with more powerful receivers are underway now. Coasting through the Solar System prior to entering the abyss of interstellar space, two Pioneers and two Voyagers carry plaques and discs bearing mankind's greetings. On these audio and video records are the sounds and sights of Earth, designed to show the human race as a species endowed with hope and perseverence, a little intelligence and some generosity, a desire to make contact with the universe. On them are the gasping roar of the youthful Earth's volcanoes, the whisper of the wind, the murmur of a river, the choruses of frogs and birds, the eerily lovely song of the whale in its ocean home and the sounds of people, heartbeats, footsteps, children laughing. The records also carry music: Included are Bach's *Prelude and Fugue in C* and *Brandenberg Concerto No. 2 in F*, first movement; Mozart's *Queen of the Night Aria* from *The Magic Flute*; Stravinsky's *Rite of Spring*; folk songs and melodies from all parts of the Earth—a Navajo night chant, Louis Armstrong's *Melancholy Blues* and Blind Willie Johnston's *Dark was the Night.*

And the pictures of mankind's world. The Sun's position in space; the varied lives of human beings—the birth of a child; infancy and growth and the complex loom of life; cities, houses, tools, plants and animals and landscapes; oceans, a sunset. Men hunting with spears and astronauts walking in space; Olympic sprinters, a traffic jam and Man speaking in the many tongues of Earth, sending greetings to the stars, even in ancient Sumerian—"Silima khemen: May all be well".

The coded information on these artifacts of the human race's faith in a non-empty universe should enable intelligent beings to find the Earth even if the spacecraft were picked up a million years hence. What these intelligent beings who followed up our invitation would find after such a period of time is anyone's guess—perhaps an Earth inhabited by the remote descendents of mankind, as different mentally and physically from us as we are from the little animals that were our ancestors; or perhaps a gutted, polluted, poisoned Earth on which no life is to be found.

In the meantime the astronomer continues his study, painstakingly attempting to improve his knowledge of our position in and relevance to the universe.

In his George Darwin lecture to the Royal Astronomical Society in 1953, Edwin Hubble expressed the astronomer's philosophy very well.

> "As for the future, it is possible to penetrate still further into space—to follow the red-shifts still further back into time—but we are already in the region of diminishing returns; instruments will be increasingly expensive, and progress increasingly slow . . .

. . . From our home on the Earth, we look out into the distances and strive to imagine the sort of world into which we are born. Today we have reached far out into space. Our immediate neighbourhood we know rather intimately. But with increasing distance our knowledge fades, and fades rapidly, until at the last dim horizon we search among ghostly errors of observation for landmarks that are scarcely more substantial. The search will continue. The urge is older than history. It is not satisfied and it will not be suppressed."

Problems: Parts 2 and 3

1. True or false?
   (*a*) Some supergiant stars are older than the Sun.
   (*b*) The spiral arms of our Galaxy in the neighbourhood of the Sun can be traced only by radio astronomical observations.
   (*c*) The Crab Nebula is the remains of the supernova observed by Tycho Brahé in 1572.
   (*d*) In the evening skies of late winter in middle northern latitudes, the Milky Way passes nearly overhead from northwest to southeast.

2. Insert the missing numbers.
   (*a*) RR Lyrae stars are members of Population . . . .
   (*b*) There are . . . . types of binary system.
   (*c*) In our Galaxy there are about $10^?$ stars.

3. In the following statements which is the correct alternative?
   A. The method of trigonometrical parallax gives reliable distances of stars up to about:

      (*a*) 10 pc, (*b*) 100 pc, (*c*) 1 kpc, (*d*) 10 kpc, (*e*) 100 kpc.

   B. The apex of the solar motion determines the direction of:
      (*a*) galactic rotation,
      (*b*) the pole of the ecliptic,
      (*c*) the Sun's peculiar motion.

   C. A G-type star of apparent visual magnitude $+6^m$ has an annual proper motion of $1''\!.0$. The star is a:
      (*a*) white dwarf,
      (*b*) main sequence star,
      (*c*) giant,
      (*d*) supergiant.

   D. If the estimated absolute magnitude of a star is half a magnitude greater than the true absolute magnitude, the derived distance of the star is (approximately),
      (*a*) 25% too large,
      (*b*) 25% too small,
      (*c*) 10% too large,
      (*d*) 10% too small.

   E. The youngest stars in our galaxy are found in
      (*a*) globular clusters,
      (*b*) galactic clusters,
      (*c*) stellar associations.

4. The largest red shift in an extra-galactic object so far measured optically corresponds to a radial velocity of about
   (*a*) $1000\ \mathrm{km\ s^{-1}}$,
   (*b*) $10^4\ \mathrm{km\ s^{-1}}$,
   (*c*) $10^5\ \mathrm{km\ s^{-1}}$,
   (*d*) $10^6\ \mathrm{km\ s^{-1}}$,
   (*e*) $10^7\ \mathrm{km\ s^{-1}}$.
   Which of these statements is correct?

5. True or false?
   (*a*) The gas in intergalactic space is too cold to emit the 21 cm line.
   (*b*) Our galaxy is larger than M31 (the Great Galaxy in Andromeda).
   (*c*) In general, spiral galaxies are larger than ellipticals.
   (*d*) The stars in elliptical galaxies mostly belong to Population II.

6. A nova in a galaxy at a distance of 600 kpc has an apparent visual magnitude of $16^{m}\!.5$. What is the ratio of its intrinsic brightness to that of the Sun? (Absolute visual magnitude of Sun $= 4^{m}\!.83$.)

7. A Cepheid in the M31 galaxy is found to be 5·64 magnitudes fainter than a Cepheid of the same period in the Lesser Magellanic Cloud. Assuming the distance of the Cloud to be 47 kpc, find the distance of M31.

Appendix: Astronomical and Related Constants

1. Physical Constants

Velocity of light $c = 299\,792{\cdot}5 \times 10^3$ m s^{-1}
Constant of gravitation $G = 6{\cdot}668 \times 10^{-11}$ N m^2 kg^{-2}
Planck constant $h = 6{\cdot}625 \times 10^{-34}$ J s
Boltzmann constant $k = 1{\cdot}380 \times 10^{-23}$ J K^{-1}
Gas constant $\mathcal{R} = 8{\cdot}317$ J K^{-1} mol^{-1} or $\mathcal{R}_{(SI)} = 8{\cdot}317 \times 10^3$ J K^{-1} kg^{-1}
Avogadro number $N_0 = 6{\cdot}025 \times 10^{23}$ mol^{-1} or $N_{0(SI)} = N_0 \times 10^3$
Stefan–Boltzmann constant $\sigma = 5{\cdot}669 \times 10^{-8}$ W m^{-2} K^{-4}
Wien displacement law constant $\lambda_{max} T = 2{\cdot}90 \times 10^{-3}$ m K
Mass of electron $m_e = 9{\cdot}108 \times 10^{-31}$ kg
Electron charge $e = 4{\cdot}803 \times 10^{-10}$ esu
Solar parallax $P_\odot = 8''79405$
Constant of aberration $= 20''49$
Constant of nutation $= 9''207$
Obliquity of ecliptic $\varepsilon = 23° \, 27' \, 08''26 - 46''84 T$†
Constant of general precession $= 50''2564 + 0''0222 T$†
Permittivity of free space $\varepsilon_0 = 8{\cdot}85424 \times 10^{-12}$ F m^{-1}
Solar constant $= 1{\cdot}360 \times 10^3$ W m^{-2}

2. Time

Length of the Second
1 ephemeris second = 1/31 556 925·975 of length of tropical year at 1900·0
1 mean solar second (smoothed) = 1 ephemeris second to about 1 part in 10^8
1 mean solar second = 1·0027379093 mean sidereal seconds

Length of the Day
1 mean solar day = 1·0027379093 sidereal days
= $24^h \, 03^m \, 56^s5554$ mean sidereal time
= 86 636·5554 mean sidereal seconds
1 sidereal day = 0·9972695664 mean solar days
= $23^h \, 56^m \, 04^s0905$ mean solar time
= 86 164·0905 mean solar seconds

Length of the Month

Synodic	29^d53059	$29^d \, 12^h \, 44^m \, 03^s$
Sidereal	27·32166	27 07 43 12
Anomalistic	27·55455	27 13 18 33
Nodical	27·21222	27 05 05 36
Tropical	27·32158	27 07 43 05

Length of the Year

Tropical (♈ to ♈)	365^d24220	$365^d \, 05^h \, 48^m \, 46^s$
Sidereal (fixed star to fixed star)	365·25636	365 06 09 10

† T is measured in Julian centuries from 1900.

Anomalistic (perigee to perigee)	365·25964	365 06 13 53
Julian	365·25	365 06 00 00

3. Mathematical Constants, Systems of Units and Conversion Factors

1 radian = 57°·29578 = 3437′·747 = 206 264″·8
Square degrees in a steradian = 3282·806
A sphere at its centre subtends 4π steradians
$\pi = 3{\cdot}14159$
e = 2·71828
$\log_{10} e = 0{\cdot}43429$
$\log_e 10 = 2{\cdot}30259$
1 km = 0·62137 miles = 0·53996 nautical miles = 3280·8 ft
1 mile = 1·6093 km
1 International nautical mile = 1·8520 km = 6076·11 ft
1 Å = 10^{-7} mm
1 AU = 149 600 000 km = 92 960 000 miles
1 parsec = 206 264·8 AU
1 km s^{-1} = 2236·9 miles per hour
1 foot per second = 0·30480 metres per second
1 AU per year = 2·9456 miles per second = 4·7404 km s^{-1}
1 kilogram = 2·204622 lb
1 lb = 0·453592 kg
1 electron volt (eV) = $1{\cdot}60206 \times 10^{-19}$ J

Basic SI Units

Physical Quantity	Name of Unit	Symbol
length	metre	m
mass	kilogram	kg
time	second	s
electric current	ampere	A
thermodynamic temperature	degree Kelvin	K

Derived SI units for special quantities

Physical Quantity	Name of Unit	Symbol	Dimensions
energy	joule	J	kg m^2 s^{-2}
power	watt	W	kg m^2 s^{-3}
force	newton	N	kg m s^{-2}
electric charge	coulomb	C	A s
magnetic flux density	tesla	T	kg s^{-2} A^{-1}
frequency	hertz	Hz	s^{-1}

Conversion of cgs units to SI units

1 ångström = 10^{-10} m
1 dyne = 10^{-5} N
1 erg = 10^{-7} J
1 millibar = 10^{2} GN m^{-2}
1 gauss = 10^{4} T
1 erg s^{-1} = 10^{-7} W

Bibliography

1 Source Books and Materials

Astronomy—Selected Readings, ed. M. A. Seeds, Benjamin/Cummings Publishing Company, 1980

Astrophysical Quantities, by C. W. Allen, Athlone Press, 3rd Edition, 1973

The Astronomical Almanac, Her Majesty's Stationery Office, published yearly

The Handbook of the British Astronomical Association, British Astronomical Association, Burlington House, Piccadilly, London, W1V ONL, published yearly

Norton's Star Atlas, by A. P. Norton and J. G. Inglis, Gall and Inglis, Edinburgh, 16th Edition, 1973

The Observer's Handbook of the Royal Astronomical Society of Canada, University of Toronto Press. Available from The Royal Astronomical Society of Canada, 124 Merton Street, Toronto, Canada M4S 2Z2

Outline of Astronomy (two volumes), by H. H. Voigt, Noordhoof Publishers, Holland, 1974

Philips Planisphere, A revolving chart of the heavens showing, for a given latitude and season, the naked eye stars visible at night

The Cambridge Atlas of Astronomy, ed. J. Audouze and G. Israël, Cambridge University Press, 1985

Colours of the Stars, by D. Malin and P. Murdin, Cambridge University Press, 1984

Astronomy: A Handbook, ed. G. D. Roth, Springer-Verlag, Berlin, 1975

In addition, current discussion on the recent developments of the subject, on forthcoming, predicted events and on historical topics can be found in the excellent journal, *Sky and Telescope*, Sky Publishing Corporation, Cambridge, Mass., U.S.A.

2 Historical Books

A History of Astronomy, by A. Pannekoek, Interscience Publishers, New York, 1961

The History of the Telescope, by C. H. King, Charles Griffin and Company Ltd., London, 1955

The Sleepwalkers, by A. Koestler, The Macmillan Company, New York, 1969

Megalithic Sites in Britain, by A. Thom, Oxford University Press, 1967

Megalithic Lunar Observatories, by A. Thom, Oxford University Press, 1969

Echoes of the Ancient Skies, by E. C. Krupp, Harper & Row, New York, 1983

3 Books on Specific Topics

Practical Work in Elementary Astronomy, by M. C. J. Minnaert, D. Reidel Publishing Company, 1969

Astronomy: Observational Activities and Experiments, by M. K. Gainer, Allyn and Bacon Inc, Boston, 1974

An Introduction to Experimental Astronomy, by R. B. Culver, W. H. Freeman and Co, 1974

Introduction to Radio Astronomy, by R. C. Jennison, Newnes, London, 1966

The Early Years of Radio Astronomy, ed. W. T. Sullivan III, Cambridge University Press, 1984

The Practical Astronomer, by C. A. Ronan, Pan Books Ltd, London, and Macmillan, London, 1981

X-ray Astronomy, by J. L. Culhane and P. W. Somford, Faber and Faber, London, 1981

The Universe and Life, by G. S. Kutter, Jones & Bartlett, 1987

The Atmosphere of the Sun, by C. J. Durrant, Adam Hilger, 1988

Planets Beyond, by M. Littmann, Wiley, 1988

The New Physics, by P. C. W. Davies, Cambridge University Press, 1988

Answers to Problems

Part 1

1. (c)
2. (a) false, (b) false, (c) true, (d) false (until the 21st century!), (e) false
3. (a) Neil Armstrong, (b) Neptune, (c) spring, (d) Jupiter
4. (a) Neptune, (b) Saturn, (c) Mercury, (d) Earth, (e) Saturn, (f) Venus
5. $1{\cdot}55 \times 10^3$ kg m^{-3}
6. 2914 metres
7. 28° 36′ N; about 22° 48′
8. 4·89
10. 0·2343
11. 346·8 days
13. 7° 52′; 290° 52′

Chapter 7

1. 12·59 kpc
2. (i) +5·33 (ii) +0·25 (iii) −2.57
3. (i) 2·05 (ii) 0·016 (iii) 100
4. (i) 0·586 (ii) 63·10 (iii) 847·2
5. (i) $5{\cdot}33 \times 10^5$ km (ii) $5{\cdot}53 \times 10^6$ km (iii) $2{\cdot}03 \times 10^7$ km
6. (i) 2·25 (ii) 4·27 (iii) 10·85
7. $M_A - M_B = 0{\cdot}6065$
8. 4·266 and 1·422 solar masses
9. 0″181
10. $0^m 238$, $-0^m 262$
11. (a) $6^m 64$ (b) $7^m 1$ (c) $6^m 65$ (Hint: brightnesses can be added but magnitudes cannot)
12. (i) 2·396 (ii) 0″0719
13. 0·6281
14. $+22^m 51$ (Hint: remember to take account of the phase)
15. $m_A = 6^m 82$, $m_B = 4^m 47$ $B_A/B_B = 0{\cdot}1145$
16. $m_B = 4^m 64$, $m_C = 6^m 15$, $m = 3^m 07$; $\sqrt{A} = 13{\cdot}86$, $\sqrt{B} = 3{\cdot}56$

Chapter 9

1. Mass of each component = 3·21 solar masses
   Radius of each component = 2·67 solar radii
   Parallax = 0″0093
   (Hint: use Kepler's 3rd law $P = \dfrac{\alpha}{T^{2/3}} \dfrac{1}{(2M)^{1/3}}$, the distance modulus, Stefan's law expressed in terms of absolute magnitude and the proportionality $T_c \propto M/R$)
2. $1{\cdot}67 \times 10^8$ years (Hint: use first three pieces of data to get mass of each component. Then find absolute bolometric magnitude of component and obtain ratio of star's and solar luminosities, i.e. the ratio of the rates of energy production.
3. $6{\cdot}84 \times 10^8$ years

Chapter 10

1. $\mu = 0''139$; $u = 5{\cdot}07$ km s^{-1}; $V = 22{\cdot}58$ km s^{-1} at an angle of 12° 59′ to the line of sight
2. 0″0416
3. (i) 39·63 km s^{-1} (ii) 31·46 pc
4. 1″278; 17·81 km s^{-1} at an angle 115° 15′ to the line of sight; 1″562
5. 89·52 km s^{-1} in a direction making an angle of 46° 41′ to the line of sight

Chapter 12

1. 11·0 kpc
2. $0''.00274$
3. 50·12
4. 451·1 kpc
5. 603·6 kpc

Chapter 15

1. $5{\cdot}26 \times 10^{41}$ kg. (Hint: the maximum line-of-sight velocity in the 30° direction must occur when that line from the Sun is tangential to a particular orbit about the galactic centre. Having found the Sun's circular velocity, equate centrifugal and gravitational forces)

Parts 2 and 3

1. (*a*) false, (*b*) false, (*c*) false, (*d*) true
2. (*a*) two, (*b*) three, (*c*) 11
3. A (*b*)
   B (*c*)
   C (*b*)
   D (*b*)
   E (*c*)
4. (*c*)
5. (*a*) false, (*b*) false, (*c*) true, (*d*) true
6. $7{\cdot}73 \times 10^{4}$ times brighter than the Sun
7. 631·2 kpc

Index

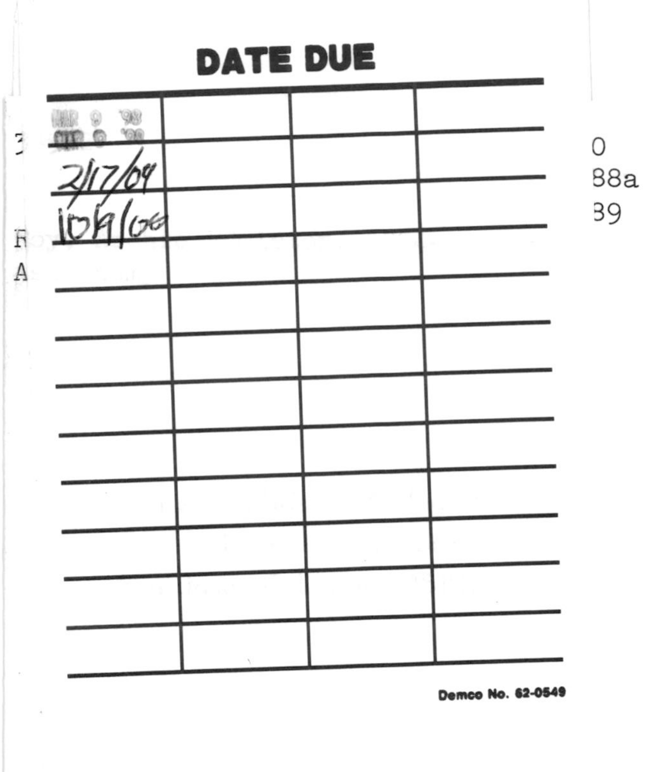
DATE DUE
2/17/04
10/9/06
Demco No. 62-0549
DEMCO